Methods in Enzymology

Volume 287
CHEMOKINES

METHODS IN ENZYMOLOGY

EDITORS-IN-CHIEF

John N. Abelson   Melvin I. Simon

DIVISION OF BIOLOGY
CALIFORNIA INSTITUTE OF TECHNOLOGY
PASADENA, CALIFORNIA

FOUNDING EDITORS

Sidney P. Colowick and Nathan O. Kaplan

Methods in Enzymology

Volume 287

Chemokines

EDITED BY

Richard Horuk

DEPARTMENT OF IMMUNOLOGY
BERLEX BIOSCIENCES
RICHMOND, CALIFORNIA

ACADEMIC PRESS

San Diego   London   Boston   New York   Sydney   Tokyo   Toronto

Academic Press
15 East 26th Street, 15th Floor, New York, New York 10010, USA
http://www.apnet.com

1001149369

Academic Press Limited
24-28 Oval Road, London NW1 7DX, UK
http://www.hbuk.co.uk/ap/

International Standard Book Number: 0-12-182188-9

PRINTED IN THE UNITED STATES OF AMERICA
97 98 99 00 01 02 MM 9 8 7 6 5 4 3 2 1

Table of Contents

Section I. C-X-C Chemokines

Section II. C-C and C Chemokines

Section III. Other Methods

Contributors to Volume 287

Article numbers are in parentheses following the names of contributors.
Affiliations listed are current.

PAOLA ALLAVENA (8), *Department of Immunology and Cell Biology, Istituto di Recherche Farmacologiche "Mario Negri," 20157 Milan, Italy*

DAVID ALLISON (6), *Department of Bioanalytical Technology, Genentech, Inc., South San Francisco, California 94080*

JENNIFER ANDERSON (16), *Biomedical Research Centre and Department of Biochemistry, University of British Columbia, Vancouver, British Columbia V6T 1Z3, Canada*

DEBORAH BALY (6), *Department of Bioanalytical Technology, Genentech, Inc., South San Francisco, California 94080*

DEBRA A. BARNES (19), *Department of Immunology, Berlex Biosciences, Richmond, California 94804-0099*

ALFONS BILLIAU (2), *Laboratory for Immunobiology, Rega Institute for Medical Research, University of Leuven, B-3000 Leuven, Belgium*

ANNE-MARIE BUCKLE (9), *Leukemia Research Fund Cellular Development Unit, Department of Biochemistry and Applied Molecular Biology, University of Manchester Institute of Science and Technology, Manchester M60 1QD, United Kingdom*

ANN CHERNOSKY (21), *Department of Neurosciences, Research Institute, Cleveland Clinic Foundation, Cleveland, Ohio 44195*

HSIUNG-FEI CHIEN (21), *Department of Neurology, Johns Hopkins University School of Medicine, Baltimore, Maryland 21287; and Department of Surgery, National Taiwan University Hospital, Taipei, Taiwan, Republic of China*

IAN CLARK-LEWIS (7, 16), *Biomedical Research Centre and Department of Biochemistry, University of British Columbia, Vancouver, British Columbia V6T 1Z3, Canada*

MARYROSE J. CONKLYN (24), *Department of Cancer, Immunology and Infectious Diseases, Central Research Division, Pfizer, Inc., Groton, Connecticut 06340*

DONALD N. COOK (13), *Department of Immunology, Schering-Plough Research Institute, Kenilworth, New Jersey 07033*

STEWART CRAIG (9), *Osiris Therapeutics, Inc., Baltimore, Maryland 21231-2001*

LLOYD G. CZAPLEWSKI (9), *British Biotech Pharmaceuticals Ltd., Cowley, Oxford OX4 5LY, United Kingdom*

PHILIP E. DAWSON (3), *Department of Chemistry, Caltech, Pasadena, California 91125*

LAURA DeFORGE (6), *Department of Bioanalytical Technology, Genentech, Inc., South San Francisco, California 94080*

STEPHEN K. DURHAM (20), *Experimental Pathology, Bristol-Myers Squibb Pharmaceutical Research Institute, Princeton, New Jersey 08543*

WAYNE J. FAIRBROTHER (4), *Department of Protein Engineering, Genentech, Inc., South San Francisco, California 94080*

GUY FROYEN (2), *Laboratory for Immunobiology, Rega Institute for Medical Research, University of Leuven, B-3000 Leuven, Belgium*

M. ELENA FUENTES (20), *Inflammatory Diseases Unit, Roche Bioscience, Palo Alto, California 94304*

HUBERT F. GAERTNER (22), *Département de Biochimie Médicale, Centre Médicale Universitaire, 1211 Geneva 4, Switzerland*

URSULA GIBSON (6), *Department of Bioanalytical Technology, Genentech, Inc., South San Francisco, California 94080*

ANDRZEJ R. GLABINSKI (21), *Department of Neurosciences, Research Institute, Cleveland Clinic Foundation, Cleveland, Ohio 44195*

ANNEMIE HAELENS (2), *Laboratory of Molecular Immunology, Rega Institute for Medical Research, University of Leuven, B-3000 Leuven, Belgium*

JOSEPH A. HEDRICK (14), *DNAX Research Institute of Molecular and Cellular Biology, Palo Alto, California 94304-1104*

HARVEY R. HERSCHMANN (17), *Departments of Biological Chemistry and Molecular and Medical Pharmacology, UCLA School of Medicine, Los Angeles, California 90095*

JOSEPH HESSELGESSER (5), *Department of Immunology, Berlex Biosciences, Richmond, California 94804*

ARLENE J. HOOGEWERF (23), *Geneva Biomedical Research Institute, Glaxo-Wellcome Research and Development, 1228 Plan-les-Ouates, Geneva, Switzerland*

RICHARD HORUK (1, 5), *Department of Immunology, Berlex Biosciences, Richmond, California 94804*

ROD E. HUBBARD (23), *Department of Chemistry, University of York, Heslington, York YO1 5DD, England*

STEPHEN W. JONES (19), *Department of Medicinal Chemistry, Berlex Biosciences, Richmond, California 94804*

CYRIL M. KAY (7), *Protein Engineering Network of Centres of Excellence (PENCE) and Department of Biochemistry, University of Alberta, Edmonton, Alberta T6G 2S2, Canada*

MICHAEL L. KEY (18), *SAIC-Frederick, Frederick, Maryland 21702*

ALAN M. KRENSKY (10, 11), *Division of Immunology and Transplantation Biology, Department of Pediatrics, Stanford University Medical Center, Stanford, California 94305-5119*

KIMBERLY KRIVACIC (21), *Department of Neurosciences, Research Institute, Cleveland Clinic Foundation, Cleveland, Ohio 44195*

GABRIELE S. V. KUSCHERT (23), *Department of Chemistry, University of York, Heslington, York YO1 5DD, England*

SERGIO A. LIRA (20), *Department of Immunology, Schering-Plough Research Institute, Kenilworth, New Jersey 07033-0539*

DAN L. LONGO (18), *National Institute on Aging, National Institutes of Health, Baltimore, Maryland 21224*

HENRY B. LOWMAN (4), *Department of Protein Engineering, Genentech, Inc., South San Francisco, California 94080*

ALBERTO MANTOVANI (8), *Department of Immunology and Cell Biology, Istituto di Recherche Farmacologiche "Mario Negri," 20157 Milan, Italy*

WILLIAM J. MURPHY (18), *SAIC-Frederick, Frederick, Maryland 21702*

PETER J. NELSON (10, 11), *Clinical Biochemistry Group, Department of Internal Medicine, Ludwig-Maximilians-University of Munich, 80336 Munich, Germany*

ROBERT C. NEWTON (12), *Department of Immunology, Inflammatory Diseases Research, The DuPont Merck Pharmaceutical Company, Wilmington, Delaware 19880-0400*

ROBIN E. OFFORD (22), *Département de Biochimie Médicale, Centre Médicale Universitaire, 1211 Geneva 4, Switzerland*

GHISLAIN OPDENAKKER (2, 8), *Laboratory of Molecular Immunology, Rega Institute for Medical Research, University of Leuven, B-3000 Leuven, Belgium*

PHILLIP OWEN (16), *Biomedical Research Centre and Department of Biochemistry, University of British Columbia, Vancouver, British Columbia V6T 1Z3, Canada*

J. M. PATTISON (10), *Renal Unit, Guy's Hospital, London SE1 9RT, United Kingdom*

H. DANIEL PEREZ (19), *Department of Immunology, Berlex Biosciences, Richmond, California 94804-0099*

JOHN W. PETERSON (21), *Department of Neurosciences, Research Institute, Cleveland Clinic Foundation, Cleveland, Ohio 44195*

CHRISTINE A. POWER (23), *Geneva Biomedical Research Institute, Glaxo-Wellcome Research and Development, 1228 Plan-les-Ouates, Geneva, Switzerland*

PAUL PROOST (2, 8), *Laboratory of Molecular Immunology, Rega Institute for Medical Research, University of Leuven, B-3000 Leuven, Belgium*

AMANDA E. I. PROUDFOOT (22), *Geneva Biomedical Research Institute, Glaxo-Wellcome Research and Development, 1228 Plan-les-Ouates, Geneva, Switzerland*

KRISHNAKUMAR RAJARATHNAM (7), *Protein Engineering Network of Centres of Excellence (PENCE) and Department of Biochemistry, University of Alberta, Edmonton, Alberta T6G 2S2, Canada*

RICHARD M. RANSOHOFF (21), *Department of Neurosciences, Research Institute, Mellen Center for Multiple Sclerosis Treatment and Research, Cleveland Clinic Foundation, Cleveland, Ohio 44195*

DOROTHEA REILLY (1), *Department of Cell Culture and Fermentation, Genentech, Inc., South San Francisco, California 94080*

JENS-MICHAEL SCHRÖDER (15), *Department of Dermatology, Clinical Research Unit, University of Kiel, D-24105 Kiel, Germany*

HENRY J. SHOWELL (24), *Department of Cancer, Immunology and Infectious Diseases, Central Research Division, Pfizer, Inc., Groton, Connecticut 06340*

JEFFREY B. SMITH (17), *Division of Neonatology, Department of Pediatrics, UCLA School of Medicine and UCLA Children's Hospital, Los Angeles, California 90095-1752*

SILVANO SOZZANI (8), *Department of Immunology and Cell Biology, Istituto di Recherche Farmacologichie "Mario Negri," 20157 Milan, Italy*

ROBERT M. STRIETER (20), *Department of Internal Medicine, University of Michigan, Ann Arbor, Michigan 48109*

SOFIE STRUYF (8), *Laboratory of Molecular Immunology, Rega Institute for Medical Research, University of Leuven, B-3000 Leuven, Belgium*

BRIAN D. SYKES (7), *Protein Engineering Network of Centres of Excellence (PENCE) and Department of Biochemistry, University of Alberta, Edmonton, Alberta T6G 2S2, Canada*

MARIE TANI (21), *Department of Neurosciences, Research Institute, Mellen Center for Multiple Sclerosis Treatment and Research, Cleveland Clinic Foundation, Cleveland, Ohio 44195*

DENNIS D. TAUB (18), *National Institute of Aging, National Institutes of Health, Baltimore, Maryland 21224*

BRUCE D. TRAPP (21), *Department of Neurosciences, Research Institute, Cleveland Clinic Foundation, Cleveland, Ohio 44195*

KRISHNA VADDI (12), *Genzyme Corporation, Framingham, Massachusetts 01701*

JO VAN DAMME (2, 8), *Laboratory of Molecular Immunology, Rega Institute for Medical Research, University of Leuven, B-3000 Leuven, Belgium*

LUAN VO (16), *Biomedical Research Centre and Department of Biochemistry, University of British Columbia, Vancouver, British Columbia V6T 1Z3, Canada*

TIMOTHY N. C. WELLS (22, 23), *Geneva Biomedical Research Institute, Glaxo-Wellcome Research and Development, 1228 Plan-les-Ouates, Geneva, Switzerland*

ANJA WUYTS (2, 8), *Laboratory of Molecular Immunology, Rega Institute for Medical Research, University of Leuven, B-3000 Leuven, Belgium*

DAN YANSURA (1), *Department of Cell Genetics, Genentech, Inc., South San Francisco, California 94080*

ALBERT ZLOTNIK (14), *DNAX Research Institute of Molecular and Cellular Biology, Palo Alto, California 94304-1104*

Preface

Chemokines play an important role in inducing the directed migration of blood leukocytes in the body. When this process goes awry and immune cells turn on and attack their own tissues, autoimmune diseases such as rheumatoid arthritis and multiple sclerosis result. In light of these proinflammatory properties, chemokines and their cellular receptors have become therapeutic targets for drug intervention by major pharmaceutical companies. In addition, chemokine receptors have been in the scientific spotlight recently because of the finding that they are coreceptors, along with CD4, for pathogenic organisms such as HIV-1, which use them to gain entry into, and infect, mammalian cells. These and other related findings have placed chemokines in the limelight and exposed them to intense scrutiny by an increasingly broad population of the scientific community.

Given this increased interest in chemokines there was a real need for a practical bench guide that gives detailed protocols and methods that can be used by researchers and advanced students as a step-by-step guide for studying these molecules. With this in mind I assembled a series of comprehensive articles from acknowledged experts in the chemokine field. They are presented in Volumes 287 and 288 of *Methods in Enzymology*. Volume 287 deals with methods in chemokine research; Volume 288 covers chemokine receptor protocols. These volumes provide a detailed compendium of laboratory methods that will appeal both to the novice and to the more experienced researcher wanting to enter this field.

I would like to express my sincere thanks to all the contributing authors for their outstanding efforts and patience during the production of this work. Also, I would like to thank Shirley Light of Academic Press for providing guidance, encouragement, and advice in the preparation of these volumes.

RICHARD HORUK

METHODS IN ENZYMOLOGY

Section I

C-X-C Chemokines

[1] Expression, Purification, and Characterization of *Escherichia coli*-Derived Recombinant Human Melanoma Growth Stimulating Activity

By RICHARD HORUK, DOROTHEA REILLY, and DAN YANSURA

Introduction

Over the past few years rapid progress has been made in relating chemokine structure to function and in understanding the complex biological and physiological roles of these proteins. These advances have been greatly aided by the bacterial expression of many chemokines in sufficient quantities to allow their complete molecular characterization. This chapter outlines the procedures and methods required to express, isolate, and purify chemokines in a biologically active form and in high yield in an *Escherichia coli* expression system. These methods are illustrated with reference to the expression, purification, and characterization of an *E. coli*-derived recombinant human chemokine known as melanoma growth stimulating activity (MGSA).

The polypeptide MGSA is a hormone of 73 amino acid residues that was first identified and assayed from conditioned medium of the human melanoma cell line Hs294T.[1,2] It has been established that the polypeptide sequence of MGSA is identical to the protein product of the human *gro* gene.[3] Structurally MGSA belongs to the C-X-C group of chemokines, which include interleukin-8 (IL-8), neutrophil activating peptide-2 (NAP-2), and platelet factor 4 (PF4).[4] MGSA has a variety of biological effects including the ability to stimulate the proliferation of melanoma cells,[2,5,6] to act as a neutrophil chemotactic factor,[5] to induce calcium mobilization in a variety of different target cells,[6,7] and to regulate collagen expression

[1] A. Richmond, D. H. Lawson, D. W. Nixon, J. S. Stevens, and R. K. Chawla, *Cancer Res.* **43,** 2106 (1983).

[2] A. Richmond and H. G. Thomas, *J. Cell Physiol.* **129,** 375 (1986).

[3] A. Anisowicz, L. Bardwell, and R. Sager, *Proc. Natl. Acad. Sci. U.S.A.* **84,** 7188 (1987).

[4] T. Schall, *in* "The Cytokine Handbook" (A. Thompson, ed.), 2nd ed., p. 419. Academic Press, San Diego, 1994.

[5] R. Derynck, E. Balentien, J. H. Han, H. G. Thomas, D. Wen, A. J. Samantha, C. O. Zachariae, P. R. Griffin, W. L. Wong, K. Matsushima, and A. Richmond, *Biochemistry* **29,** 10225 (1990).

[6] R. Horuk, D. G. Yansura, D. Reilly, S. Spencer, J. Bourell, W. Henzel, G. Rice, and E. Unemori, *J. Biol. Chem.* **268,** 541 (1993).

[7] B. Moser, C. Schumacher, V. von Tscharner, I. Clark-Lewis, and M. Baggiolini, *J. Biol. Chem.* **266,** 10666 (1991).

in fibroblasts.[8] These studies suggest that MGSA participates in a number of biological processes including tumor growth, wound healing, and inflammation. An understanding of the mechanism of action of MGSA, its overall structure, and its cellular regulation should provide insight into these important physiological processes.

Until recently lack of sufficient MGSA had hampered progress in the elucidation of the biological role of the chemokine and identification of its putative receptor(s).[2,5,7,9] Expression of MGSA in Chinese hamster ovary (CHO) cells was described[5] but yielded material missing two C-terminal amino acids at relatively low expression levels (60 μg from an initial 5 liters of conditioned media). Given the relative unavailability of MGSA we decided to express the protein in *E. coli*.[6]

In this chapter we provide in detail the methods and protocols that have enabled us to express, isolate, and purify several hundred milligrams of recombinant MGSA in *E. coli*.[6] With this expression system the recombinant MGSA is secreted into the extracellular medium, which greatly simplifies its isolation. Furthermore, the recombinant chemokine is intact and has the same mass and N-terminal sequence as authentic MGSA. Finally, we give detailed protocols for the analysis and characterization of recombinant MGSA that should greatly aid the investigator embarking on similar projects with other chemokines. The availability of large amounts of chemokines should facilitate progress in the identification of specific receptors for these molecules and help to provide an understanding of their physiological role.

Plasmid Construction

The expression plasmid pMG34 was designed for the secretion of mature MGSA[10] into the *E. coli* periplasmic space (Fig. 1). Secretion offers two distinct advantages over the more common mode of expression in the bacterial cytoplasm. The first is the formation *in vivo* of the correct N-terminal amino acid on the secreted protein, rather than having the usual methionine following cytoplasmic expression. The second is the ability to form disulfide bonds in the periplasm to maintain correct protein conformation[11]; in the reducing environment of the cytoplasm disulfides generally do not form. Although the plasmid pMG34 direct MGSA expression primarily to the *E. coli* periplasm, the small size of the protein results in significant leakage into the outside growth medium.

[8] E. N. Unemori, E. P. Amento, E. A. Bauer, and R. Horuk, *J. Biol. Chem.* **268,** 1338 (1993).
[9] M. P. Beckmann, W. E. Munger, C. Kozlosky, T. VandenBos, V. Price, S. Lyman, N. P. Gerard, C. Gerard, and D. P. Cerretti, *Biochem. Biophys. Res. Commun.* **179,** 784 (1991).
[10] A. Richmond, E. Balentien, H. G. Thomas, G. Flaggs, D. E. Barton, J. Spiess, R. Bordoni, U. Francke, and R. Derynck, *EMBO J.* **7,** 2025 (1988).
[11] J. C. A. Bardwell, J.-O. Lee, G. Jander, N. Martin, D. Belin, and J. Beckwith, *Proc. Natl. Acad. Sci. U.S.A.* **90,** 1038 (1993).

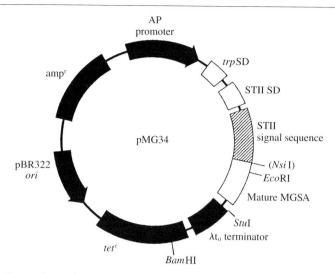

FIG. 1. Expression of human MGSA in *E. coli*. Schematic representation of the pMG34 expression vector. The alkaline phosphatase (AP) promoter and origin of replication (pBR322) are shown as black arrows, the *amp*[r] and *tet*[r] antibiotic resistance markers and λt$_o$ terminator are shown as black boxes, the STII signal sequence is shown as a hatched box, and the *trp* and STII Shine–Dalgarno (SD) sequences and coding sequence for MGSA are shown as open boxes. (Reprinted with permission from Horuk *et al.*[6])

Transcription and translation controls for MGSA expression in the plasmid pMG34 are provided by the alkaline phosphatase promoter and tandem *trp* and heat-stable enterotoxin II (STII) Shine–Dalgarno sequences as previously described.[12] Secretion of MGSA into the periplasm is facilitated with the bacterial STII signal sequence,[13,14] a sequence used successfully with several other chemokines including regulated upon activation normal T expressed and secreted (RANTES), macrophage inflammatory protein-1α (MIP-1α), MIP-1β, and IL-8.[15–17] A convenient *Nsi*I restric-

[12] C. N. Chang, M. Rey, B. Bochner, H. Heyneker, and G. Gray, *Gene* **55,** 189 (1987).
[13] C. H. Lee, S. L. Moseley, H. W. Moon, S. C. Whipp, C. L. Gyles, and M. So, *Infect. Immun.* **42,** 264 (1983).
[14] R. N. Picken, A. J. Mazaitis, W. K. Maas, M. Rey, and H. Heyneker, *Infect. Immunol.* **42,** 269 (1983).
[15] P. Kuna, S. R. Reddigari, T. J. Schall, D. Rucinski, M. Y. Viksman, and A. P. Kaplan, *J. Immunol.* **149,** 636 (1992).
[16] A. Rot, M. Krieger, T. Brunner, S. C. Bischoff, T. J. Schall, and C. A. Dahinden, *J. Exp. Med.* **176,** 1489 (1992).
[17] P. P. Massion, C. A. Hebert, S. Leong, B. Chan, H. Inoue, K. Grattan, D. Sheppard, and J. A. Nadel, *Am. J. Physiol.* **268,** L85 (1995).

tion site near the end of the STII signal sequence in the plasmid pTF2A12[18] provided for precise in-frame fusion of the MGSA coding sequence using synthetic DNA. The major part of the MGSA coding sequence was obtained from the mammalian expression vector pCMV-M23.[5] Finally, the λt_o transcription terminator[19] was placed approximately 45 nucleotides downstream of the MGSA translation termination codon.

Transformation and Fermentation

Competent *E. coli* K12 strains were transformed with pMG34 DNA using the $CaCl_2$–heat shock method.[20] Fermentations were performed in 15-liter Biolafitte fermentors (Biolafitte, France) as described previously.[18]

Purification

Although chemokines have been successfully produced in a variety of mammalian expression systems, a major disadvantage has been their low yield, typically 10 to 20 μg per liter.[5] In contrast, expression of the proteins in *E. coli* and in yeast has generated yields of around 2 to 20 mg per liter.[6,21–23] However, the recombinant expression of chemokines in non-mammalian expression systems, although it offers much higher chemokine yields, suffers from some major disadvantages compared to mammalian expression. First, the chemokines, even though they are small proteins, are not always efficiently secreted. This necessitates their extraction from the cell paste using either 8 M urea or 6 M guanidine. We have found that chemokines extracted in these chaotropic agents will often quite naturally refold correctly during the course of the purification. In many cases, however, the chemokines have to be carefully refolded to recover their biological activity. Second, bacterial expression of chemokines can often be accompanied by the retention of the initiating methionine at the N-terminal end of the protein. The N-terminal extension of chemokines can be problematic

[18] L. R. Paborsky, K. M. Tate, R. J. Harris, D. G. Yansura, L. Band, G. McCray, C. M. Gorman, D. P. O'Brien, J. Y. Chang, J. R. Schwartz, V. P. Fung, J. N. Thomas, and G. A. Vehar, *Biochemistry* **28,** 8072 (1989).

[19] S. Scholtissek and F. Grosse, *Nucleic Acids Res.* **15,** 3185 (1987).

[20] M. Mandel and A. Higa, *J. Mol. Biol.* **53,** 159 (1970).

[21] I. Lindley, H. Aschauer, J.-M. Seifert, C. Lam, W. Brunowsky, E. Kownatzki, M. Thelen, P. Peveri, B. Dewald, V. von Tscharner, A. Walz, and M. Baggiolini, *Proc. Natl. Acad. Sci. U.S.A.* **85,** 9199 (1988).

[22] J. M. Clements, S. Craig, A. J. H. Gearing, M. G. Hunter, C. M. Heyworth, T. M. Dexter, and B. I. Lord, *Cytokine* **4,** 76 (1992).

[23] K. E. Driscoll, D. G. Hassenbein, B. W. Howard, R. J. Isfort, D. Cody, M. H. Tindal, M. Suchanek, and J. M. Carter, *J. Leukocyte Biol.* **58,** 359 (1995).

because, as described for RANTES,[24] it can convert the protein into an antagonist. We discuss some of these problems in more detail later and also describe some generalized procedures for the extraction and purification of chemokines from an *E. coli* expression system, illustrated with reference to MGSA.

Isolation and Purification of Melanoma Growth Stimulating Activity from Cell Supernatants

The *E. coli* expression system for the production of MGSA was engineered so that the protein is efficiently excreted into the supernatant. Thus, the first step in the isolation of the protein is to concentrate the cell supernatants (usually 10 liters or more) to a small volume using tangential flow ultrafiltration (Filtron Minisette, Filtron, Northborough, MA). With this system 10 liters of cell culture supernatant can be rapidly concentrated to 500 ml. Given the small molecular mass of the chemokines, a 670 cm^2 alpha membrane screen-channel cassette with a 3-kDa cutoff is used. The concentrated supernatant is then diafiltered against 5 liters of 1 m*M* Tris-HCl buffer, pH 7.8, to reduce its conductivity.

The purification of MGSA takes advantage of the fact that, like most chemokines, it is a very basic protein with an isoelectric point above pH 9. Thus, the initial purification is achieved by passing the diafiltered supernatant over a column (10 × 5 cm) of S-Sepharose Fast Flow at a flow rate of 400 ml/hr. The column is developed with a step gradient of NaCl (125 and 350 m*M*) in 50 m*M* Tris-HCl buffer, pH 7.8. The ion-exchange column efficiently concentrates the chemokine and effectively removes most of the *E. coli* proteins, which are acidic and thus pass through the column in the flow-through. The MGSA is recovered in the 350 m*M* NaCl eluate (if the column is developed in the presence of 8 *M* urea the MGSA is recovered in the 175 m*M* NaCl eluate) and can be directly subjected to reversed-phase high-performance liquid chromatography (HPLC) on a Vydak Hi-Pore C_{18} column. We use a 50-min gradient from 10 to 60% (v/v) acetonitrile in 0.1% (v/v) trifluoroacetic acid (TFA). This step usually produces pure MGSA. However, in some cases, we rerun the MGSA-containing peak on the same C_{18} column using a shallower, 40-min gradient from 30 to 40% acetonitrile in 0.1% (v/v) TFA.

If the combination of ion-exchange chromatography and reversed-phase HPLC do not produce a pure protein, the investigator can take advantage of the fact that most chemokines have a heparin binding site, which appears to be associated with residues in the α-helical region of the molecule near

[24] A. E. I. Proudfoot, C. A. Power, A. J. Hoogewerf, M. O. Montjovent, F. Borlat, R. E. Offord, and T. N. C. Wells, *J. Biol. Chem.* **271**, 2599 (1996).

the carboxyl terminus.[25–27] Thus, the chemokines can be further purified by interposing a heparin affinity chromatography step in between the ion-exchange and reversed-phase steps. If this is required, then proteins from the 350 mM NaCl pool are diafiltered against 1 mM Tris-HCl buffer, pH 7.8, to reduce their molarity to 75 mM NaCl or less. The diafiltered proteins are then applied to a heparin-Sepharose affinity column equilibrated in 75 mM Tris-HCl buffer, pH 7.8. The column is washed with 2 bed volumes of the same buffer and is then developed with a step gradient of NaCl (250 mM, 500 mM, and 1 M) in 50 mM Tris-HCl buffer, pH 7.8. Under these conditions the MGSA elutes in the 500 mM NaCl and can then be directly applied to a reversed-phase HPLC column as described earlier. It is important to optimize the heparin column with each chemokine in turn, as some, like RANTES and PF4, require higher NaCl concentrations to be eluted. Figure 2 shows the purification of MGSA through the various chromatographic steps.

Isolation and Extraction from Cell Pastes

In many cases, bacterial expression of chemokines does not result in the secretion of the protein into the cell supernatant, and the protein has to be extracted from the cell paste.[21,28,29] A variety of methods to disrupt bacterial cells have been described, including sonication, the nitrogen cavitation bomb, and the French press.[21,29] The disrupted cells are then usually extracted with a chaotropic agent such as urea or guanidine hydrochloride. The procedure currently in use in our laboratory combines cellular disruption and extraction in one step. Typically the combined cell pastes from a 10-liter fermentor are vigorously vortexed in 1500 ml of 8 M urea/10 mM Tris, pH 8.0. This solution is then placed in an end-over-end rotator and extracted for 2 to 4 hr. The disruption of the bacterial cells results in the release of bacterial DNA, and this is removed by precipitation with polyethyleneimine [PEI, final concentration 0.1% (v/v)]. The extract is then centrifuged at 10,000 g for 20 min at 4°, and the supernatant is then carefully removed by vacuum aspiration. The cell pastes are discarded, and the supernatant is further clarified by capsule filtration to yield a clear, straw-colored solution. The chemokines are then purified from the extracted supernatants by the methods described above.

[25] R. St. Charles, D. A. Walz, and B. F. P. Edwards, *J. Biol. Chem.* **264,** 2092 (1989).

[26] A. D. Luster, S. M. Greenberg, and P. Leder, *J. Exp. Med.* **182,** 219 (1995).

[27] G. M. Clore and A. M. Gronenborn, *FASEB J.* **9,** 57 (1995).

[28] J. Hesselgesser, C. Chitnis, L. Miller, D. J. Yansura, L. Simmons, W. Fairbrother, C. Kotts, C. Wirth, B. Gillece-Castro, and R. Horuk, *J. Biol. Chem.* **270,** 11472 (1995).

[29] I. von Luettichau, P. J. Nelson, J. M. Pattison, M. van de Rijn, P. Huie, R. Warnke, C. J. Wiedermann, R. A. Stahl, R. K. Sibley, and A. M. Krensky, *Cytokine* **8,** 89 (1996).

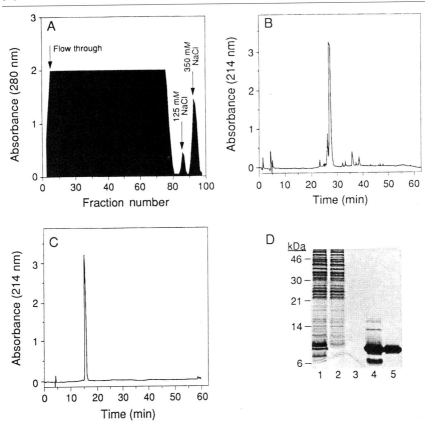

FIG. 2. Purification of MGSA. (A) Elution from a S-Sepharose ion-exchange column with a step NaCl gradient as shown. (B) Reversed-phase HPLC of the 350 mM NaCl peak on a Vydak C_{18} column. (C) Reversed-phase HPLC of the major protein peak from B above. (D) Silver stain of SDS–PAGE analysis of MGSA fractions: lane 1, starting material from *E. coli;* lane 2, flow-through fraction from ion-exchange column; lane 3, blank; lane 4, 350 mM NaCl pool from ion-exchange column; lane 5, purified MGSA from reversed-phase HPLC pool shown in C above. Numbers represent molecular mass derived from molecular mass marker proteins. (Reprinted with permission from Horuk *et al.*[6])

We have had great success extracting chemokines from *E. coli* cell pastes using the methods described. However, with some chemokines, notably RANTES, we have experienced difficulties both with obtaining properly refolded protein and also in producing recombinant protein with an extra methionine residue at the N-terminal end. Methods to overcome both of these problems are discussed next.

Refolding of Recombinant Chemokines

In the vast majority of cases, chemokines produced by recombinant technology are properly refolded without the need to employ any special procedures.[21,28] Occasionally, however, the recombinant chemokine needs to be treated further to facilitate the correct, disulfide bond formation. In such cases we take supernatants eluted from the S-Sepharose column, dialyze them against 10 mM Tris-HCl, pH 8, and then add the refolding solution to yield final concentrations of 1.5 mM 2,2'-dipyridyl disulfide, 50 mM Tris-HCl, pH 8, 1 mM EDTA, 12 mM reduced glutathione, 750 mM arginine, and 1.2 mM oxidized glutathione. The chemokine is stirred vigorously in the refolding solution overnight at 4° and then dialyzed against 20 mM acetic acid containing 50 mM NaCl, pH 2. The chemokine is then filtered and directly applied to an HPLC column as described above.

Recombinant Chemokines with N-Terminal Extension

Occasionally, production of chemokines in *E. coli* yields proteins having extra residues at the N-terminal end. We have found evidence of extension of RANTES by addition of a methionine residue at the N terminus using the *E. coli* expression system described by von Luettichau *et al.*[29] Others

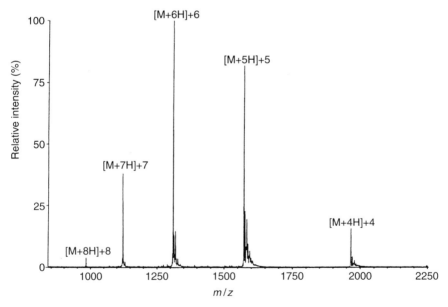

Fig. 3. Electrospray mass spectrum of intact MGSA from the reversed-phase HPLC pool (Fig. 2C). (Reprinted with permission from Horuk *et al.*[6])

have described similar N-terminal extensions of RANTES,[24] MIP-2,[23] and MCP-1 (monocyte chemotactic protein, P. Domaille, personal communication). We have found that the following procedure, adapted from a method of Domaille (P. Domaille, personal communication), works effectively to remove extra N-terminal residues from recombinant chemokines. Ten milligrams of lyophilized protein are dissolved in 500 μl of 40 mM Tris-HCl, pH 8.0. Fifty microliters of aminopeptidase (PeproTech, Inc., Rocky Hill, NJ; 1.7 mg/ml) are added to 100 μl of 40 mM Tris-HCl, pH 8.0, and incubated at 70° for 10 min. The two solutions are combined and left at room temperature overnight. The chemokine–enzyme mixture is applied to a C_{18} analytical reversed-phase HPLC column, and the column is developed with a linear gradient of acetonitrile in 0.1% (v/v) TFA (20–50% B 10–50% (v/v) acetonitrile in 0.1% (v/v) TFA over 60 min) at a flow rate of 1 ml/min. Depending on the efficiency of the enzymatic cleavage, two peaks may be observed; the first peak to elute is the cleaved material, and the second peak represents the starting material. If the reaction is complete, however, only one peak is observed.

Sequence Analysis and Mass Spectrometry of Melanoma Growth Stimulating Activity

To verify the authenticity of the primary sequence of recombinant chemokines we usually carry out N-terminal sequencing and mass determinations using mass spectrometry. Figure 3 shows mass spectrometric analysis of HPLC-purified MGSA.

Verification of Biological Activity of Melanoma Growth Stimulating Activity

An important consideration in the molecular characterization of recombinant MGSA is the determination of its biological activity. A number of bioassays based on the biological effects of MGSA in a variety of different target cells are available. We discuss one of these in more detail below, namely, the ability of MGSA to stimulate the proliferation of melanoma cells.[2,5,6] In addition to these biological effects, MGSA, in common with a number of other C-X-C chemokines, also activates neutrophils and stimulates their degranulation, upregulates CD11/CD18 adhesion proteins, and induces chemotaxis and calcium mobilization. These assays are covered in more detail in the chapters by Baly *et al.*[30] and McColl and Naccache[31] and are not mentioned further here.

[30] D. Baly, U. Gibson, D. Allison, and L. DeForge, *Methods Enzymol.* **287**, [6], 1997 (this volume).
[31] S. R. McColl and P. H. Naccache, *Methods Enzymol.* **288**, [18] (1997).

TABLE I

DOSE–RESPONSE MITOGENIC EFFECT OF
RECOMBINANT HUMAN MGSA ON HS294T
MELANOMA CELLS[a]

Addition	Counts (cpm ± SE)[b]
None	
MGSA	4,111 ± 100
1.2 fM	5,471 ± 632
12 fM	15,300 ± 77
120 fM	16,382 ± 46
1.2 pM	17,453 ± 1,117
12 pM	17,427 ± 2,580
12 pM	16,102 ± 112
1.2 nM	10,649 ± 1,278
12 nM	11,834 ± 429

[a] Reprinted with permission from Horuk et al.[6]
[b] Counts represent the average of six replicate
wells per dose. The addition of 10% serum resulted in the incorporation of 33,440 ± 1,639
cpm of [³H]thymidine into the cells.

[³H]Thymidine Incorporation Assay for Mitogenesis

The [³H]thymidine proliferation assay, which measures [³H]thymidine
incorporation into DNA, is a simple assay to measure the biological activity
of MGSA. A typical assay procedure is given.[6] Basically, around 10^3 to 10^4
human melanoma (HST) cells are plated out in 96-well plates and are
labeled with 1 μCi of [methyl-³H]thymidine (5 Ci/mmol, Amersham, Arlington Heights, IL) in 0.2 ml of serum-free medium for the last 18 hr of a 24-
hr MGSA treatment period. Cells are harvested from the 96-well plates
with a cell harvester, and incorporated [³H]thymidine is quantified in a
scintillation counter. Dulbecco's modified Eagle's medium (DMEM) supplemented with 10% (v/v) fetal bovine serum is a good positive control for
melanoma cell proliferation. Typical data are shown in Table I.

[2] Purification and Identification of Human and Mouse Granulocyte Chemotactic Protein-2 Isoforms

By ANJA WUYTS, PAUL PROOST, GUY FROYEN, ANNEMIE HAELENS, ALFONS BILLIAU, GHISLAIN OPDENAKKER, and JO VAN DAMME

Introduction

Chemokines are produced by a variety of normal and transformed cells in low amounts (micrograms/liter) after stimulation with different endogenous (cytokines) or exogenous inducers (viral, bacterial, or plant products). This chapter describes the methods to purify novel chemokines to homogeneity from natural cellular sources by a multistep chromatographic procedure. After each purification step, fractions are analyzed for purity [by sodium dodecyl sulfate–polyacrylamide gel electrophoresis (SDS–PAGE)] and for biological activity (chemotaxis). This allows one to determine which fractions need to be further purified. Finally, pure proteins are identified by NH_2-terminal amino acid sequence analysis of both intact proteins and peptide fragments. With this purification strategy, human and mouse granulocyte chemotactic protein-2 (GCP-2) were isolated from human MG-63 osteosarcoma cells and from the murine fibroblastoid cell line MO, respectively. The primary structure was identified completely by sequence analysis. The biological activity of these novel chemokines was tested in the microchamber chemotaxis assay and in the gelatinase B release assay. As an alternative to the time-consuming and expensive purification of novel chemokines, human GCP-2, once identified, was chemically synthesized and biochemically and biologically compared with the natural chemokine. Finally, after cDNA cloning, the regulation of human GCP-2 transcription was studied in various cell types by quantitating mRNA levels using reverse transcription–polymerase chain reaction (RT–PCR).

Production of Natural Chemokine Forms

To obtain microgram amounts of homogeneous chemokine, liters of conditioned medium from stimulated cells are recommended as a starting material. For production of chemotactic activity on this scale, 50 or more confluent tissue cultures are usually supplemented with medium containing a low serum concentration and appropriate inducers. The low serum concentration facilitates subsequent purification of chemokines to homogeneity. If serum is totally omitted, the production levels are often very low. Supernatants are routinely harvested 48 to 72 hr after induction. Unless cell detachment occurs, cultures can be refed with fresh induction medium and harvested a second and third time. As an example, the production of neutrophil chemotactic factors by human osteosarcoma cells[1-3] and by murine fibroblastoid cells[4,5] is described.

Human MG-63 osteosarcoma cells are grown in Eagle's minimum essential medium with Earle's salts (EMEM; GIBCO, Paisley, Scotland) supplemented with 10% (v/v) fetal calf serum (FCS; GIBCO) for 7 days. Confluent monolayers (175-cm^2 flasks; Nunc, Roskilde, Denmark) are induced with a semipurified cytokine mixture (derived from mitogen-stimulated mononuclear cells) in medium containing 2% (v/v) FCS. Supernatants are harvested after 48 hr, and the cultures are induced two more times in the same way.

The murine fibroblastoid cell line MO is grown in EMEM supplemented with 10% FCS for 4 days. Confluent monolayers are induced with a mixture of 50 μg/ml of the double-stranded RNA polyriboinosinic : polyribocytidylic acid [poly(rI) : poly(rC); P.-L. Biochemicals, Milwaukee, WI] and 10 μg/ml lipopolysaccharide (LPS, *Escherichia coli* 0.111.B4; Difco, Detroit, MI) in medium containing 2% FCS. Supernatants are harvested after 72 hr.

Concentration and Partial Purification of Natural Chemokines from Cell Culture Conditioned Medium

Before the purification of chemokines, a large volume of conditioned medium has to be concentrated. Two methods that are routinely used in our laboratory for simultaneous concentration and partial purification of chemokines are adsorption to controlled pore glass beads[6] (particle size 120–200 mesh, pore size 350 Å; CPG-10-350; Serva, Heidelberg, Germany) and adsorption to silicic acid[7] (particle size 35–70 μm, pore size 100 Å; Matrex; Amicon, Beverly, MA).

For adsorption to CPG, 1 to 3 liters of conditioned medium at neutral pH is magnetically stirred with CPG (1/30, v/v) for 2 hr at 4° in spinner flasks (3 to 10 liters). After sedimentation, the CPG beads are washed

[1] A. Billiau, V. G. Edy, H. Heremans, J. Van Damme, J. Desmyter, J. A. Georgiades, and P. De Somer, *Antimicrob. Agents Chemother.* **12**, 11 (1977).

[2] P. Proost, C. De Wolf-Peeters, R. Conings, G. Opdenakker, A. Billiau, and J. Van Damme, *J. Immunol.* **150**, 1000 (1993).

[3] P. Proost, A. Wuyts, R. Conings, J.-P. Lenaerts, A. Billiau, G. Opdenakker, and J. Van Damme, *Biochemistry* **32**, 10170 (1993).

[4] A. Billiau, H. Sobis, H. Eyssen, and H. van den Berghe, *Arch. Virol.* **43**, 345 (1973).

[5] A. Wuyts, A. Haelens, P. Proost, J.-P. Lenaerts, R. Conings, G. Opdenakker, and J. Van Damme, *J. Immunol.* **157**, 1736 (1996).

[6] J. Van Damme and A. Billiau, *Methods Enzymol.* **78**, 101 (1981).

[7] M. De Ley, J. Van Damme, H. Claeys, H. Weening, J. W. Heine, A. Billiau, C. Vermylen, and P. De Somer, *Eur. J. Immunol.* **10**, 877 (1980).

consecutively with phosphate-buffered saline (PBS; 500–1000 ml) and with 10 mM glycine-hydrochloride, pH 3.5 (250–500 ml), by manual stirring of the beads over 5 min. To elute proteins from the beads, CPG is stirred twice in 0.3 M glycine-hydrochloride, pH 2.0 (25–50 ml), for 5 min (manually) and twice over 30 min (magnetically at 4°) to further increase recovery. The eluates are pooled and stored at −20° until further purification.

If adsorption to silicic acid is used as an alternative, 1 to 3 liters of cell culture supernatant (neutral pH) is magnetically stirred with silicic acid (10 g/liter) for 2 hr at 4°. After centrifugation to sediment the silicic acid, adsorbed chemokines are washed (4°, 10 min) once with PBS at pH 7.4 (500–1000 ml) and once with PBS, pH 7.4, containing 1 M NaCl (250–500 ml). Elution of proteins is done by magnetically stirring in PBS containing 1.4 M NaCl and 50% ethylene glycol (50–100 ml) over 30 min at 4° (four times to improve recovery). A second elution with 0.3 M glycine-hydrochloride, pH 2.0 (four times with 50–100 ml for 15 min at 4°) can be done to further remove remaining chemokines from the silicic acid. The bulk of chemokines is normally eluted from the silicic acid in the first elution step. The eluates can further be concentrated by dialysis (3.5-kDa cutoff membranes; Spectra/Por; Spectrum Medical Industries, Houston, TX) against 20 mM Tris-HCl, pH 7.5, supplemented with 15% polyethylene glycol (PEG 20,000; Fluka Chemie, Buchs, Switzerland).

The recovery of chemokines by adsorption to CPG is usually better than that by adsorption to silicic acid. Therefore, adsorption to CPG is preferred for the purification and concentration of conditioned medium from cultured cells containing low serum concentrations. A recovery of the expensive CPG used for such purification is possible by cleaning the glass beads with 70% (w/w) nitric acid to remove the noneluted proteins. Adsorption to silicic acid, however, is often used for concentrating supernatants from stimulated buffy coats containing higher protein concentrations. The inexpensive silicic acid is discarded after a single use.

On average, adsorption to CPG results in a 10-fold reduction of the initial volume and a 50-fold decrease of the protein content. Because a recovery of chemokine activity of 75% can be reached, the procedure yields a 40-fold increase of specific chemokine activity. This indicates that the concentration step also results in a partial purification of chemotactic proteins.

Purification of Novel Chemokines to Homogeneity

Affinity Chromatography

Chemokines possess a high affinity for heparin. This characteristic can be used for further purification by heparin affinity chromatography. The

CPG eluate or concentrated silicic acid eluate is extensively dialyzed (3.5-kDa cutoff membranes) against 50 mM Tris-HCl, 50 mM NaCl, pH 7.4. The dialyzed eluate is loaded on a heparin-Sepharose column (1.6 × 40 cm; 10 ml/hr; CL-6B; Pharmacia, Uppsala, Sweden) equilibrated with 50 mM Tris-HCl, 50 mM NaCl, pH 7.4. After washing of the column with equilibration buffer (100 ml), proteins are eluted with a linear NaCl gradient (0.05–2 M) in the same buffer (20 ml/hr; 5-ml fractions). The protein concentration in the fractions is determined by a Coomassie Blue G-250 binding assay[8] using the Bio-Rad commercial kit (Bio-Rad Laboratories, Richmond, CA) with bovine serum albumin as the standard. The bulk of proteins elute at lower NaCl concentration, whereas chemokines elute at higher salt concentration. Thus, the method gives a good separation of chemokines from the bulk of contaminating proteins. Depending on their degree of affinity for heparin, chemokines can be partially fractionated. Because platelet factor 4 (PF-4) has a significantly higher affinity for heparin than does interleukin-8 (IL-8), these proteins are separated by heparin affinity chromatography. However, several chemokines are only poorly separated during the purification step.

If available, antibody affinity chromatography can be used as an alternative for heparin-Sepharose chromatography to isolate chemokines. A polyclonal antiserum[9] against CPG-purified cytokines from serum-free fibroblasts has been developed by repeated injection in goat. The immunoglobulin (Ig) fraction, obtained by Protein A affinity chromatography (Pharmacia) of 20 ml of the antiserum, has been coupled to 6 g of CNBr-activated Sepharose 4B (Pharmacia) to make an antibody column (1.5 × 15 cm). The CPG or silicic acid eluate is neutralized with NaOH until neutral pH before loading on the column (20 ml/hr), equilibrated with PBS, pH 7.4. After washing with PBS (50 ml), proteins are eluted with 0.5 M NaCl, 0.1 M citrate-hydrochloride, pH 2.0 (20 ml/hr, 5-ml fractions). This purification step removes contaminating proteins and allows further concentration of chemokines. If only a particular chemokine needs to be purified, antibody affinity chromatography using a specific antibody (monoclonal antibodies if available) can be sufficient to yield pure chemokine.

Fractions derived from heparin-Sepharose or antibody affinity chromatography are tested for biological activity (e.g., chemotaxis) and purity (SDS–PAGE) as described later in this chapter. Each fraction is assayed at several dilutions in the biotest in order to indicate the peak of activity.

---

[8] M. M. Bradford, *Anal. Biochem.* **72**, 248 (1976).
[9] J. Van Damme, S. Cayphas, J. Van Snick, R. Conings, W. Put, J.-P. Lenaerts, R. J. Simpson, and A. Billiau, *Eur. J. Biochem.* **168**, 543 (1987).

Low molecular weight proteins (5000–15,000) with chemotactic activity for neutrophilic granulocytes are subjected to the next purification step.

Cation-Exchange Chromatography

Chemokines are basic proteins with a high isoelectric point. Cation-exchange chromatography at low pH can therefore be used to purify chemokines from other proteins and to separate chemokines from each other. Fractions (5 to 20 ml) derived from heparin-Sepharose or antibody affinity chromatography containing the peak of chemotactic activity are first dialyzed against 50 mM formate, 0.01% Tween 20, pH 4.0, to remove salt and to adjust the pH. Dialyzed fractions are injected on a Mono S column (Pharmacia) equilibrated with 50 mM formate, 0.01% Tween 20, pH 4.0. After washing with 20 ml of equilibration buffer, proteins are eluted with a linear NaCl gradient (0–1 M) in buffer (1 ml/min; 1-ml fractions). Absorbance at 220 nm is measured as a parameter for protein concentration. Proteins purified by cation-exchange fast protein liquid chromatography (FPLC) are analyzed for purity and relative molecular mass by SDS–PAGE (see later) and for biological activity in the microchamber chemotaxis assay (*vide infra*). Cation-exchange FPLC separates most chemokines from each other; for example, human GCP-2 is separated from GRO and IL-8 during this purification step. To purify chemokines to homogeneity, a fourth purification step is often necessary.

Reversed-Phase High-Performance Liquid Chromatography

Reversed-phase high-performance liquid chromatography (RP-HPLC) on an analytical C_8 column is used as a final purification step. The organic solvents used allow the chemokines to retain their biological activity and are compatible with subsequent amino acid sequence analysis and mass spectrometry. The FPLC fractions (1 to 5 ml) containing chemotactic proteins are injected on a 220 × 2.1 mm C_8 Aquapore RP-300 column (Applied Biosystems, Foster City, CA) equilibrated with 0.1% (v/v) trifluoroacetic acid (TFA) in water. Chemokines are eluted with an acetonitrile gradient [0–80% (v/v) acetonitrile in 0.1% (v/v) TFA; 0.4 ml/min; 0.4-ml fractions]. Proteins derived from HPLC are analyzed for purity and relative molecular weight by SDS–PAGE and are tested for chemotactic activity in the microchamber chemotaxis assay (*vide infra*). RP-HPLC does not yield only homogeneous protein, but often also results in separation of different chemokine isoforms. For example, semipurified natural GCP-2, isolated from human MG-63 cells (purified by CPG, heparin-Sepharose affinity chromatography, and FPLC), was separated by C_8 RP-HPLC into four peaks of neutrophil

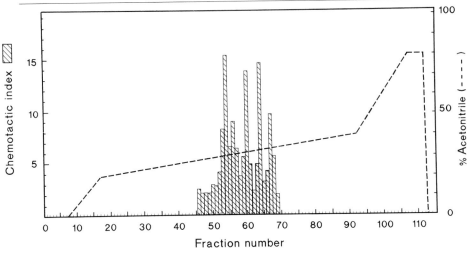

FIG. 1. Purification of human GCP-2 isoforms to homogeneity by RP-HPLC. GCP-2, isolated from human osteosarcoma cells, was further purified (after adsorption to CPG, heparin affinity chromatography, and FPLC) by RP-HPLC on a C_8 Aquapore RP-300 column. Proteins were eluted with an acetonitrile gradient. Fractions were tested (dilution 1/100) for neutrophil chemotactic activity (expressed as chemotactic index) in the microchamber assay.

chemotactic activity (fractions 54, 56, 60, and 67; Fig. 1), which corresponded to 6-kDa proteins on a polyacrylamide gel under reducing conditions. Figure 2 shows the protein band obtained by SDS–PAGE of fraction 67 under reducing conditions (Fig. 1). The four 6-kDa proteins were identified as differently NH_2-terminally truncated forms of GCP-2 (see Fig. 3A). In a

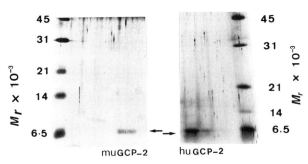

FIG. 2. SDS–PAGE analysis of mouse GCP-2 (more truncated forms; left-hand side) and of human GCP-2 (fraction 67, Fig. 1; right-hand side) purified by RP-HPLC. Samples (20 μl) were loaded on Tris–Tricine gels under reducing conditions. Molecular weight markers: ovalbumin (M_r 45,000), carbonate dehydratase (M_r 31,000), soybean trypsin inhibitor (M_r 21,500), lysozyme (M_r 14,400), and aprotinin (M_r 6500).

similar way, RP-HPLC of mouse GCP-2, isolated from the fibroblastoid cell line MO, yielded two protein bands (7 and 8 kDa) on SDS–PAGE, corresponding to differently NH$_2$-terminally truncated forms of mouse GCP-2 (see Fig. 3B). Figure 2 shows the protein band corresponding to the GCP-2 form eluting first in the acetonitrile gradient (26.5% acetonitrile) and containing the more truncated forms [GCP-2(S); *vide infra*].

Sodium Dodecyl Sulfate–Polyacrylamide Gel Electrophoresis

Proteins obtained by heparin-Sepharose chromatography, FPLC, and HPLC are checked for purity and relative molecular weight by SDS–PAGE under reducing or nonreducing conditions on Tris–Tricine gels.[10] The stacking, spacer, and separating gels contain 5% (w/v) T (total percentage concentration of acrylamide and bisacrylamide) and 5% (w/w) C (percentage concentration of cross-linker relative to T), 10% T and 3.3% C, and 13% T and 5% C, respectively. Proteins are visualized by silver staining. Relative molecular weight markers (Bio-Rad Laboratories) are ovalbumin (M_r 45,000), carbonate dehydratase (M_r 31,000), soybean trypsin inhibitor (M_r 21,500), and lysozyme (M_r 14,400) and the low molecular weight marker (Pierce, Rockford, IL) aprotinin (M_r 6500).

Identification of GCP-2 Isoforms by NH$_2$-Terminal Amino Acid Sequence Analysis

The NH$_2$-terminal amino acid sequence of pure chemotactic proteins is identified by Edman degradation on a pulsed liquid-phase protein sequencer (477A/120A; Applied Biosystems) with on-line detection of phenylthiohydantoin amino acids. Because of the high sensitivity (low picomole level) of currently available protein sequencers, 10 pmol of protein is sufficient for the identification by NH$_2$-terminal sequence analysis. Low molar concentrations of contaminants compared to chemokine will not interfere with the identification of the amino acid sequence, but such contaminants have to be completely removed if the chemokine needs to be digested for internal sequence analysis or if biological activity is tested.

The presence of cysteine residues in the primary structure is often obvious from the absence of any detectable signal after Edman degradation. Their presence can be confirmed after modification of cysteine residues. This can be done in solution or *in situ* on the sequencing reaction cartridge

---

[10] H. Schägger and G. von Jagow, *Anal. Biochem.* **166,** 368 (1987).

filter. For such *in situ* alkylation[11] of cysteine residues, the protein is loaded on the sequencer as usual. Twenty microliters of a reduction/alkylation mixture [2 μl of tri-*n*-butylphosphine, 5 μl of 4-vinylpyridine (Aldrich, Milwaukee, WI), and 100 μl of acetonitrile] is added to the sample, and the cartridge is immediately reassembled. The reaction cartridge is equilibrated with *N*-methylpiperidine, and excess reagents are removed by washing with *n*-butyl chloride and ethyl acetate. This reaction cycle is followed by the normal reaction cycles. Cysteine residues can also be alkylated in solution with acrylamide.[12] In a first step, the protein is reduced by incubation (total volume of 10 μl) for 30 min at 70° in 0.2 M Tris, pH 8.4, 100 mM dithiothreitol (DTT), and 1% SDS (from a 20% stock solution). After incubation, the sample is diluted with 4 volumes of MilliQ (Millipore, Milford, MA) water. The protein is alkylated by incubation at 37° for 45 min in 2 M acrylamide (from a 6 M stock solution in water) in the dark. After alkylation, methanol is added to a final concentration of 10%, and the protein is desalted by spinning in a ProSpin cartridge (Applied Biosystems). The ProBlott membrane is washed by adding 50 μl of 20% methanol and by spinning this solution through the membrane. The membrane is punched out, vortexed once in 0.1% TFA and three times in MilliQ water, and dried before transfer to the protein sequencer.

Amino-terminal amino acid sequence analysis of the four human GCP-2 protein bands obtained after RP-HPLC of the conditioned medium from osteosarcoma cells (*vide supra*) revealed the presence of four differently NH$_2$-terminally truncated forms of a single chemokine (Fig. 3A). For mouse GCP-2, the two proteins separated by RP-HPLC similarly corresponded to different NH$_2$-terminal forms of the same molecule (Fig. 3B). In contrast to human GCP-2, where each RP-HPLC fraction contained a single NH$_2$-terminal form, differently NH$_2$-terminally truncated forms of murine GCP-2 were present in single fractions. The fractions containing short and long NH$_2$-terminal forms of mouse GCP-2 were called GCP-2(S) and GCP-2(L), respectively.

Maximally about 50 residues can be determined during one sequencing run using 500 pmol of pure chemokine. Thus, chemical digestion or enzymatic cleavage is necessary to determine the complete amino acid sequence of proteins consisting of about 75 amino acids. Several chemokines contain the sequence Asp-Pro in the COOH-terminal tail. This Asp-Pro bond can be chemically cleaved by formic acid. Therefore, protein is incubated with 75% (v/v) formic acid at 37° for 50 hr. Because human and mouse GCP-2 possess only a single cleavage site for formic acid, the chemokine can be

[11] P. C. Andrews and J. E. Dixon, *Anal. Biochem.* **161,** 524 (1987).
[12] D. C. Brune, *Anal. Biochem.* **207,** 285 (1992).

	Fraction	% Acetonitrile	NH$_2$-terminal amino acid sequence	Relative amount
A				
	67	32.5	GPVSAVLTELRXTXLRVTLR	100%
	60	31	VSAVLTELRXTXLRVTLRVNPKTIGK	100%
	56	30	VLTELRXTXLRVTLRVNP	100%
	54	29.5	ELRXTXLRVTLRVNPKTIGKLQVFPAG	100%
B				
	48	28	APSSVIAATELRXVXLTVTPKINPXLXANXXV	78%
			SSVIAATELR	16%
			SXXXATELRXVXXXXT	6%
	42	26.5	AATELRXVXLTVTPK	15%
			ATELRXVXLTVTP	25%
			TELRXVXLTVTPKINPKLIANXEVIPXGXQ	44%
			ELRXVXLXXTPKXNXXXXA	16%

FIG. 3. Identification of neutrophil chemotactic proteins as GCP-2 isoforms by NH$_2$-terminal amino acid sequence analysis. (A) Four 6-kDa proteins purified by RP-HPLC from the conditioned medium of human osteosarcoma cells and corresponding to neutrophil chemotactic activity (fractions 54, 56, 60, and 67, see Fig. 1) were analyzed by amino acid sequence analysis. Unidentified amino acids are marked with an X. (B) GCP-2 proteins (S and L forms), corresponding to bands of 7 (Fig. 2) and 8 kDa on SDS–PAGE, were obtained after RP-HPLC (eluting at 26.5 and 28% acetonitrile, respectively) of the conditioned medium from murine fibroblasts and were identified by amino acid sequence analysis. Unidentified amino acids are marked with an X.

loaded directly on the protein sequencer after cleavage (without the need of RP-HPLC separation of the peptide fragments) to identify the COOH terminus. Indeed, to obtain only the sequence of the COOH-terminal part, the original NH$_2$-terminus of the protein can be blocked with o-phthalaldehyde[13] (Fluoropa, Pierce). This reagent will block all sequenceable amino acids except NH$_2$-terminal prolines. To that purpose, 20 mg o-phthalaldehyde and 50 μl of β-mercaptoethanol is added to 10 ml of acetonitrile; the reaction cartridge of the sequencer is equilibrated with N-methylpiperidine and wetted with the o-phthalaldehyde reaction mixture. After several washing steps with ethyl acetate and n-butyl chloride, the sequencing proceeds with the normal reaction cycles using double coupling times in the cycle immediately following the o-phthalaldehyde treatment. The o-phthalaldehyde reaction can also be used to extend the length of the readable sequence. Indeed, if the presence of a proline residue in a sequence is known, treatment with o-phthalaldehyde just before cleavage of the proline can be used to remove background signals.

[13] A. W. Brauer, C. L. Oman, and M. N. Margolies, *Anal. Biochem.* **137,** 134 (1984).

Different enzymes have been used for proteolytic cleavage of natural chemokines[14]: trypsin (sequencing grade, Boehringer Mannheim, Mannheim, Germany), which cleaves the protein after arginine and lysine residues (poor cleavage of Arg-Pro and Lys-Pro bonds), and endoproteinase Asp-N, Glu-C, Arg-C, or Lys-C (all sequencing grade; Boehringer Mannheim), which cleave the protein before Asp and after Glu, Arg, or Lys, respectively. After proteolytic digestion, chemokine fragments need to be separated by RP-HPLC on a C_8 Aquapore RP-300 column (50 × 1 mm, Applied Biosystems) before they are subjected to amino acid sequence analysis. The complete amino acid sequence (Fig. 4) of natural human GCP-2 was identified by the alignment of sequences of peptide fragments. These fragments were obtained after chemical digestion (1.5 μg of GCP-2) with formic acid and enzymatic cleavage of 4 μg of GCP-2 with 0.2 μg of endoproteinase Glu-C (25°, 18 hr in 25 mM ammonium carbonate buffer, pH 7.8), Lys-C (37°, 18 hr in 25 mM Tris-HCl buffer, pH 8.5, 1 mM EDTA), or Asp-N (37°, 18 hr in 50 mM sodium phosphate buffer, pH 8.0). The presence of cysteines was confirmed by *in situ* alkylation. The complete sequence of murine GCP-2 was identified by NH_2-terminal sequence analysis of the natural purified protein (3.5 μg) and sequence analysis after formic acid digestion of 3.5 μg GCP-2 (Fig. 4). The cysteines were detected after alkylation with acrylamide in solution.

Synthesis of Human Chemokines

As an alternative for the expensive and time-consuming purification of natural chemokines from the conditioned medium of cells, these small proteins (70–80 residues) can be chemically synthesized within 1 week. This method allows rapid generation of medium-size (milligram level) quantities of chemokine. This scaling-up alternative broadens the research spectrum to include, for example, detailed evaluation of all possible target cells, development of immunotests, and experiments in animals. As another alternative, natural chemokine can be substituted at an unlimited scale by recombinant material.

Both IL-8 and neutrophil-activating protein-2 (NAP-2) have been synthesized successfully using Boc (*tert*-butoxycarbonyl) synthesis.[15] Using this method, extensive safety precautions are necessary because of the use of highly corrosive and toxic HF. This can be avoided by the use of Fmoc (9-fluorenylmethoxycarbonyl) chemistry.

[14] J. Van Damme, P. Proost, J.-P. Lenaerts, and G. Opdenakker, *J. Exp. Med.* **176,** 59 (1992).
[15] I. Clark-Lewis, B. Moser, A. Walz, M. Baggiolini, G. J. Scott, and R. Aebersold, *Biochemistry* **30,** 3128 (1991).

NH₂-terminal

GPVSAVLTELRCTCLRVTLRVNPKTIGKLQVFPAG
VSAVLTELRCTCLRVTLRVNPKTIGKLQVFPAG
VLTELRCTCLRVTLRVNPKTIGKLQVFPA
ELRCTCLRVTLRVNPKTIGKLQVFPAG

Glu-C digest

VSAVLTELRXTXLRVTLRVN   VVASLKNGKQVCLDPE

Lys-C digest

LQVFPAGPQCSK
LQVFPAGPQCSKVEVVA
VEVVASLK
QVCLDPEAPFLK

Asp-N digest

ELRXTXLRVTLRVNPKTIGKLQVFPAGPQXSKVEVV

HCOOH digest

PEAPFLKKVIQKILDSGNK

---

huGCP-2   GPVSAVLTELRCTCLRVTLRVNPKTIGKLQVFPAGPQCSKVEVVASLKNGKQVCLDPEAPFLKKVIQKILDSGNK

muGCP-2   APSSVIAATELRCVCLTVTPKINPKLIANLEVIPAGPQCPTVEVIAKLKNQKEVCLDPEAPVIKKIIQKILGSDKKKA

NH₂-terminal

APSSVIAATELRXVXLTVTPKINPXLXANXXV
PSSVIAATELRXVXLTVTPKIN
SSVIAATELRXVXLTVTPKINPXLIAXLEVXP
SVIAATELRXVXLTVTPKINXXLXAXLEV
VIAATELRXVXLTVTPKINPXL
IAATELRXVXLTVXXK
AATELRXVXLTVTPKINPXXXAN
ATELRCVCLTVTPKINPKLIANLEVIPAGPQCPTVEVIAKLKNQKEVCLXXEA
TELRXVXLTVTPKINPKLIANLEVIPAGPQXXTVVXA
ELRXVXLTVTPKINXXLXXXXEV
LRXVXLTVTXXIXP

HCOOH digest

PEAPVIKKIIQKILGSDKKKA

FIG. 4. Complete amino acid sequence of human and murine GCP-2. The complete primary protein structure of human GCP-2 was aligned after NH₂-terminal sequence analysis and sequencing of internal peptide fragments obtained by chemical digestion with formic acid or enzymatic cleavage with endoproteinase Glu-C, Lys-C, or Asp-N. The complete amino acid sequence of murine GCP-2 was obtained after NH₂-terminal sequence analysis and by sequencing after formic acid digestion. Unidentified amino acids are marked with an X.

Human GCP-2 was synthesized by solid-phase peptide synthesis on a model 431A peptide synthesizer (Applied Biosystems) using Fmoc chemistry. Solid-phase peptide synthesis involves five steps: chain assembly, cleavage of the peptide from the resin and removal of side-chain protecting groups, purification of crude peptide, additional chemical modification (e.g., formation of disulfide bonds), and finally biochemical and biological characterization of the folded and purified chemokine.

Chemokine Chain Assembly

Fmoc chemistry was used to synthesize the primary structure of GCP-2 on a 0.25-mmol scale. The peptide is assembled from the COOH terminal toward the NH$_2$-terminal side. All amino acids used for synthesis contain an Fmoc group to protect the α-amino group. Reactive side chains of amino acids are protected: trityl (Trt) for Asn, Cys, and Gln; *tert*-butyl (tBu) for Ser and Thr; *tert*-butyl ester (OtBu) for Asp and Glu; *tert*-butyloxycarbonyl (Boc) for Lys; and 2,2,5,7,8-pentamethylchroman-6-sulfonyl (Pmc) for Arg.

The COOH-terminal Fmoc-protected amino acid is first linked to an HMP (4-hydroxymethyl-phenoxymethyl-copolystyrene–1% divinylbenzene) resin. Therefore, 2 equivalents of protected amino acid react with 1 equivalent of dicyclohexylcarbodiimide (DCC) to form a symmetric anhydride and dicyclohexylurea (DCU) precipitate. Before the activated amino acid is transferred to the reaction cartridge, 0.1 equivalent dimethylaminopyridine (DMAP) is added to the resin as a coupling catalyst. After coupling of the COOH-terminal amino acid, remaining free hydroxyl functions of the resin are capped with benzoic anhydride in the presence of DMAP. The Fmoc group of the coupled amino acid is removed (deprotection) by 20% piperidine in *N*-methylpyrrolidone (NMP), and the resin is washed with NMP. The next Fmoc-protected amino acid is activated with HBTU [2-(1*H*-benzotriazol-1-yl)-1,1,3,3-tetramethyluronium hexafluorophosphate], HOBt (1-hydroxybenzotriazole), and diisopropylethylamine (DIEA): the amino acid is dissolved in NMP and a solution of HBTU (1 equivalent) and HOBt (1 equivalent) in *N,N*-dimethylformamide (DMF). This solution is transferred to the reaction vessel, where 1.5 equivalents DIEA in NMP is added to initiate the activation. In this coupling step, the activated amino acid reacts with the NH$_2$-terminal amino acid of the growing chemokine peptide chain. Four equivalents of the activated amino acid is added for one equivalent of the peptide chain. After each coupling step, the free NH$_2$-termini of nonreacted peptides are capped with acetic anhydride to avoid formation of peptides that lack internal amino acids, and the resin is washed with NMP. The Fmoc removal, coupling, and capping steps are repeated for each amino acid. After the last coupling step, the Fmoc group is removed from the NH_2-terminal amino acid, and the resin is washed with NMP and dichloromethane (DCM). The resin is dried on the synthesizer.

A 22-min-long coupling program and a deprotection of 15 min were used for the synthesis of human GCP-2. Coupling efficiency is measured by monitoring Fmoc deprotection using a UV spectrophotometer (301 nm). High coupling and deprotection yields are essential to generate a peptide of 75 residues. Assuming that every combined coupling and deprotection step is performed with a yield of 99%, only 47.5% of the generated chemokine is correctly assembled. Whenever troublesome coupling reactions are expected (e.g., for the coupling of successive hydrophobic residues), double coupling steps can be programmed. For example, for the synthesis of GCP-2, double coupling steps were used for cycles 9 to 13 (coupling of I, V, K, K, and L) and cycles 32 to 35 (coupling of V, E, V, and K).

Cleavage of Chemokine Peptide from Resin and Removal of Side-Chain Protecting Groups

Trifluoroacetic acid is used to cleave the synthesized chemokine from the resin and to remove side-chain protecting groups. Therefore, the resin-bound peptide is stirred in a cleavage mixture containing 0.75 g of phenol, 250 μl of ethanedithiol, 500 μl of thioanisole, 500 μl of bidistilled water, and 10 ml of TFA for 90 min at room temperature.[16] Thioanisole is added to accelerate the cleavage of Pmc groups from Arg residues. Ethanedithiol, water, and phenol are added as scavengers to minimize side-chain reactions. After cleavage, synthesized chemokine is separated from the resin by filtration through a medium porosity glass filter, and protein is precipitated in 30 ml cold methyl *tert*-butyl ether (MTBE). The peptide is washed four times with MTBE and dissolved in bidistilled water. After lyophilization of the protein, it can be stored at 4° until purification.

Purification of Crude Synthetic Chemokine

After cleavage of the peptide from the resin and removal of side-chain protecting groups, crude chemokine has to be purified. Intact protein can be purified from incompletely synthesized peptides by RP-HPLC. The peptide is dissolved in 0.1% TFA and injected on a 100×8 mm C_{18} Deltapak (30 nm) column (Waters, Millipore). Peptides are eluted with an acetonitrile gradient [0–80% (v/v) acetonitrile in 0.1% (v/v) TFA; 2 ml/min, 2-ml fractions]. Absorbance at 220 nm is measured as a parameter for protein concentration. Finally, intact chemokine is identified by SDS–PAGE and by NH_2-terminal amino acid sequence analysis.

---

[16] D. S. King, C. G. Fields, and G. B. Fields, *Int. J. Pept. Protein Res.* **36,** 255 (1990).

Folding of Chemokines

Disulfide bridges can be formed by incubation of purified unfolded chemokine (100 μg/ml) in 150 mM Tris-HCl, pH 8.7, containing 1 mM EDTA, 0.3 mM oxidized glutathione, 3 mM reduced glutathione, and 1 M guanidinium chloride for 90 min at room temperature.[17] Folded peptide is purified to homogeneity by RP-HPLC as described above: after incubation, the reaction mixture is injected on a 220 $\times$ 2.1 mm C_8 Aquapore RP-300 column (Applied Biosystems), and proteins are eluted with an acetonitrile gradient [0–80% (v/v) acetonitrile in 0.1% (v/v) TFA; 0.4 ml/min, 0.4-ml fractions]. Absorbance at 220 nm is measured.

Characterization of Synthetic GCP-2 Protein

The fractions obtained after RP-HPLC of folded protein are analyzed by SDS–PAGE under reducing and nonreducing conditions for purity and for the presence of monomeric or multimeric forms of synthetic protein. The relative molecular masses of synthetic GCP-2 and natural GCP-2 (*vide supra*) were identical (6 kDa on SDS–PAGE under reducing conditions). Monomeric forms of the synthetic protein eluting at the same acetonitrile concentration as natural GCP-2 protein on HPLC were further analyzed by NH_2-terminal amino acid sequence analysis and molecular mass analysis using a matrix-assisted laser desorption/ionization (MALDI) mass spectrometer. Synthetic GCP-2 showed a molecular mass of 8076 ± 16 Da, which corresponds to the theoretical molecular mass of 8070 Da. The intact synthetic chemokine was compared with natural GCP-2 for chemotactic activity on neutrophilic granulocytes. No difference in specific activity was observed between synthetic and natural GCP-2. Thus, the synthetic protein is a good alternative to the natural one for further biological characterization.

Assays for Neutrophil Chemotaxis and Activation by C-X-C Chemokines

Isolation of Human and Murine Neutrophilic Granulocytes

Human and murine neutrophilic granulocytes are isolated from fresh heparinized peripheral blood from a single human donor or from pooled mouse blood (obtained by cardiac punctures), respectively. Blood is mixed with 1 volume of PBS, pH 7.4, and 1 volume of hydroxyethyl-starch solution

---

[17] R. Jaenicke and R. Rudolph, *in* "Protein Structure, A Practical Approach" (T. E. Creighton, ed.), p. 206. IRL Press, Oxford, 1989.

(Plasmasteril, Fresenius, Bad Homburg, Germany) and placed at 37° for 30 min to allow sedimentation of erythrocytes. The supernatant is centrifuged at 400 g, and leukocytes are washed with PBS. To separate the granulocytes and mononuclear cells, the leukocytes, suspended in PBS, are loaded on 3 volumes of Ficoll-sodium metrizoate (Lymphoprep, Nycomed Pharma, Oslo, Norway) or Nycodenz solution (Nycoprep Animal, Nycomed Pharma) for human and murine cells, respectively, and centrifuged for 30 min at 400 g. Remaining erythrocytes are removed from the granulocyte pellet by lysis in bidistilled water for 30 sec, and the medium is made isotonic by adding one-third volume of 3.6% (w/v) NaCl solution. After centrifugation, the granulocyte pellet is washed twice with Hanks' balanced salt solution (HBSS; GIBCO) supplemented with 1 mg/ml of human serum albumin (HSA). The granulocytes are used as a source for neutrophils (95% purity) in the microchamber assay at a concentration of 10^6 cells/ml in HBSS supplemented with HSA (1 mg/ml).

Neutrophil Chemotaxis Assay

The neutrophil chemotactic activity of proteins can be tested in the 48-well microchamber chemotaxis assay,[18,19] which is a sensitive but labor-intensive method. To test the chemotactic activity of murine chemokines, human neutrophilic granulocytes may be used. The data obtained for murine chemokines using human cells were confirmed by chemotaxis assays with murine neutrophils.

Crude conditioned media and column fractions containing purified chemokine are tested for neutrophil chemotactic activity in the 48-well microchamber assay[18,19] (Neuro Probe, Cabin John, MD). A 5-μm pore size polyvinylpyrrolidone-free filter (Nuclepore, Pleasanton, CA) is placed in the chamber to separate the lower and the upper compartment. The lower compartment of the microchamber is filled with 27 μl of test sample dilutions (in HBSS + 1 mg/ml HSA), positive control (e.g., the chemokine IL-8), or negative control (dilution buffer). After mounting the micropore filter in the chamber, the upper compartment is filled with 50 μl of cell suspension. The microchamber is incubated for 45 min at 37° in a 5% (v/v) CO_2 incubator. The filter is removed from the chamber, fixed for 2 min in a fixative solution (Hemacolor 1; Merck, Darmstadt, Germany), and stained with Diff-Quick [2 min in color reagent red, Hemacolor 2 (Merck), and 2 min in color reagent blue, Hemacolor 3 (Merck)]. The cells that have migrated through the filter are counted in 10 microscopic fields (magnifica-

---

[18] W. Falk, R. H. Goodwin, Jr., and E. J. Leonard, *J. Immunol. Methods* **33**, 239 (1980).

[19] J. Van Damme and R. Conings, *in* "Cytokines, A Practical Approach" (F. R. Balkwill, ed.), p. 215. IRL Press, Oxford, 1995.

tion: ×500) per well. If single sample dilutions (e.g., for multiple column fractions) are tested, chemotactic activity is expressed as the chemotactic index, being the number of cells that migrated toward the sample divided by the number of cells that migrated to the negative control. In case a dilution series of the sample is tested, bioactivity can also be expressed in units/ml (U/ml), with 1 U/ml corresponding to the half-maximal migration index. Subsequently, a specific activity (U/mg protein) can be calculated.

Using this assay, neutrophil chemotactic activity of different NH_2-terminally truncated forms of natural human GCP-2 were compared (Fig. 5A). No difference in neutrophil chemotactic activity was observed for the different truncated forms of human GCP-2. Fractions of mouse GCP-2 containing shorter forms (S) and longer forms (L) were tested in parallel with synthetic GCP-2 (75 amino acids) (Fig. 5B). The data show that NH_2-terminal truncation of mouse GCP-2 results in enhanced neutrophil chemotactic activity. Mouse GCP-2(S) was also more potent than human GCP-2. This indicates that the truncated forms of mouse GCP-2 could function as a major neutrophil chemotactic factor in the mouse.[5]

Release of Gelatinase B

Release of gelatinase B[20] can be used as a parameter for neutrophil activation. Purified neutrophils (135 μl), suspended at 5×10^6 cells/ml in HBSS plus 1 mg/ml HSA, are stimulated with chemokine samples (15 μl) at different dilutions for 30 min at 37° in microtiter plates. After incubation, 100 μl of cell supernatant is centrifuged to remove suspended cells, and 70 μl of clarified supernatant is frozen at $-20°$ until determination of gelatinase B activity. Gelatinase B activity can be determined by SDS–PAGE zymography with copolymerized gelatin as the substrate. Twenty test samples (5 μl) are diluted with 5 μl of distilled water and 10 μl of sample buffer containing 125 mM Tris-HCl, pH 6.8, 4% (w/v) SDS, 0.1% (w/v) bromphenol blue, and 0.25 M sucrose. Samples are run on polyacrylamide gels [7.5% T, 3.3% C, 1% (w/v) SDS] in which 0.1% (w/v) of gelatin substrate is copolymerized. Stacking gels (3% T, 3.3% C, 0.5% SDS) do not contain gelatin. After electrophoresis at 4° for 16 hr at 80 to 85 V (Protean II system, Bio-Rad Laboratories), the gel is washed twice for 20 min in a 50 mM Tris-HCl, 10 mM CaCl$_2$, 0.02% (w/v) NaN$_3$, pH 7.5, buffer (washing buffer) containing 2.5% (v/v) Triton X-100 to remove SDS and to reactivate enzymes. After incubation for 24 hr at 37° in washing buffer containing 1% (v/v) Triton X-100, zones of gelatin degradation are visualized by staining for 30 min with 0.25% (w/v) Coomassie Brilliant Blue R-250 in 45% (v/v)

[20] S. Masure, A. Billiau, J. Van Damme, and G. Opdenakker, *Biochim. Biophys. Acta* **1054,** 317 (1990).

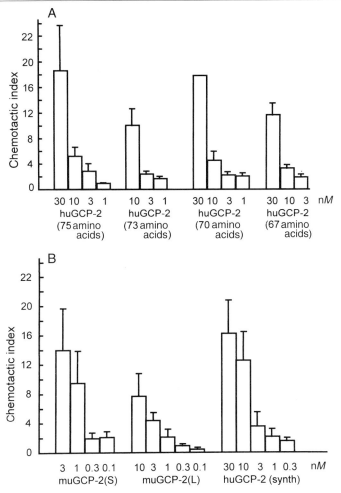

FIG. 5. Chemotactic activity of truncated forms of human and murine GCP-2. (A) Four truncated forms (Fig. 3A) of natural human GCP-2 (75, 73, 70, and 67 amino acids) were compared for chemotactic activity on neutrophilic granulocytes in the microchamber chemotaxis assay. (B) Fractions of murine GCP-2 containing shorter (S) and longer (L) NH$_2$- terminal forms (Fig. 3B) were tested for neutrophil chemotactic activity in parallel with synthetic human GCP-2 (75 residues) in the microchamber assay. Chemotactic indices are derived from five independent experiments (mean ± SEM). Within each experiment, the chemotactic index is calculated from triplicate cultures.

methanol/10% (v/v) acetic acid and destaining with 30% (v/v) methanol/
10% (v/v) acetic acid. Coomassie Blue stains gelatin, and gelatinase activity
appears as unstained zones on a blue background.

Quantification of gelatinase activity is achieved by scanning of negatively
stained gels, for example, with a CAMAG TLC Scanner 88 densitometer
(Muttenz, Switzerland) equipped with a transmission option 76 760 and on-
line data acquisition and processing with the appropriate software package.
Scanning speed is usually 5 mm/sec. The scanning wavelength is set at 600
nm. Gelatinase activity is expressed as scanning units corresponding to the
area under the curve, which is an integration ratio that takes into account
both brightness and width of the substrate lysis zone.

Human synthetic GCP-2, natural IL-8 and natural GROα, and the
natural murine chemokines GCP-2(S) and GCP-2(L) were compared in
the gelatinase B release assay. The data obtained show that IL-8 is more
potent than GROα and GCP-2 for the induction of gelatinase B release
in human neutrophils. Mouse GCP-2(S) showed a higher specific activity
than mouse GCP-2(L) and was also more potent than human GCP-2 (Fig.
6). The data correspond to the results obtained in the chemotaxis assay.

Regulation of Human GCP-2 Transcription in Cells: Quantitation of mRNA Levels by Reverse Transcription–Polymerase Chain Reaction

Amplification of individual RNA molecules can be achieved by a
method that combines reverse transcription (RT) and the polymerase chain
reaction (PCR), called RT-PCR. This method has been demonstrated to
be extremely sensitive for mRNA analysis, and quantitative information
can be obtained. Therefore, this technique is a powerful tool to study the
regulation of transcription.[21] In the PCR step the mRNA of interest is
specifically amplified by using two gene-specific oligonucleotide primers
that anneal in two different exons to discriminate between genomic DNA.
Relative mRNA levels can be measured by relating the amplicons derived
from the mRNA of interest to those of a housekeeping gene. The latter
functions as an external control because its transcription is not affected
by inducers.

As an example, the lung carcinoma cell line A549 (ATCC, Rockville,
MD, CCL 185) and the amnion-derived Wish cell line (ATCC CCL 25),
in which GCP-2 mRNA induction has been studied, were cultured for 4
days. Cell cultures were induced for 6 hr at 37° with 10 μg/ml LPS, 40
U/ml interleukin-1β (IL-1β), or 50 ng/ml phorbol 12-myristate 13-acetate
(PMA). As a control for basal mRNA induction, cells were left untreated.

[21] A. M. Wang, M. V. Doyle, and D. F. Mark, *Proc. Natl. Acad. Sci. U.S.A.* **86,** 9717 (1989).

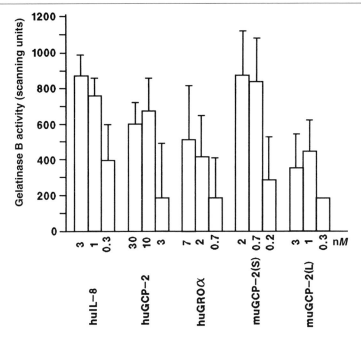

Fig. 6. Induction of gelatinase B release by human C-X-C chemokines and murine GCP-2 from neutrophils. Purified neutrophils were stimulated with chemokines at different concentrations. Gelatinase B activity was measured by SDS–PAGE zymography. Enzyme activity is expressed in scanning units after subtraction of values obtained with control unstimulated neutrophils (mean of four independent experiments ± SEM).

Total RNA was extracted with TRIzol Reagent (Life Technologies, Gaithersburg, MD), and the concentration was measured with the GeneQuant RNA/DNA Calculator (Pharmacia). One microgram total RNA was reverse transcribed in $1\times$ RT buffer (0.14 M KCl, 8 mM MgCl$_2$, 50 mM Tris-HCl, pH 8.1), 5 mM DTT, 0.15 mM dNTPs, 1 μg random hexamer primers (Life Technologies), 50 U human placental ribonuclease inhibitor, and 4 U RAV-2 reverse transcriptase (both from Amersham, Buckinghamshire, UK) in a final volume of 50 μl. The mixtures were kept at 23° for 10 min and then incubated for 1 hr at 42°. The enzyme was denatured at 95° for 5 min. In a first PCR, the amount (volume) of first-strand cDNA that was necessary to yield equal amounts of amplicons of the housekeeping gene β-actin [primers Huβ-ACTINF (5'-TACAATGAGC TGCGTGTGGC TCCCG-3') and Huβ-ACTINB (5'-AATGGTGATG ACCTGGCCGT CAGGC-3')] was determined. Subsequently, the relative amounts of human GCP-2 [primers HuGCP2FOR (5'-GCACTTGTTT ACGCGTTACG CTG-3') and HuGCP2BACK (5'-CCACTGTCCA

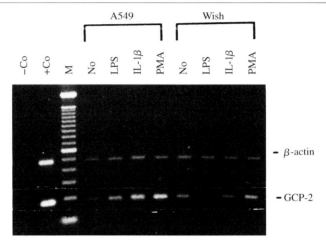

FIG. 7. Semiquantitative determination of GCP-2 mRNA induction in human epithelial cell lines. A549 and Wish cells were induced with 10 μg/ml LPS, 40 U/ml IL-1β, or 50 ng/ml PMA or were left untreated (No). Six hours later total RNA was extracted, and a semiquantitative analysis of GCP-2 mRNA was performed using β-actin-specific external control primers. The number of PCR cycles performed was 32 for β-actin and 38 for GCP-2 cDNAs. In each PCR, a negative (−Co) and positive plasmid (+Co) control were included. Lane M, 100 bp marker (GIBCO).

AAATTTTCTG GATGAC-3′) based on the cDNA sequence] mRNA molecules were quantified. The conditions applied in all PCRs were 30 sec at 95°, 40 sec at 62°, and 40 sec at 72° with a final extension of 5 min at 72°, done on a Gene Amp PCR System 2400 (Perkin-Elmer, Foster City, CA) for 32 (β-actin) or 38 (GCP-2) cycles.

From Fig. 7, it is clear that the amount of RNA used in the RT-PCR was the same in all samples because the β-actin amplicons (479 bp) showed equal intensities on an ethidium bromide-stained agarose gel. From the GCP-2 amplicons (183 bp) it can be seen that all three inducers caused an increase in the GCP-2 mRNA levels in A549 cells, with PMA and IL-1β being better inducers than LPS. In Wish cells, LPS decreased and PMA increased the mRNA levels, whereas IL-1β did not show any effect. In several other studies, results obtained by this new technique were compared with those of more classic quantitation methods such as Northern and slot-blot analyses. The quantitative capability of RT-PCR was confirmed in these experiments.[22,23] Furthermore, the reproducibility of the RT-PCR

[22] K. E. Noonan, C. Beck, T. A. Holzmayer, J. E. Chin, J. S. Wunder, I. L. Andrulis, A. F. Gazdar, C. L. Willman, B. Griffith, D. D. Von Hoff, and I. B. Roninson, *Proc. Natl. Acad. Sci. U.S.A.* **87,** 7160 (1990).

[23] L. D. Murphy, C. E. Herzog, J. B. Rudick, A. T. Fojo, and S. E. Bates, *Biochemistry* **29,** 10351 (1990).

has been demonstrated in our laboratory for the quantitation of other cytokine mRNAs.[24] In the latter study, the quantitation was even performed with an internal control to obtain absolute values of mRNA molecules. The method described here is especially useful for the quantitation of cytokine mRNA molecules because these are present at low abundance and have a short half-life. The technology can be applied when the number of cells obtained from primary *in vivo* sources is limited.

Conclusion

Novel chemokines can be isolated and identified from natural cellular sources by a four-step chromatographic procedure. The described purification procedure yields not only several homogeneous chemokines (e.g., IL-8, GCP-2, GRO)[2] at the same time, but also results in the separation of different isoforms of the C-X-C chemokine GCP-2, which is chemotactic for neutrophils. Isolated proteins are identified by amino acid sequence analysis and are biologically characterized by *in vitro* tests (microchamber chemotaxis assay, induction of gelatinase B release). Under similar conditions but using monocytes as responder cells, different monocyte chemotactic proteins (MCP-1, -2, -3) were discovered (see Van Damme *et al.*[25]). The isolation procedure is further discussed and illustrated in Section II of this volume.

For a detailed evaluation of novel chemokines *in vitro* (study of all possible target cells, gene regulation, receptor usage) and to perform *in vivo* experiments, sufficient pure protein can be obtained by chemical synthesis as an alternative for purification of natural chemokine. The labor-intensive nonspecific biotests (chemotaxis, enzyme release) do not allow one to study chemokine gene regulation. However, after cDNA cloning, this can be done by quantitation of mRNA levels by RT-PCR. Finally, as soon as antibodies are available, the production of chemokine can be measured *in vitro* and *in vivo* using specific immunotests as described for MCPs (see Van Damme *et al.*[25]).

Acknowledgments

This work was supported by the Concerted Research Actions of the Regional Government of Flanders, the Interuniversity Attraction Pole of the Belgian Federal Government (IUAP), the Belgian National Fund for Scientific Research (NFWO; including "Levenslijn"), the General Savings and Retirement Fund (ASLK), and the European Community. A. Wuyts is a research assistant of the Belgian NFWO.

[24] N.-M. Zhou, P. Matthys, C. Polacek, P. Fiten, A. Sato, A. Billiau, and G. Froyen, *Cytokine* **9,** 212 (1997).
[25] J. Van Damme, S. Struyf, A. Wuyts, G. Opdenakker, P. Proost, P. Allavena, S. Sozzani, and A. Mantovani, *Methods Enzymol.* **287** [8] 1997 (this volume).

[3] Synthesis of Chemokines by Native Chemical Ligation

By PHILIP E. DAWSON

Introduction

The chemical synthesis of proteins has been developed to complement the recombinant expression of proteins for structure–function studies. Solid-phase peptide synthesis[1] has been used to produce both natural and unnatural analogs of several proteins to investigate specific aspects of protein function.[2] In particular, several members of the chemokine family have been synthesized,[3] resulting in many insights into the molecular basis of chemokine function.[4] Despite these successes, the application of chemical synthesis to the study of proteins has been limited by the difficulty of synthesizing polypeptides that are greater than approximately 50 amino acids in length.

To overcome this limitation, chemical ligation techniques have been introduced that couple unprotected peptides in denaturing aqueous solution.[5] Several ligation strategies have been developed, each exploiting a different chemoselective reaction at the site of ligation. These reactions produce a nonnative thioester,[5] oxime,[6] thiazolidine,[7] nonnative amide,[8] or thioether[9] bond at the site of ligation, and they have been used to synthesize a wide variety of protein analogs up to 200 amino acids in length.[10,11] A method for synthesizing polypeptides with only native peptide bonds has also been introduced.[12] In this chapter, the native chemical ligation tech-

[1] R. B. Merrifield, *J. Am. Chem. Soc.* **85,** 2149 (1963).

[2] T. W. Muir and S. B. H. Kent, *Curr. Opin. Biotechnol.* **4,** 420 (1992).

[3] I. Clark-Lewis, L. Vo, P. Owen, and J. Anderson, *Methods Enzymol.* **287** [16], 1997 (this volume).

[4] I. Clark-Lewis, B. Dewald, M. Loetscher, B. Moser, and M. Baggiolini, *J. Biol. Chem.* **269,** 16075 (1994).

[5] M. Schnölzer and S. B. H. Kent, *Science* **256,** 221 (1992).

[6] K. Rose, *J. Am. Chem. Soc.* **116,** 30 (1994).

[7] J. Shao and J. P. Tam, *J. Am. Chem. Soc.* **117,** 3893 (1995).

[8] C.-F. Liu, C. Rao, and J. P. Tam, *J. Am. Chem. Soc.* **118,** 307 (1996).

[9] D. R. Englebretsen, B. G. Garnham, D. A. Bergman, and P. F. Alewood, *Tetrahedron Lett.* **36,** 8871 (1995).

[10] M. Baca, T. W. Muir, M. Schnölzer, and S. B. H. Kent, *J. Am. Chem. Soc.* **117,** 1881 (1995).

[11] L. E. Canne, A. R. Ferre-D'Amare, S. K. Burley, and S. B. H. Kent, *J. Am. Chem. Soc.* **117,** 2998 (1992).

[12] P. E. Dawson, T. W. Muir, I. Clark-Lewis, and S. B. H. Kent, *Science* **266,** 776 (1994).

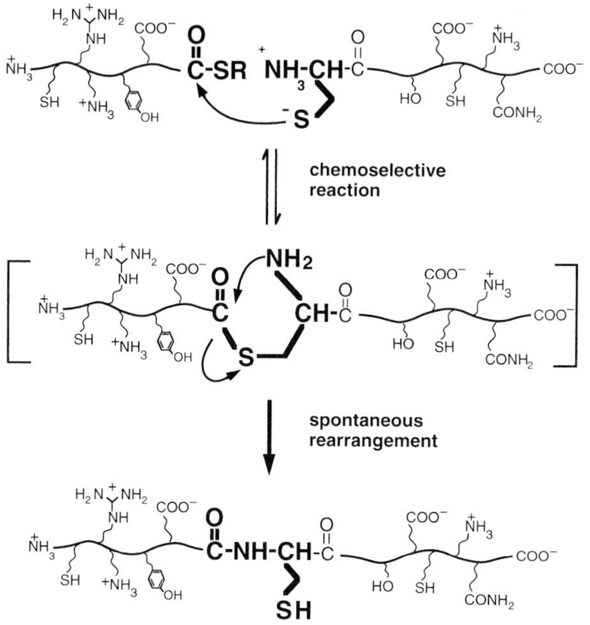

FIG. 1. The native chemical ligation reaction. [Reprinted with permission from P. E. Dawson, T. W. Muir, I. Clark-Lewis, and S. B. H. Kent, *Science* **266,** 776 (1994). Copyright 1994 American Association for the Advancement of Science.]

nique is described, focusing on its application to the synthesis of chemokines.

Description of Method

Strategy of Native Chemical Ligation

Native chemical ligation allows two completely unprotected peptides to be joined through formation of a native peptide bond at the ligation site.[12] This strategy is schematically illustrated in Fig. 1. The ligation is initiated by a chemoselective reaction between the N-terminal cysteine thiol of one peptide and a thioester functionality at the C-terminal residue (Xaa) of a second peptide. Exchange of the peptide thioester with the thiol side chain gives a thioester-linked intermediate as the initial covalent product. This thioester-linked intermediate then undergoes a spontaneous rearrangement, involving an *S*- to *N*-acyl transfer,[13] to generate a native

[13] T. Wieland, E. Bokelmann, L. Bauer, H. U. Lang, and H. Lau, *Liebigs Ann. Chem.* **583,** 129 (1953).

peptide bond at an Xaa-Cys site in the polypeptide sequence. Native chemical ligation is compatible with all side-chain functionalities found in proteins, including multiple cysteine side chains, and can be performed in denaturing aqueous solutions such as 6 M guanidine hydrochloride, permitting high peptide concentrations ($\sim$1 mM) to be used.

Thiol Additives

In the first step of native chemical ligation, a reversible thiol exchange reaction takes place to form a new thioester bond (Fig. 1). Peptides in a typical ligation reaction often contain several cysteine residues, each of which has a free thiol capable of thioester exchange. Because only the N-terminal cysteine residue can rearrange to form an amide bond, attack by another thiol in the ligation reaction simply produces a new thioester that is still capable of further thiol exchange. To facilitate this reversible exchange, a thiol additive is added to the reaction mixture, typically a mixture of thiophenol and benzyl mercaptan.[14] The thiophenol additive maintains a population of highly reactive phenyl thioester peptides, whereas the benzyl mercaptan prevents the cysteine side chains from forming unreactive disulfide bonds during ligation.[15,16] Another advantage of using this thiol additive is that weakly activated thioester peptides can be synthesized and stored for several months until they are converted to more reactive phenyl thioester peptides during the ligation reaction.[14]

Selection of Ligation Site

The native chemical ligation reaction requires a cysteine residue at the site of ligation.[17] For this reason, proteins that contain a cysteine in the middle of the polypeptide sequence are well suited for this technique. If a protein does not contain a cysteine in the middle of the sequence, one must be introduced into the polypeptide sequence. Proteins are extremely tolerant to individual substitutions in their sequence. As a result, cysteine residues can often be introduced into a synthetically convenient position

[14] P. E. Dawson, M. Churchill, R. M. Ghadiri, and S. B. H. Kent, *J. Am. Chem. Soc.* **119**, 4325 (1997).

[15] Phosphines such as [Tris(2-carboxyethyl)phosphine, hydrochloride] TCEP (Strem Chemicals, Newburyport, MA) can be added to reduce any disulfide bonds after ligation as discussed by Dawson *et al.*[12]

[16] Phosphines have also been added during ligation as a reducing agent. J. P. Tam, *Proc. Natl. Acad. Sci. U.S.A.* **92**, 12485 (1995).

[17] A promising extension has been reported that permits Gly-Xaa and Xaa-Gly sites to be utilized. This extension has not yet been applied to proteins [L. E. Canne, S. J. Bark, and S. B. H. Kent, *J. Am. Chem. Soc.* **118**, 55891 (1996)].

in the polypeptide chain. Knowledge of the three-dimensional structure of the protein or sequence conservation among similar proteins can aid in the selection of a nonperturbing substitution. This approach was utilized in the synthesis of the 110-amino acid polypeptide chain of the ribonuclease barnase.[14] A Gly[49]-Lys[50] to Gly[49]-Cys[50] substitution was chosen because residue 50 had been structurally characterized as remote from the active site of the ribonuclease.[18] In cases where a free cysteine side chain is undesirable, the thiol can be modified after the ligation reaction using a variety of alkylating agents to produce both hydrophobic and hydrophilic side chains.[19]

Another consideration in the selection of a ligation site concerns the synthesis of an α-carboxyl thioester peptide. The optimum choice for the C-terminal amino acid of this peptide is a nonfunctionalized and nonhindered amino acid such as Gly,[14] Ala,[12] or Leu.[20] If one of these amino acids is not suitable for a particular peptide sequence, a variety of amino acid side chains have been used at this position, including Asn and Lys.[15,21] Even the β-branched amino acid Val has been used in model systems although longer reaction times were required, presumably owing to the steric hindrance of the branched side chain. Not all side chains have been examined, but there appears to be considerable synthetic flexibility at this position of the polypeptide sequence.

Application to Chemokines

Chemokines are especially well suited for synthesis by native chemical ligation. Chemokines are 70- to 80-amino acid proteins with two disulfide bonds that are separated into two families based on the relative positions of the first two cysteine residues. These cysteines are either separated by one amino acid (C-X-C or α family) or are adjacent (C-C or β family). The third cysteine in both families is located roughly in the middle of the sequence; in interleukin-8, for example, it is located at residue 34 out of a total of 72 amino acids. Because of its central location, this cysteine can be utilized in native chemical ligation. The ligation of interleukin-8 is schematically represented in Fig. 2. This strategy should be applicable for both α- and β-chemokines.

[18] D. Sali, M. Bycroft, and A. R. Fersht, *J. Mol. Biol.* **220**, 779 (1991).
[19] S. S. Wong, "Chemistry of Protein Conjugation and Cross-Linking." CRC Press, Boca Raton, Florida, 1991.
[20] W. Lu, M. A. Qasim, and S. B. H. Kent, *J. Am. Chem. Soc.* **118**, 8518 (1996).
[21] M. A. Siani, D. A. Thompson, L. E. Canne, G. M. Figliozzi, B. Robson, S. B. H. Kent, and R. J. Simon, poster presented at the IB3 Chemokines 3 Conference, San Francisco, California, October 1996.

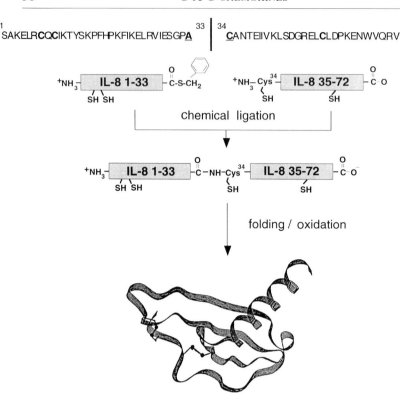

SAKELRCQCIKTYSKPFHPKFIKELRVIESGPA | CANTEIIVKLSDGRELCLDPKENWVQRVVEKFLKRAENS

FIG. 2. Synthetic strategy for the chemical ligation of interleukin-8. The peptide corresponding to [Ala33]IL-8(1–33)COSbenzyl was reacted with IL-8(34–72), which contained a cysteine
residue at the N terminus. The resulting product contained a native peptide bond at the site
of ligation. The H33A IL-8 polypeptide was folded by air oxidation to produce the correct
disulfide-linked product. [Reprinted with permission from P. E. Dawson, T. W. Muir, I. Clark-
Lewis, and S. B. H. Kent, *Science* **266,** 776 (1994). Copyright 1994 American Association for
the Advancement of Science.]

Method

General Equipment

Manual peptide synthesis reaction vessel[22]
HF apparatus
Analytical/semipreparative high-performance liquid chromatography
    (HPLC) setup

[22] M. Schnolzer, P. Alewood, D. Alewood, and S. B. H. Kent, *Int. J. Pept. Protein Res.* **85,**
180 (1992).

Electrospray or matrix-assisted laser desorption/ionization (MALDI) mass spectrometer

Lyophilizer

Magnetic stir plate and stir bar

Reaction vessels, typically an Eppendorf tube or glass scintillation vial depending on scale

Synthesis of Amino-Terminal Cysteine Peptides

The peptide corresponding to the C-terminal half of the desired polypeptide product can be synthesized by a variety of standard procedures. In the syntheses performed to date, these peptides have all been synthesized using highly optimized *in situ* neutralization protocols for Boc (*tert*-butoxycarbonyl) chemistry solid-phase peptide synthesis. Procedures for both manual and machine-assisted peptide synthesis have been described in detail elsewhere.[22] Crude peptide products are purified by reversed-phase HPLC and characterized by electrospray mass spectrometry. In addition to these protocols, peptides derived from Fmoc (9-fluorenylmethoxycarbonyl) chemistry or even from expression in genetically engineered microorganisms should be applicable to synthesis of peptides containing an N-terminal cysteine.

Notes: N-terminal cysteine residues are susceptible to thiazolidine formation. As shown in Fig. 3, the N-terminal cysteine can react with aldehydes such as formaldehyde, and ketones such as acetone, to form a thiazolidine ring that is no longer competent for native chemical ligation. This side reaction can be largely prevented by taking the following precautions: (1) Avoid the use of acetone when washing *all* glassware. (2) Do not use acetone/dry ice to freeze peptide solutions for lyophilization. (3) Avoid the *N*-π-benzyloxymethyl (Bom) protecting group for the histidine side chain. During HF cleavage an equivalent of formaldehyde is released, which is inefficiently scavenged by additives such as resorcinol. In addition, this protecting group should not be used when synthesizing the thioester-containing peptide because peptides synthesized using the Bom group have been observed to retain a reactive equivalent of formaldehyde even after

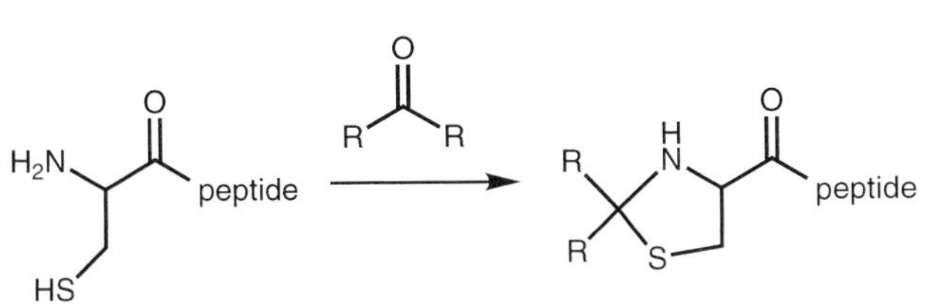

FIG. 3. Thiazolidine formation.

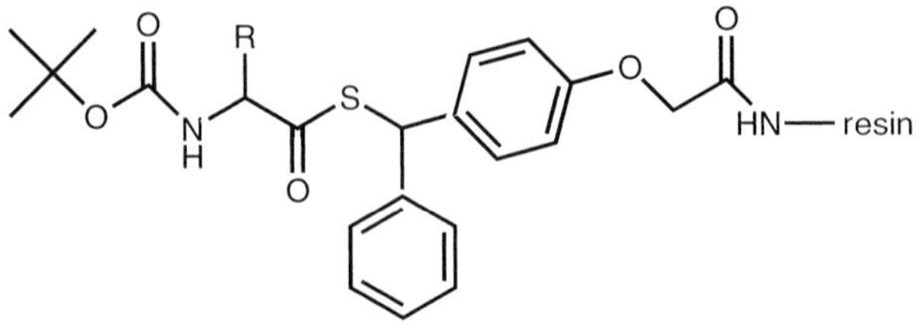

FIG. 4. Thio acid linker.

HPLC purification. (4) Use high-purity solvents and reagents throughout the handling of the peptide. This is particularly important for denaturants such as urea and guanidine hydrochloride because they are used in high concentration.

Synthesis of Thioester Peptides: General Considerations

The α-carboxyl thioester peptides are synthesized from thio acid peptides using the same highly optimized *in situ* neutralization Boc chemistry protocols used for the N-terminal cysteine-containing peptide.[23] One exception to the synthetic procedure regards selection of side-chain-protecting groups. Thioester bonds are labile to nucleophiles during synthesis and deprotection, which prevents histidine dinitrophenyl (DNP) and tryptophan formyl protecting groups from being removed by nucleophiles before cleavage from the resin. As a result, tryptophan residues are left unprotected[24] and the DNP group on the histidine side chain is left on during HF cleavage. The His(DNP) protecting group is removed efficiently from the peptide under native ligation conditions.[25] The histidine Bom-protecting group, His(Bom), should be avoided as a result of side reactions with N-terminal cysteine residues (see above).

Synthesis of Thioester Peptides through Alkylation of Thio Acid Peptide

Peptides with an α-carboxy thio acid can be alkylated in aqueous buffer to give the desired thioester product. These peptides can be synthesized using a Boc-amino acid thio acid linker (Fig. 4).[26] This Boc-amino acid

---

[23] Fmoc chemistry has not been utilized because of the susceptibility of thioester peptides to aminolysis by piperidine. Use of a resin designed to be cleaved by nucleophiles following synthesis may allow thioester peptides to be synthesized by Fmoc protocols.

[24] Small amounts of alkylation of the tryptophan side chain are observed during synthesis. In principle, the formyl protecting group could be removed either following or during ligation.

[25] T. Hackeng, personal communication (1996).

[26] J. Blake, *Int. J. Pep. Protein Res.* **17,** 273 (1981).

linker must be synthesized for each amino acid desired at the C terminus of the thioester peptide. The procedure has been described in detail elsewhere.[27] The Boc-amino acid linker is then coupled to aminomethyl resin[28,29] to produce the peptide–thio acid yielding resin.

Boc-Amino Acid Thio Acid Resin

Reagents

Boc-amino acid thio acid linker
Aminomethyl resin (0.8–1.0 mmol/g)
2-(1-*H*-benzotriazol-1-yl)-1,1,3,3-tetramethyluronium hexafluorophosphate (HBTU) (Richeleau, Montreal, Canada)
Dimethylformamide (DMF) (spectrophotometric grade)
Ninhydrin reagents[29,30]

Procedure

1. Dissolve 1.5 equivalents of Boc-amino acid thioester linker · dicyclohexylamine and 1.3 equivalents of HBTU in a minimal volume of DMF.
2. Add to 1 equivalent of aminomethyl resin (0.8–1.0 mmol/g) preswollen in DMF.
3. After 1 hr, check the coupling by ninhydrin analysis[29,30] and continue the reaction until greater than 99.5% complete.

Peptide Synthesis

Reagents

Trifluoroacetic acid (TFA) (Biograde, Halocarbon)
Dichloromethane (spectrophotometric grade)
Methanol (spectrophotometric grade)
Diisopropylethylamine (DIEA) (Applied Biosystems, Foster City, CA)

Procedure

1. Swell the Boc-amino acid thio acid resin in DMF for 10 min.
2. Remove the Boc group with 100% TFA (two times, 1 min each time). Flow wash resin twice for 20 sec each time with DMF.

[27] L. E. Canne, S. M. Walker, and S. B. H. Kent, *Tetrahedron Lett.* **36,** 1217 (1995).
[28] A. R. Mitchell, S. B. H. Kent, B. W. Erickson, and R. B. Merifield, *Tetrahedron Lett.* **42,** 3795 (1976).
[29] J. Tam, *Methods Enzymol.* **146,** 127 (1987).
[30] V. K. Sarin, S. B. H. Kent, J. P. Tam, and R. B. Merrifield, *Anal. Biochem.* **117,** 147 (1981).

3. Synthesize the peptide using either manual or machine-assisted *in situ* neutralization synthetic cycles for Boc chemistry.[22]
4. Remove the final Boc group with 100% TFA (two times, 1 min each). Flow wash resin twice for 20 sec with DMF.
5. Neutralize with 10% DIEA in DMF (two times, 1 min each).
6. Wash the peptide resin using a 20-sec flow wash of DMF followed by a 20-sec flow wash of 1:1 (v/v) dichloromethane:methanol.
7. Dry the resin for at least 1 hr by vacuum desiccation.
8. Deprotect the amino acid side chains and cleave from the resin using HF (see below).

Notes on Side-Chain Deprotection and Cleavage from Resin. During HF cleavage thiol scavengers should be avoided; typically, 5–10% (v/v) *p*-cresol or anisole is used as a scavenger. Following treatment with HF for 1 hr in an ice bath, the HF should be removed as quickly as possible, ideally in less than 10 min. The peptide should then be precipitated in cold ether and filtered through a clean fritted funnel. Because thio acid peptides can be prone to hydrolysis under oxidizing conditions, it is important to avoid suction of air through the funnel following filtration of the peptide. Finally, the peptide should be dissolved in 40% (v/v) acetonitrile in water containing 0.1% TFA and lyophilized. Thio acid peptides should be stored at $-20°$ until they are converted to more stable thioester peptides for prolonged storage.

Alkylation of Thio Acid Peptides

Reagents

Thio acid peptide
Alkylation buffer: 6 *M* guanidine hydrochloride (ultrapure grade, ICN, Costa Mesa, CA), 100 m*M* sodium acetate, pH 4.0
Bromoacetic acid (Aldrich, Milwaukee, WI)

Procedure

1. Dissolve the crude thio acid peptide in alkylation buffer to give an approximately 5 mg/ml peptide solution.
2. Add 5 equivalents of bromoacetic acid and mix with a pipette.[31]
3. Monitor the reaction at 30-min intervals by analytical HPLC and mass spectrometry if possible.

---

[31] The alkylation reaction can also be performed using benzyl bromide as the alkylating agent, producing a benzyl thioester peptide. Complications can arise owing to the poor solubility of benzyl mercaptan in aqueous solutions and its greater reactivity toward the peptide side chains.

4. Purify the crude thioester peptide by semipreparative or preparative HPLC and lyophilize.

Notes. The alkylation reaction can be monitored by analytical reversed-phase HPLC at 30-min intervals and is usually complete within an hour. The thioester product typically elutes later than the thio acid peptide, but the product should also be characterized by mass spectrometry. It is advisable to perform a trial reaction with approximately 1 mg of the crude thio acid peptide. The alkylation reaction can be scaled up using the same peptide and bromoacetic acid concentrations, followed by purification by semipreparative or preparative reversed-phase HPLC. Lyophilized thioester peptides can typically be stored for at least a month at $-20°$.

Direct Synthesis of Thioester Peptides. Hojo and colleagues[32] developed a procedure for synthesizing thioester peptides directly from a modified peptide resin. The resin is formed by attaching the preformed dipeptide analog $BocXaa-SC(CH_3)_2CH_2COOH$ to norleucine-loaded 4-(methyl) benzhydrylamine (MBHA) resin, where Xaa is the desired C-terminal amino acid. Thioesters generated by this process should be exchangeable with thiols such as thiophenol to generate more reactive thioester peptides.

Procedure for Ligation of Peptides

Reagents

1 equivalent of purified and lyophilized thioester peptide
1 equivalent of purified and lyophilized N-terminal cysteine-containing peptide
Ligation buffer: 6 M guanidine hydrochloride (ultrapure grade, ICN), 100 mM sodium phosphate, pH 7.5
Thiol additives: thiophenol and benzyl mercaptan (Aldrich)

Procedure

1. Weigh the two lyophilized peptides into the same reaction vessel.
2. Add the ligation buffer to a final concentration of >1 mM (5–10 mg/ml) in each peptide.
3. Immediately add 3% thiophenol and 1% benzyl mercaptan by volume.
4. Vortex for 2 sec to fully dissolve peptides and to saturate the buffer with the thiol additive.
5. Remove 2 μl for analytical reversed-phase HPLC.
6. Close reaction vessel tightly.

---

[32] H. Hojo, Y. Kwon, Y. Kakuta, S. Tsuda, I. Tanaka, K. Hikichi, and S. Aimoto, *Bull. Chem. Soc. Jpn.* **66,** 2700 (1993).

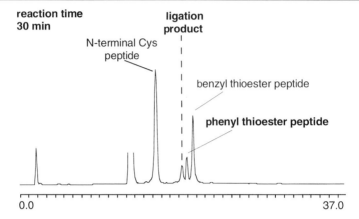

FIG. 5. Ligation reaction in progress, as shown by an HPLC trace of the ligation of barnase(1–48)COSbenzyl with [Cys^{49}]barnase(49–110) in the presence of thiophenol after just 30 min. In addition to the starting materials, the polypeptide product [Cys^{49}]barnase(1–110) as well as the phenyl thioester peptide barnase(1–48)COSphenyl were present as characterized by electrospray mass spectrometry. This particular ligation reaction was complete in approximately 5 hr.

7. Store in a ventilated area at room temperature. (For ligation volumes >1 ml, insert a stir bar, and place on a magnetic stir plate.)

8. Continue ligation until >90% complete as monitored by analytical HPLC. Ligation times vary from 5 hr to 36 hr, depending on the C-terminal amino acid of the thioester peptide.[33]

9. The full-length polypeptide product can be purified by gel filtration or HPLC and then oxidized to form the folded chemokine as described elsewhere in this volume.[3]

Monitoring Ligation Reaction. The ligation reaction can be monitored by reversed-phase HPLC of a few microliters (10–20 μg of each peptide) of the reaction mixture. A gradient should be selected that allows separation of the peptide components from each other as well as from the thiol additives. In most cases lengthening the gradient will allow elution of thiol additives before the polypeptides. To reduce the amount of thiol additive present during analysis, a quick centrifugation of the solution can separate the undissolved thiol additive and its oxidation products from the polypeptides in solution. A typical HPLC trace of an incomplete ligation reaction is shown in Fig. 5.

---

[33] During ligation, a white precipitate may form in the ligation mixture. This is a result of oxidation of thiophenol and benzyl mercaptan and can be removed after ligation by filtration.

To be confident of the ligation products, the mass of each peptide HPLC fraction should be determined by electrospray mass spectrometry. Matrix-assisted laser desorption mass spectrometry (MALDI) can also be used on the HPLC fractions and even the crude ligation mixture using a 1:10 dilution into the matrix solution.[34] In the absence of ready access to mass spectrometry, a small-scale reaction containing only the thioester peptide and the thiol additives can be used to determine the HPLC elution times of the various thioester peptide intermediates.

Acknowledgment

The author thanks Stephen Kent, Reyna Simon, and Manuel Baca for helpful discussions during the preparation of this manuscript.

[34] M. C. Fitzgerald, personal communication (1996).

[4] Molecular Approaches to Structure–Function Analysis of Interleukin-8

By Wayne J. Fairbrother and Henry B. Lowman

Introduction

Chemokines in general bind promiscuously to certain subclasses of chemokine receptors. This cross-talk, involving the stimulation of different receptors by the same chemokine [e.g., interleukin-8 (IL-8) binding to neutrophil receptors CXCR1 and CXCR2] as well as the activation of a given receptor by multiple chemokines, may in a normal animal provide a relatively fail-safe set of redundant controls for appropriate immunological responses. However, from the standpoint of rational drug design, an understanding of specific ligand and receptor roles is necessary if predictable therapeutic effects are to be obtained.

We have studied the receptor-binding specificity of IL-8 to its two known neutrophil receptors, CXCR1 and CXCR2 (also known as IL-8 receptors A and B, respectively)[1,2] as a model of chemokine receptor specificity.[3]

[1] W. E. Holmes, J. Lee, W.-J. Kuang, G. C. Rice, and W. I. Wood, *Science* **253,** 1278 (1991).
[2] P. M. Murphy and H. L. Tiffany, *Science* **253,** 1280 (1991).
[3] H. B. Lowman, P. H. Slagle, L. E. DeForge, C. M. Wirth, B. L. Gillece-Castro, J. H. Bourell, and W. J. Fairbrother, *J. Biol. Chem.* **271,** 14344 (1996).

0076-6879/97 $25.00

FIG. 1. Primary structure alignment of IL-8, MGSA, and NAP-2. Conserved residues are shaded. For the alignment shown, 31 residues of MGSA and 32 residues of NAP-2 are identical to the equivalent residues in IL-8.

The two receptors bind IL-8 with similar high affinities ($\sim$2 nM) but differ considerably in their affinities for the homologous chemokines melanoma growth stimulating activity (MGSA) and neutrophil activating peptide-2 (NAP-2); CXCR2 binds both MGSA and NAP-2 with similar affinities to IL-8, whereas CXCR1 interacts only weakly with these chemokines.[4–9] The Glu-Leu-Arg (ELR) motif in the N-terminal region of IL-8 (Fig. 1) was originally identified by alanine scanning mutagenesis as a major determinant for neutrophil binding.[10] However, as both MGSA and NAP-2 have corresponding ELR motifs (Fig. 1) but do not bind significantly to the neutrophil receptor CXCR1, other binding determinants giving rise to the observed receptor specificities must exist. We used an iterative approach of comparing homologous protein structures (IL-8 and MGSA), constructing variants that exchange differing features of the two structures, and analyzing the receptor (CXCR1 vs CXCR2) binding specificities of such variants to determine the region of IL-8 responsible for its observed receptor specificity. These variants demonstrate the degree to which a small number of residues can dramatically alter specificity. The approach may be useful for determining which portions of other chemokine structures direct specific ligand–receptor interactions, and for engineering monospecific or multispecific reagents for binding to these receptors.

[4] B. Moser, C. Schumacher, V. van Tscharner, I. Clark-Lewis, and M. Baggiolini, *J. Biol. Chem.* **266,** 10666 (1991).

[5] J. Lee, R. Horuk, G. C. Rice, G. L. Bennett, T. Camerato, and W. I. Wood, *J. Biol. Chem.* **267,** 16283 (1992).

[6] G. J. LaRosa, K. M. Thomas, M. E. Kaufmann, R. Mark, M. White, L. Taylor, G. Gray, D. Witt, and J. Navarro, *J. Biol. Chem.* **267,** 25402 (1992).

[7] C. Schumacher, I. Clark-Lewis, M. Baggiolini, and B. Moser, *Proc. Natl. Acad. Sci. U.S.A.* **89,** 10542 (1992).

[8] D. P. Cerretti, C. J. Kozlosky, T. V. Bos, N. Nelson, D. P. Gearing, and M. P. Beckman, *Mol. Immunol.* **30,** 359 (1993).

[9] F. Petersen, H.-D. Flad, and E. Brandt, *J. Immunol.* **152,** 2467 (1994).

[10] C. A. Hébert, R. V. Vitangcòl, and J. B. Baker, *J. Biol. Chem.* **266,** 18989 (1991).

Structural Analysis for Design of Variants

Atomic resolution structures have been determined for IL-8, MGSA, and NAP-2 using either solution nuclear magnetic resonance (NMR) or X-ray crystallographic techniques.[11–15] The coordinates of the different structures were compared using the program ALIGN, which uses an iterative procedure to identify structurally equivalent residues in a pair of structures; atoms in these residues then form the basis of a best-fit superposition.[16] The program uses the algorithm of Needleman and Wunsch to identify structural homology and to account for amino acid insertions and deletions.[17] The best-fit superposition of the IL-8 crystal structure[12] and the MGSA solution structure[13] is illustrated in Fig. 2A. Deviations between the Cα coordinates of the individual monomer structures are plotted in Fig. 3. The most significant difference found between the monomer subunits of these structures is in the loop region preceding the first β strand, which is shorter by one residue in MGSA relative to IL-8 (Fig. 1). This region, which we have designated the N-loop, includes IL-8 residues 13–17 and the corresponding MGSA residues 15–18. Observation of this conformational difference, and a similar difference between IL-8 and NAP-2 (Fig. 3), led to the suggestion that this region may be responsible for the receptor specificity differences observed among IL-8, MGSA, and NAP-2.[13,15] To test whether this structural difference accounts for the observed differences in the receptor-binding specificities of these chemokines, we constructed variants of IL-8 and MGSA in which portions of the N-loop regions were swapped between the two proteins and tested them for binding to the receptors CXCR1 and CXCR2.

Expression of Mutants

The analysis of a large number of protein variants is facilitated by the use of an integrated mutagenesis and protein expression vector. Such a

[11] G. M. Clore, E. Appella, M. Yamada, K. Matsushima, and A. M. Gronenborn, *Biochemistry* **29,** 1689 (1990).

[12] E. T. Baldwin, I. T. Weber, R. St. Charles, J.-C. Xuan, E. Appella, M. Yamada, K. Matsushima, B. F. P. Edwards, G. M. Clore, A. M. Gronenborn, and A. Wlodawer, *Proc. Natl. Acad. Sci. U.S.A.* **88,** 502 (1991).

[13] W. J. Fairbrother, D. Reilly, T. J. Colby, J. Hesselgesser, and R. Horuk, *J. Mol. Biol.* **242,** 252 (1994).

[14] K.-S. Kim, I. Clark-Lewis, and B. D. Sykes, *J. Biol. Chem.* **269,** 32909 (1994).

[15] M. G. Malkowski, J. Y. Wu, J. B. Lazar, P. H. Johnson, and B. F. P. Edwards, *J. Biol. Chem.* **270,** 7077 (1995).

[16] Y. Satow, G. H. Cohen, E. A. Padlan, and D. R. Davies, *J. Mol. Biol.* **190,** 593 (1986).

[17] S. B. Needleman and C. D. Wunsch, *J. Mol. Biol.* **48,** 443 (1970).

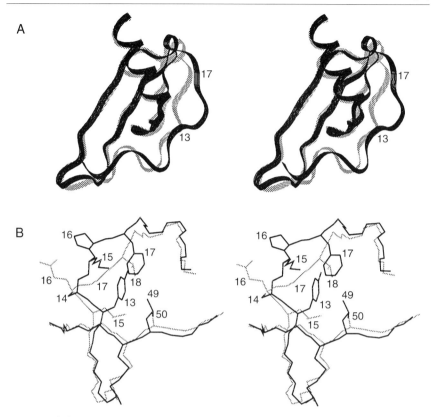

FIG. 2. (A) Stereo view showing ribbon representations of the best-fit superposition of a single subunit of the IL-8 crystal structure (black) and the solution structure of MGSA (gray). The Cα alignment was performed using the program ALIGN using IL-8 residues 7 to 68 and MGSA residues 9 to 69 as initial input. (B) Overlay of the N-loop regions and strand β3 of the IL-8 crystal structure (black) and the MGSA solution structure (gray). IL-8 residues 9 to 12, 18 to 21, and 48 to 51 and the corresponding residues from MGSA (11 to 14, 19 to 22, and 49 to 52) were optimally aligned for this comparison.

system has been used both to generate site-specific variants of IL-8 and to direct the secretion of IL-8 variants into the periplasm of *Escherichia coli* through the use of the *St*II secretion signal sequence.[3] In this case, a plasmid, pPS0170, for expression of IL-8 was constructed from the parental vector phGHam-g3,[18] which contains a β-lactamase (*bla*) gene, as well as origins of replication for both double- and single-strand DNA synthesis (Fig. 4). The former allows for normal plasmid replication and preparation of

[18] H. B. Lowman, S. H. Bass, N. Simpson, and J. A. Wells, *Biochemistry* **30,** 10832 (1991).

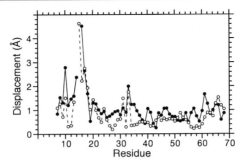

FIG. 3. Deviations between the monomer subunit Cα atomic positions of the IL-8 crystal structure and the MGSA solution (filled circles) and NAP-2 crystal structure (open circles). The Cα alignment was performed using the program ALIGN with IL-8 residues 7 to 68, MGSA residues 9 to 69, and NAP-2 residues 5 to 65 used as initial input. The optimal overlay between IL-8 and MGSA (Fig. 2A) was achieved using IL-8 residues 7 to 14 and 17 to 68, and MGSA residues 9 to 16 and 18 to 69; the resulting best-fit superposition of the Cα atoms for these 60 pairs of residues had a root mean square (RMS) deviation of 1.15 Å. The optimal overlay between IL-8 and NAP-2 was obtained using IL-8 residues 7 to 13 and 18 to 68, and NAP-2 residues 5 to 11 and 15 to 65; the RMS deviation of the resulting best-fit superposition of the Cα atoms for these 58 pairs of residues was 0.84 Å.

double-stranded DNA for long-term storage and for restriction nuclease digestions; the latter permits rapid and efficient site-directed mutagenesis using, for example, the method of Kunkel.[19] Once mutagenesis has been carried out, clones are sequenced to confirm the desired mutations(s). Thereafter, the corresponding plasmid DNA is transformed into a strain of *E. coli* suitable for expression, such as W3110 (ATCC, Rockville, MD). MGSA variants were constructed from a similar starting plasmid pMG34[20] or from another phGHam-g3 derivative, plasmid pH1817.[3]

Protein expression is directed in this vector by the alkaline phosphatase promoter (*PphoA*), for which induction is accomplished by starving cells of phosphate.[21] A 5-ml starting culture is grown in enriched medium, such as YT broth[22] containing the appropriate antibiotic, for 12–16 hr at 37° with shaking. The starting culture is then added to 500 ml of minimal medium such as AP5 [2.2 g/liter casamino acids, 0.3 g/liter yeast extract, 1.5 g/liter glucose, 20 mM NH_4Cl, 1 mM KCl, 1.6 mM $MgSO_4$, 70 mM NaCl, 120 mM triethanolamine (pH 7.4)], and incubated for 20–24 hr, with

[19] T. A. Kunkel, *Proc. Natl. Acad. Sci. U.S.A.* **82,** 488 (1985).
[20] R. Horuk, D. G. Yansura, D. Reilly, S. Spencer, J. Bourell, W. Henzel, G. Rice, and E. Unemori, *J. Biol. Chem.* **268,** 541 (1993).
[21] C. N. Chang, M. Rey, B. Bochner, H. Heynecker, and G. Gray, *Gene* **55,** 189 (1987).
[22] J. Sambrook, E. F. Fritsch, and T. Maniatis, *in* "Molecular Cloning: A Laboratory Manual," 2nd ed., p. A.3. Cold Spring Harbor Laboratory, Cold Spring Harbor, New York, 1989.

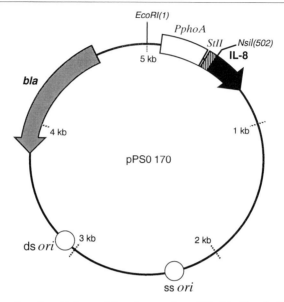

FIG. 4. Expression plasmid for periplasmic secretion of IL-8 in *E. coli*. Plasmid pPS0170 contains a copy of the IL-8 gene (beginning at the *Nsi*I site), with the *St*II secretion signal sequence, under control of the promoter *PphoA*. The plasmid contains origins for single-stranded (ss) as well as double-stranded (ds) DNA replication, and a marker (*bla*) for ampicillin resistance.

shaking, in a 4-liter flask. For IL-8 and many variants, expression is enhanced by inducing at a temperature of 25–30°, although some variants express at higher levels using a temperature of 37°. Poorly expressing variants should be tested for expression levels over a range of temperatures. At the end of the induction period, cells are pelleted and frozen at −20°. If they are not to be processed within 24 hr, cell pellets can be transferred to −70° for storage over periods of months.

Protein Purification

Several protein purification schemes have been described for IL-8 and homologous proteins.[20,23,24] The approach detailed here allows for rapid purification of protein using an *E. coli* periplasmic secretion system. All

[23] C. A. Hébert, F. W. Luscinskas, J.-M. Kiely, E. A. Luis, W. C. Darbonne, G. L. Bennett, C. C. Liu, M. S. Obin, M. A. Gimbrone, Jr., and J. B. Baker, *J. Immunol.* **145,** 3033 (1990).
[24] G. Williams, N. Borkakoti, G. A. Bottomley, I. Cowan, A. G. Fallowfield, P. S. Jones, S. J. Kirtland, G. J. Price, and L. Price, *J. Biol. Chem.* **271,** 9579 (1996).

steps of the purification utilize aqueous buffers; this is a distinct advantage for purifying proteins that may irreversibly unfold in organic solvents. Indeed, attempts to purify certain monomeric IL-8 variants by reversed-phase chromatography led to loss of functional activity, loss of protein during chromatography because of irreversible column binding, or protein aggregation as judged by size-exclusion chromatography (P. H. Slagle and H. B. Lowman, unpublished observations, 1993).

The frozen cell pellet is thawed at room temperature and resuspended thoroughly in 25 ml cold TE buffer [10 mM Tris, pH 7.6, 1 mM EDTA, 0.1 mM phenylmethylsulfonyl fluoride (PMSF), 0.5 mM benzamidine]. After incubation on ice for 30 min, the cells are pelleted and the supernatant discarded. The cell pellet is suspended in high-salt solution (300 mM NaCl, 50 mM Tris, pH 8.0, 10 mM EDTA, 25% (w/v) sucrose, 0.1 mM PMSF, 0.5 mM benzamidine) and incubated on ice for an additional 30 min. The cells are again pelleted, and a 5% (v/v) solution of polyethyleneimine (PEI) is added, with stirring, to the supernatant to give a final PEI concentration of 0.25%. After 10 min on ice, the debris is pelleted, and the supernatant is sterile filtered.

As the first step in purification, a 5-ml DEAE-Sepharose (Pharmacia, Piscataway, NJ) column is connected in tandem to a 5-ml S-Sepharose (Pharmacia, Piscataway, NJ) column and equilibrated with 0.2 M NaCl in 10 mM MES (2-[N-Morpholino]ethane-sulfonic acid) buffer, pH 6.2. The PEI supernatant is diluted two-fold with 10 mM MES buffer to a final salt concentration of about 150 mM. Note that in some cases the concentration of IL-8 will be high enough to cause precipitation, in which case the sample must be diluted further with 0.2 M NaCl. The sample is loaded onto the tandem columns, and then the S-Sepharose column is disconnected and washed with 0.2 M NaCl. Partially purified IL-8 is then eluted with two column volumes of 1 M NaCl in MES buffer, pH 6.2. α_2-Macroglobulin is added to 1 μg/mg of total protein to further reduce proteolysis. The sample may be stored overnight at 4°.

The eluted protein is then diluted seven-fold with 10 mM phosphate buffer, pH 7.4, and loaded onto a 1-ml HiTrap heparin column (Pharmacia, Piscataway, NJ). A linear gradient of 0–1 M NaCl in phosphate buffer can be used for elution, with wild-type IL-8 eluting at approximately 0.5 M NaCl. An example chromatogram is shown in Fig. 5A.

An equal volume of saturated $(NH_4)_2SO_4$ is added, with mixing, to the pooled peak fraction(s). If precipitation occurs, 10 mM phosphate buffer should be added to bring the protein into solution. The protein is then loaded onto a hydrophobic interaction chromatography (HIC) phenyl-Superose column (Pharmacia) and eluted using a reverse gradient from 60–65% saturated $(NH_4)_2SO_4$ to 0% in 10 mM phosphate buffer,

A

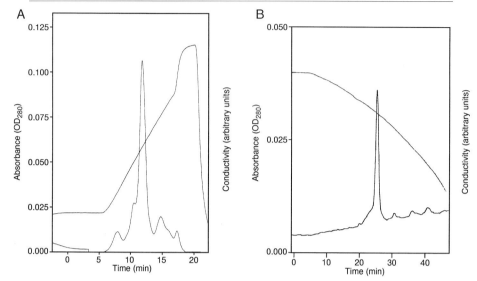

Fig. 5. (A) Typical chromatogram of an IL-8 variant partially purified on a 1-ml HiTrap heparin column, following tandem DEAE- and S-Sepharose columns (see text). The major peak, eluting at approximately 0.5 M NaCl, corresponds to the IL-8 variant. (B) Chromatogram showing the pooled peak fractions from A further purified on a phenyl-Superose hydrophobic interaction column, with a decreasing gradient of $(NH_4)_2SO_4$.

pH 7.4 (Fig. 5B). The protein is desalted into phosphate-buffered saline (PBS; 137 mM NaCl, 3 mM KCl, 8 mM Na_2HPO_4, 1.5 mM KH_2PO_4, pH 7.2 adjusted with HCl) using a NAP-5 gel-filtration column (Pharmacia) and may be concentrated using Centricon-3 concentrators (Amicon, Beverly, MA).

This purification procedure has been successful for a number of IL-8 and MGSA variants; however, certain variants, even those containing single point mutations, may require modifications. For example, for variants having reduced pI, lower pH and lower NaCl concentration in the loading buffer may be required for binding to the S-Sepharose and HiTrap heparin columns. In addition, certain monomer variants of IL-8 appear to bind irreversibly to phenyl-Superose but can be successfully purified on the less hydrophobic alkyl-Superose HIC column (Pharmacia).

Receptor-Specific Binding Assays

As noted above, many chemokines are promiscuous with respect to receptor binding. In the case of IL-8, at least two known receptors, CXCR1 and CXCR2, are known to bind IL-8 on neutrophils. To understand how the binding of each receptor is affected by mutations in IL-8, the binding experiments must be carried out under conditions where only one type of receptor is available. Transiently or stably transfected mammalian cells provide such a system, and such cell lines expressing chemokine receptors are commercially available. We have made use of stably transfected 293 cells (ATCC, Rockville, MD), expressing 10^5 to 10^6 copies of CXCR1 or CXCR2 per cell to determine the binding affinities of IL-8 variants to each receptor type.

Stably transfected 293 cells are maintained in 150-cm^2 dishes in Ham's F12 medium[25] containing 10% (v/v) fetal calf serum, 2 mM glutamine, 20 mM HEPES, pH 7.2, penicillin G (100 units/ml), streptomycin (100 μg/ml), and an appropriate selection antibiotic. Cells are washed by aspirating the medium, gently washing with 10 ml PBS per dish, and again aspirating. The cells are then harvested by "jetting" 10 ml of PBS per dish to dislodge the cells, which are then gently pelleted and resuspended in cold binding buffer. The concentration of cells should be about 10^7 cells/ml, with 50 μl of cell suspension needed per concentration of competitor to be tested.

The binding assay described is an adaptation of a procedure used for neutrophils and other cell lines, which makes use of competitive displacement of ^{125}I-labeled IL-8.[10] A 25-μCi vial of [^{125}I]IL-8 (New England Nuclear, Boston, MA) is dissolved in 0.5 ml of sterile distilled water to give a stock solution of about 23 nM labeled IL-8, which can be stored at 4°. A working solution of labeled IL-8 (25 μl per competitor concentration to be tested) is prepared by making a 10-fold dilution of the stock in binding buffer [0.5% (w/v) bovine serum albumin (BSA; Sigma, St. Louis, MO, RIA grade), 25 mM HEPES (without Mg^{2+}), pH 7.2, in Hanks' balanced salt solution containing phenol red (BioWhitaker, Walkersville, MD)].

A set of eight serial dilutions of each competitor (IL-8 variant) is conveniently prepared using a multichannel pipettor and a Nunc microtiter "F" plate (Nunc, Naperville, IL), which is then used for the binding reaction. A wild-type (IL-8) control series is included with each plate of competitors. The plate is placed on ice, and 175 μl of cold binding buffer is pipetted into wells 2 through 8 of each column. Into the first well of each column is placed 263 μl of the highest concentration of competitor to be tested (e.g., 300 μM IL-8, for CXCR1 binding). Three-fold serial dilutions are then performed by transferring 87.5 μl from each well to the next and mixing by pipetting several times. An additional well for each plate should contain 175 μl of binding buffer only. Labeled IL-8 (25 μl of working solution per well) is then added to each well with mixing. It is helpful to use antiaerosol pipette tips in this and subsequent steps to avoid contamina-

[25] W. L. McKeehan, K. A. McKeehan, S. L. Hammond, and R. G. Ham, *In Vitro* **13**, 399 (1977).

tion of pipettors. Set aside one 25-μl aliquot of working labeled IL-8 solution to measure the total number of counts added per well. With the plate still on ice, 50 μl of cell suspension is added to each well, with mixing. The plate is then transferred to a 4° refrigerator or cold room to incubate for 1 hr.

For separation of cell-bound IL-8 from free IL-8, a sucrose solution [20% sucrose, 140 mM NaCl, 40 mM Tris, pH 7.6, 0.4% (w/v) BSA (Sigma RIA grade)] is prepared and stored at 4°. The sucrose solution (0.75 ml per tube) is then divided into conical polystyrene tubes, each labeled to correspond to wells on the binding plate, and stored at 4° until needed. Following incubation of the binding reactions, each well is mixed by pipetting, and the contents are layered carefully on top of the sucrose solution in the corresponding labeled tube. The tubes are spun for 5 min at 4°, 3000 rpm in a Sorvall (Newtown, CT) RC-3B centrifuge to pellet the cells. The phenol red-colored upper layer followed by the lower (sucrose) layer are removed in two steps (on ice or in a cold room). First, 0.8 ml is removed from each tube. Then, the tubes are again spun as above, and the remaining liquid is removed, leaving behind the cell pellet. Finally, the tubes are counted in a gamma counter.

The data (Fig. 6A) can be analyzed by Scatchard analysis using any of a variety of computer programs, such as LIGAND.[26] The K_d is calculated as the slope of the linearized plot, and the intercept gives the concentration of binding sites (Fig. 6B). Note that the total reaction volume is 0.25 ml in this assay. Therefore, if 300 nM competitor was added in the first well, the final concentration in that well is 210 nM.

Structural Interpretation of Affinity Assays

As discussed above, variants of IL-8 and MGSA in which portions of the N-loop regions of the two proteins were swapped were designed to test their effects on receptor-binding specificity. The initial N-loop swaps involved substituting five IL-8 residues, Tyr-13 to Phe-17, with the corresponding four residues from MGSA, Leu-15 to Ile-18 (IL-8: YSKPF→ LQGI), and vice versa (MGSA: LQGI→YSKPF). In competition binding assays with ^{125}I-labeled wild-type IL-8 in stably expressing 293 cells, as described above, the first of these variants (IL-8: YSKPF→LQGI) showed a markedly reduced affinity for CXCR1 (Table I) as expected based on the MGSA affinity. However, this variant was also reduced in CXCR2 affinity, by over 20-fold relative to wild-type IL-8 and approximately 7-fold relative to MGSA. The corresponding MGSA variant (MGSA: LQGI→

---

[26] P. Munson and D. Rodbard, *Anal. Biochem.* **107,** 220 (1980).

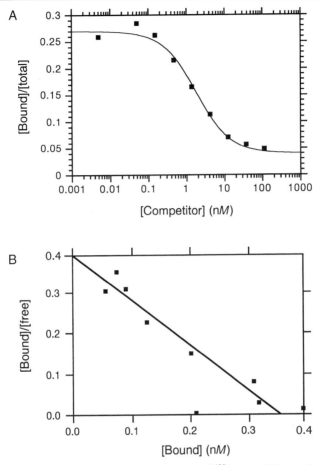

FIG. 6. (A) Displacement plot showing the fraction of ^{125}I-labeled IL-8 precipitating with 293 cells bearing CXCR2, as a function of increasing concentration of unlabeled IL-8. (B) Linearized (Scatchard) plot of the data from A, showing an approximate K_d of 0.90 ± 0.13 nM (slope of the fitted line ± standard error), with 0.35 nM binding sites (x intercept).

YSKPF) showed only a small loss in CXCR2 affinity, relative to wild-type IL-8, but instead of gaining affinity toward CXCR1 it showed little or no improvement (Table I).

More detailed examination of the protein structures revealed differences between IL-8 and MGSA in residues that pack against the N-loop region. The contributions of several of these potentially important packing residues to receptor specificity were tested by making point mutations in the context of the IL-8 or MGSA direct loop swap variants discussed above.[3] As shown

TABLE I

COMPETITIVE BINDING OF IL-8 AND MGSA VARIANTS TO NEUTROPHIL
RECEPTORS CXCR1 AND CXCR2 EXPRESSED ON STABLY
TRANSFECTED 293 CELLS[a]

| Protein background | Mutations | Relative affinity[b] | |
		CXCR1	CXCR2
IL-8	None	1	1
IL-8	YSKPF→LQGI	>260	23
YSKPF→LQGI	L49A	300	9.7
IL-8	L49A	3.6	0.7
YSKPF→LQGI	E48K/L49A	>300	2.0
MGSA	None	230	3.5
MGSA	LQGI→YSKPF	>200	4.1
LQGI→YSKPF	A50L	8.2	1.6
MGSA	A50L	40	0.88
LQGI→YSKPF	L12I/Q13K	27	0.61
LQGI→YSKPF	L12I/Q13K/A50L	0.9	1.3

[a] From Lowman et al.[3]
[b] Relative affinity $[K_d(\text{mutant})/K_d(\text{IL-8})]$, normalized for each receptor type.

in Table I, substitution of IL-8 residue Leu-49 for Ala, and MGSA residue Ala-50 for Leu, resulted in an IL-8 variant with receptor specificity similar to MGSA and an MGSA variant with high affinity for both receptors. The side chain of Leu-49 in IL-8 packs into a bulge in the N-loop, making contacts with the side chains of Tyr-13 and Phe-17 (Fig. 2B), suggesting that it may be required for packing of the N-loop residues in a conformation optimal for binding to CXCR1. The analogous position in MGSA, Ala-50, is closely packed with the side chain of Leu-15 (Fig. 2B).

To test that IL-8 residue Leu-49 is not solely responsible for the observed specificity differences, point mutants of this residue were constructed in the background of the wild-type proteins, IL-8(L49A) and MGSA(A50L). The L49A mutation in IL-8 reduced CXCR1 binding by only about four-fold and had a negligible effect on CXCR2 binding, relative to wild-type IL-8, whereas the MGSA A50L mutation resulted in a four- to six-fold improvement in binding to both receptors, relative to wild-type MGSA (Table I). The modest loss in CXCR1 affinity for IL-8(L49A) is likely due to the loss of packing interactions between Leu-49 and Tyr-13/Phe-17, whereas the slight gain in affinity for MGSA(A50L) may be due to repacking of the side chain of Leu-15 to accommodate the bulkier side chain of Leu-50.

Clearly, confirmation of these suggested packing effects will require detailed structural analysis. In this regard, Williams *et al.*[24] have demonstrated, using NMR spectroscopy, that the mutation L49A in the background of wild-type IL-8 induces small structural changes that extend beyond the immediate environment of the mutated side chain. Although the nature of these changes were not characterized, the observation is consistent with our conclusion that the role of Leu-49 in determining receptor specificity is to optimize the conformation of the N-loop region of IL-8 through specific packing interactions. Nevertheless, the receptor specificities of these point mutants do not differ greatly from the respective wild-type proteins, indicating that IL-8 residue Leu-49 is not the sole receptor specificity determinant.

The exchanged receptor specificities of the loop-swap variants discussed above are similar but not identical to the corresponding wild-type proteins (Table I). Optimization was possible, however, by the inclusion of a small number of additional mutations. In the case of the IL-8 variant IL-8: YSKPF→LQGI + L49A, inclusion of the mutation E48K, which replaces the negatively charged IL-8 residue with the corresponding positively charged MGSA residue, resulted in an improved (~five-fold) affinity for CXCR2 and gave a variant that was indistinguishable, within the experimental uncertainties, from wild-type MGSA (Table I). The choice of this additional mutation was based on previous mutagenesis results, in the backgrounds of both IL-8 and MGSA, that suggested a lysine residue at position 48 in IL-8 and position 49 in MGSA provides a favorable interaction with CXCR2.[3,27] The affinities of both MGSA mutants MGSA:LQGI→YSKPF and MGSA:LQGI→YSKPF + A50L for CXCR1 were improved by extending the loop swap to include IL-8 residues Ile-10 and Lys-11 (i.e., L12I and Q13K) (Table I). The final MGSA mutant, MGSA/LQGI→ YSKPF + L12I/Q13K/A50L binds both CXCR1 and CXCR2 with affinities indistinguishable from wild-type IL-8. These additional mutations were made as part of the original series of loop swaps and were chosen simply to extend the length of the exchanged N-loop in an empirical effort to exchange the receptor specificities.[3] The reason for the observed receptor-binding differences between the "short" and "long" loop swaps is not clear, although we note that Ile-10 was previously identified by alanine scanning mutagenesis as being important for IL-8 binding to neutrophils, whereas Lys-11 was unimportant.[10] We have not investigated the effect of the L12I mutation alone in the background of MGSA/LQGI→YSKPF + A50L.

[27] J. Hesselgesser, C. E. Chitnis, L. H. Miller, D. G. Yansura, L. C. Simmons, W. J. Fairbrother, C. Kotts, C. Wirth, B. L. Gillece-Castro, and R. Horuk, *J. Biol. Chem.* **270**, 11472 (1995).

Conclusion

By constructing and measuring receptor-binding affinities of IL-8 and MGSA variants we have shown that the N-loop region of IL-8 is required for binding to CXCR1. The conformation of the N-loop appears critical for CXCR1 binding, based on the need for Leu-49, which packs against the N-loop residues Tyr-13 and Phe-17. The N-terminal domains of CXCR1 and CXCR2 have been previously shown to interact with IL-8 and to play a role in determining chemokine-binding specificity.[6,28–31] We suggest, therefore, that the N-loop region of IL-8 interacts directly with the N-terminal domains of the IL-8 receptors. This conclusion is supported by the observation that a peptide corresponding to the 40 N-terminal residues of CXCR1 results in NMR chemical shift changes for several residues in this region when it is bound to IL-8 in solution.[32]

The N-loop region is separated by more than 20 Å from the important ELR residues, suggesting that this region of IL-8 is a secondary binding site distinct from the important ELR motif.[3,24] Based on the available data we speculate that there is a one-to-one interaction of the secondary N-loop site on IL-8 with a portion of the CXCR1 N-terminal domain and that the ELR residues interact with a region of charged residues, Arg-199, Arg-203, and Asp-265, identified by alanine scanning mutagenesis in CXCR1 extracellular loops 3 and 4.[30]

Acknowledgments

We thank Nicholas Skelton for critical reading of the manuscript, David Wood for assistance in preparing Fig. 5, and Joshua Theaker for interfacing the programs ALIGN and INSIGHT-II (Molecular Simulations, San Diego, CA).

[28] R. B. Gayle III, P. R. Sleath, S. Srinivason, C. W. Birks, K. S. Weerawarna, D. P. Cerretti, C. J. Kozlosky, N. Nelson, T. Vanden Bos, and M. P. Beckmann, *J. Biol. Chem.* **268,** 7283 (1993).

[29] C. A. Hébert, A. Chuntharapài, M. Smith, T. Colby, J. Kim, and R. Horuk, *J. Biol. Chem.* **268,** 18549 (1993).

[30] S. R. Leong, R. C. Kabakoff, and C. A. Hébert, *J. Biol. Chem.* **269,** 19343 (1994).

[31] I. U. Schraufstätter, M. Ma, Z. G. Oades, D. S. Barritt, and C. G. Cochrane, *J. Biol. Chem.* **270,** 10428 (1995).

[32] R. T. Clubb, J. G. Omichinski, G. M. Clore, and A. M. Gronenborn, *FEBS Lett.* **338,** 93 (1994).

[5] Alanine Scan Mutagenesis of Chemokines

By JOSEPH HESSELGESSER and RICHARD HORUK

Introduction

Chemokines are a family of chemotactic cytokines that play a critical role in the regulation and trafficking of immune cells.[1] Some chemokines have also been shown to be suppressive factors[2] of human immunodeficiency virus type-1 (HIV-1). There are currently three families of chemokines, grouped according to the arrangement of their invariant cysteines. The C-X-C branch in which the first two cysteines are separated by an intervening amino acid includes interleukin-8 (IL-8), melanoma growth stimulating activity (MGSA), platelet factor 4 (PF4), and stromal cell derived factor-1 (SDF-1).[3] This group is further divided by the presence (IL-8, MGSA) or absence (SDF-1, PF4) of an E-L-R motif prior to the first N-terminal cysteine, The C-X-C chemokines are generally chemoattractors and activators of neutrophils, exceptions being PF4, which also attracts fibroblasts and monocytes and SDF-1, which is a chemoattractant for T cells. The C-C chemokines are chemoattractants and cellular activators for monocytes, basophils, eosinophils, and lymphocytes. This is the largest family of chemokines and includes the macrophage inflammatory proteins MIP-1α and MIP-1β, RANTES, monocyte chemoattractant proteins MCP-1 to MCP-4, I-309, and eotaxin.[3] The third subdivision of the chemokines is the C branch, which is characterized by its only member lymphotactin, in which only two cysteines are conserved.[4]

In addition to their normal role in immune cell function, chemokines play an important part in a number of autoimmune diseases such as multiple sclerosis and rheumatoid arthritis,[3] and chemokine receptors are gateways

---

[1] T. Schall, *in* "The Cytokine Handbook" (A. Thompson, ed.), p. 419. Academic Press, San Diego, 1994.

[2] F. Cocchi, A. L. DeVico, A. Garzino-Demo, S. K. Arya, R. C. Gallo, and P. Lusso, *Science* **270,** 1811 (1995).

[3] R. Horuk, ed., "Chemoattractant Ligands and Their Receptors." CRC Press, Boca Raton, Florida, 1996.

[4] G. S. Kelner, J. Kennedy, K. B. Bacon, S. Kleyensteuber, D. A. Largaespada, N. A. Jenkins, N. G. Copeland, J. F. Bazan, K. W. Moore, T. J. Schall, and A. Zlotnik, *Science* **266,** 1395 (1994).

of infection for HIV-1 and the malaria parasite.[5-8] Thus, it is essential to understand how chemokines interact with their cellular receptors to produce their biological effects to develop therapeutic approaches of intervention.

Rationale

Structure–function relationships between ligands and their receptors can be examined in a number of ways. Site-directed mutagenesis, particularly alanine scan mutagenesis, has been a successful technique for identifying functionally important residues in proteins. This approach has been applied to a number of receptors and ligands as exemplified for human growth hormone (hGH), IL-8, MGSA, CXCR1 (IL-8RA), and CXCR2 (IL-8RB).[9-13] In a variation of alanine scan mutagenesis the charged residues of a molecule only are replaced, and thus residues forming ion pairs with charged residues in a cognate receptor can be identified. In addition, substitutions of charged amino acid residues by residues with increasing side-chain length or bulkiness can be used to determine what can be tolerated for receptor binding. Finally, replacement of hydrophobic residues by alanine gives a glimpse of important nonaqueous interactions between ligand–receptor pairs.

Site-directed mutagenesis to determine structure–function relationships for ligand–receptor pairs is preferred over ligand truncation mutants because it minimizes structural modifications to the native molecule. Another advantage of site-directed mutagenesis is that, if structural data exist, the surface residues of a protein can be substituted in a predictable pattern, one at a time or in groups. The substitution of groups of amino acids

[5] H. Choe, M. Farzan, Y. Sun, N. Sullivan, B. Rollins, P. D. Ponath, L. J. Wu, C. R. Mackay, G. LaRosa, W. Newman, N. Gerard, C. Gerard, and J. Sodroski, *Cell* (*Cambridge, Mass.*) **85**, 1135 (1996).

[6] T. Dragic, V. Litwin, G. P. Allaway, S. R. Martin, Y. X. Huang, K. A. Nagashima, C. Cayanan, P. J. Maddon, R. A. Koup, J. P. Moore, and W. A. Paxton, *Nature* (*London*) **381**, 667 (1996).

[7] B. J. Doranz, J. Rucker, Y. J. Yi, R. J. Smyth, M. Samson, S. C. Peiper, M. Parmentier, R. G. Collman, and R. W. Doms, *Cell* (*Cambridge, Mass.*) **85**, 1149 (1996).

[8] R. Horuk, *Immunol. Today* **15**, 169 (1994).

[9] B. C. Cunningham and J. A. Wells, *Science* **244**, 1081 (1989).

[10] C. A. Hébert, R. V. Vitangcol, and J. B. Baker, *J. Biol. Chem.* **266**, 18989 (1991).

[11] J. Hesselgesser, C. Chitnis, L. Miller, D. J. Yansura, L. Simmons, W. Fairbrother, C. Kotts, C. Wirth, B. Gillece-Castro, and R. Horuk, *J. Biol. Chem.* **270**, 11472 (1995).

[12] C. A. Hébert, A. Chuntharapai, M. Smith, T. Colby, J. Kim, and R. Horuk, *J. Biol. Chem.* **268**, 18549 (1993).

[13] S. R. Leong, R. C. Kabakoff, and C. A. Hébert, *J. Biol. Chem.* **269**, 19343 (1994).

from IL-8 with those from MGSA to produce a hybrid protein capable of exhibiting high-affinity binding to CXCR1 (native MGSA binds only with low affinity) provides an excellent example of the utility of this technique in helping to identify domains of IL-8 that are responsible for receptor specificity.[14] A further example of the application of alanine replacement is BB-10010, a variant of MIP-1α, that has a single substitution Asp to Ala at position 27.[15] This substitution results in a molecule that is fully active but does not form the extremely high molecular weight multimers that the wild-type chemokine does. BB-10010 has thus found favor in clinical situations in which patients undergoing chemotherapy can be dosed with high concentrations of the chemokine without fear of aggregate formation that could reduce its efficacy.

Protocols

Expression Systems

Recombinant expression in *Escherichia coli* is one of the most common methods for producing chemokines.[16–19] A number of vectors have been described including the pET-3 and PET-11 series, from Stratagene (La Jolla, CA). These are generally used in combination with a BL21(DE3) *E. coli* strain (Strategene) that contains T7 RNA polymerase under the control of *lacUV5*. This expression system allows genes placed downstream of the RNA polymerase binding site (in pET vectors) to be expressed on addition of isopropylthiogalactoside (IPTG). Upstream cloning of a signal sequence, prior to the chemokine gene, such as human growth hormone (hGH) or heat-stable enterotoxin II (hstII)[11] has been shown to aid in the expression of a soluble secreted protein.

Once an appropriate vector construct has been made with the chemokine gene of interest, site-directed mutagenesis can be performed to generate a variety of clones. The site-directed alanine substitution mutants can

[14] H. B. Lowman, P. H. Slagl, L. E. DeForge, C. M. Wirth, B. L. Gillece-Castro, J. H. Bourell, and W. J. Fairbrother, *J. Biol. Chem.* **271,** 14344 (1996).
[15] M. G. Hunter, L. Bawden, D. Brotherton, S. Craig, S. Cribbes, L. G. Czaplewski, T. M. Dexter, A. H. Drummond, A. H. Gearing, and C. M. Heyworth, *Blood* **86,** 4400 (1995).
[16] I. Lindley, H. Aschauer, J.-M. Seifert, C. Lam, W. Brunowsky, E. Kownatzki, M. Thelen, P. Peveri, B. Dewald, V. von Tscharner, A. Walz, and M. Baggiolini, *Proc. Natl. Acad. Sci. U.S.A.* **85,** 9199 (1988).
[17] R. Horuk, D. G. Yansura, D. Reilly, S. Spencer, J. Bourell, W. Henzel, G. Rice, and E. Unemori, *J. Biol. Chem.* **268,** 541 (1993).
[18] M. C. N. Johnson *et al., J. Biol. Chem.* **271,** 10853 (1996).
[19] J. Zagorski and J. E. DeLarco, *Protein Expression Purif.* **5,** 337 (1994).

be made by constructing oligonucleotide probes that cover the desired mutation followed by polymerase chain reaction (PCR) to fill in the rest of the chemokine gene sequence. There are a variety of kits available from several companies that use similar, simple, and easy-to-follow methods to insert the desired mutations. These kits include the Transformer Site-Directed Mutagenesis Kit (Clontech, Palo Alto, CA) and the Chameleon Double-Stranded Site-Directed Mutagenesis Kit (Stratagene). The basic principle behind these kits is illustrated in Fig. 1. With these bacterial expression systems it is simple to produce small amounts of proteins using 1-liter shake flasks; alternatively, this process can be scaled up to larger 10- to 20-liter fermentation runs to produce gram quantities of protein. In general, chemokines can be produced as secreted, fully folded proteins,[11,17] but sometimes problems with folding and also with the generation of N-terminal addition mutants can occur (see Horuk et al.[19a] for a discussion).

Chemokines have also been produced in mammalian expression systems.[20] A variety of cell lines, such as human kidney 293 cells, chinese hamster ovary (CHO) cells, and COS cells, can be used. Mammalian expression vectors are generally larger than E. coli vectors, with viral promoters and enhancers (pSI, pCI, and pCI-neo from Promega, Madison, WI; pcDNA3 from Invitrogen, San Diego, CA; or pPUR from Clontech). The use of an upstream signal sequence, such as hGH, CD4, or human immunoglobulin (Ig), can greatly enhance the production of a soluble secreted protein. A major advantage of mammalian expression is that the secreted protein is usually correctly folded. In addition with the introduction of defined, nonserum-containing medium (HyQ-CCM 1 to 5, Hyclone, Logan, UT, and CHO-S-SFM II, GIBCO-BRL, Grand Island, NY) downstream purification steps are easier, and there is less risk of contamination with lipopolysaccharide (LPS), which is frequently a problem with E. coli expression systems.

Peptide Synthesis

An alternative to protein expression that has found increasing use for the generation of chemokine mutants is peptide synthesis. Although this can be an expensive process ($20/amino acid plus purification), it is quick compared to expression methods. For a discussion of peptide synthesis of chemokines see Clark-Lewis et al.[20a]

[19a] R. Horuk, D. Reilly, and D. Yansura, *Methods Enzymol.* **287** [1], 1997 (this volume).
[20] E. Balentien, J. H. Han, H. G. Thomas, D. Z. Wen, A. K. Samantha, C. O. Zachariae, P. R. Griffin, R. Brachmann, W. L. Wong, K. Matsushima, and R. Derynck, *Biochemistry* **29**, 10225 (1990).
[20a] I. Clark-Lewis, L. Vo, P. Owen, and J. Anderson, *Methods Enzymol.* **287** [16], 1997 (this volume).

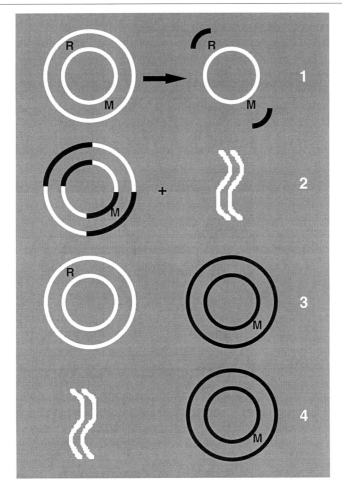

FIG. 1. Strategy for generation of alanine scan mutants. (1) Denature the double-stranded plasmid DNA. Anneal the two oligonucleotide primers, the R primer altering a unique restriction site and the M primer containing the mutation codon. (2) Perform primer extension using dNTPs and T7 or T4 DNA polymerase and ligation with T4 DNA ligase. Digest DNA with restriction enzyme for unique site. Renature DNA to result in a mixture of linear and plasmid DNA. Transform into *E. coli* strains [BMH 71-18 mutS (Clontech) or XLmutS (Stratagene)] that are unable to perform DNA mismatch repair. (3) Recover plasmid DNA (Plasmid Mini Kit, Qiagen, Chatsworth, CA). (4) Perform second digestion with restriction enzyme. Transfect into *E. coli* and recover final plasmid. Sequence the mutated gene to verify the correct sequence.

Purification

Most chemokines are very basic proteins with isoelectric points between pH 8.0 and 10.5. In contrast most *E. coli* and serum proteins from mammalian cell expression have acidic isoelectric points. Thus chemokines can be purified at neutral pH by a combination of S-Sepharose, heparin-Sepharose, and hydrophobic interaction (phenyl-Sepharose) chromatography (see Horuk *et al.*[19a]). This procedure coupled with reversed-phase high-performance liquid chromatography (HPLC) can give chemokines that are >95% pure based on silver staining sodium dodecyl sulfate–polyacrylamide gel electrophoresis (SDS–PAGE) and amino acid composition.[11] Soluble chemokine supernatants from *E. coli* or mammalian cell supernatants can be purified by this method. Alternatively, the construction of expression plasmids containing nucleotides encoding hexahistidine, FLAG peptide, glutathionine, or immunoglobulin sequences engineered into the cDNAs of specific chemokines can greatly aid in their purification.[21–23] The protein epitope can be engineered into the amino- or carboxyl-terminal region of the molecule (the position is largely dependent on the maintenance of full biological activity).

Biological Activity

Once wild-type and mutant chemokine proteins have been generated and purified they are usually analyzed by receptor-binding and bioactivity studies to determine structure–function relationships. For receptor-binding studies, a wide variety of radiolabeled chemokines are available commercially (Dupont NEN, Boston MA; Amersham, Arlington Heights, IL) or can be iodinated.[23a] Binding assays can be carried out on whole cells or membranes prepared from cells that express the appropriate chemokine receptors. Scatchard analysis of the binding data can then be used to determine the relative receptor-binding affinity of the mutant chemokine, which is defined as the K_D of the wild-type/K_D mutant $\times$ 100%. Biological assays for chemokines are numerous and are described elsewhere.[23b] However, one biological assay that is used as a standard is chemotaxis. All chemokines induce a migratory response in target cells that express functional G-protein-coupled chemokine receptors. In addition, a primary response to li-

[21] H. M. Sassenfeld, *Trends Biotechnol.* **8,** 88 (1990).
[22] A. S. Robeva, R. Woodard, D. R. Luthin, H. E. Taylor, and J. Linden, *Biochem. Pharmacol.* **51,** 545 (1996).
[23] A. Kuusinen, M. Arvola, C. Oker-Blom, and K. Keinanen, *Eur. J. Biochem.* **233,** 720 (1995).
[23a] G. L. Bennett and R. Horuk, *Methods Enzymol.* **288** [10] (1997).
[23b] D. Baly, U. Gibson, D. Allison, and L. DeForge, *Methods Enzymol.* **287** [6], 1997 (this volume); R. C. Newton and K. Vaddi, *Methods Enzymol.* **287** [12], 1997 (this volume).

gand–receptor binding is the induction of the intracellular release of calcium stores. Calcium flux assays are also popular methods of assessing chemokine function.[23c]

Applications

The following section gives an outline for the specific production of alanine scan mutants of MGSA that have been generated to probe structure–function relationships of the chemokine receptors CXCR2 and duffy antigen receptor for chemokines (DARC).[11] This approach is useful in generating a mutant of MGSA, E6A, that is able to inhibit malaria parasite invasion of human erythrocytes by receptor blockade of the receptor DARC but is a very poor agonist of the CXCR2 receptor on neutrophils.

Plasmid Construction

Plasmid construction and mutagenesis are achieved by a combination of vectors and kits described above (see also Fig. 1) and are illustrated with reference to the MGSA mutants. The pMG34 *E. coli* secretion vector has been previously described.[17] This vector contains the hstII signal sequence to aid in secretion of the chemokine and an alkaline phosphatase promoter that is induced when the *E. coli* cells are grown in a low phosphate-containing medium.[24] The MGSA gene sequence is fused to the hstII sequence as described.[17] The resulting plasmid contains an *Eco*RI site 5′ to the MGSA gene and a *Bsa*JI site 3′ to the MGSA sequence. For each mutant the entire MGSA gene is synthesized (~219 bases) to include the respective codon changes for the amino acid substitution. Double-stranded DNA is synthesized by the use of an AmpliTaq DNA Polymerase kit (Perkin-Elmer, Roche Molecular Systems, Branchburg, NJ) with appropriate amounts of DNA and polymerase per the manufacturer's specifications. DNA is purified by the use of QIAquick Nucleotide Removal Kit (Qiagen, Chatsworth, CA) per the manufacturer's specifications.

The MGSA DNA and pMG34 vector are cut separately with *Eco*RI restriction endonuclease (New England Biolabs, Beverly, MA) in 50 mM NaCl, 10 mM Tris-HCl, 10 mM MgCl$_2$, 1 mM dithiothreitol (DTT) at 37° for 2 hr. The temperature is then raised to 65° for 20 min to inactivate *Eco*RI followed by addition of *Bsa*JI endonuclease (New England Biolabs). The mixture is then incubated at 60° for 2 hr. This results in *Eco*RI overhangs with the MGSA DNA and cleaved plasmid. The DNA mixture is

[23c] S. R. McColl and P. H. Naccache, *Methods Enzymol.* **288** [18] (1997); K. B. Bacon, *Methods Enzymol.* **288** [22] (1997).
[24] C. N. Chang, M. Rey, B. Bochner, H. Heyneker, and G. Gray, *Gene* **55,** 189 (1987).

purified by the QIAquick kit as above. The mutated MGSA DNA is then ligated into the vector by the use of T4 DNA ligase (20 units, 0.3 Weiss units; New England Biolabs) in 50 mM Tris-HCl (pH 7.5), 10 mM MgCl$_2$, 10 mM DTT, 0.5 μM ATP, 25 μg/ml bovine serum albumin (BSA), 200 ng vector, and 100 ng insert at 16° for 16 hr at a final volume of 20 μl. Ligase activity is stopped by heat inactivation at 65° for 20 min.

The plasmid is transfected into *E. coli* by the CaCl$_2$/heat shock method[25] and grown in LB broth with 50 μg/ml carbenicillin. The plasmid is recovered from the *E. coli* by the use of Qiagen Plasmid Mini Kit as above. Finally, the plasmid is sequenced through the entire gene and restriction sites to verify the correct orientation and nucleotide sequence. Following plasmid recovery and sequencing, the plasmid is transformed into *E. coli*, cells are grown overnight in low phosphate medium, and the MGSA is purified as described.[11]

Studies with Melanoma Growth Stimulating Activity Mutants

Because MGSA binds with high affinity to the chemokine receptors CXCR2 (IL-8RB) and DARC,[26,27] receptor-binding studies with the MGSA mutants are carried out in cells expressing these receptors. Cells are incubated with ^{125}I-labeled MGSA in the absence and presence of unlabeled MGSA or MGSA mutants; typical competition binding curves are shown in Fig. 2. The E6A mutant of MGSA is able to bind with high affinity to DARC (K_D = 7 nM compared to K_D = 3.5 nM for MGSA), but in contrast binds with low affinity to CXCR2 (K_D = 476 nM for E6A compared to K_D = 2.3 nM for MGSA). In contrast, the R8A mutant exhibits low-affinity binding to both receptors (for CXCR2 K_D = 850 nM and for DARC K_D = 157 nM). Thus, the Arg-8 residue of MGSA appears to be crucial for expression of high-affinity binding for both receptors.

In light of the binding data obtained for the E6A and R8A mutants, functional assays are carried out both with DARC and with CXCR2. DARC is known to be a cofactor for the invasion of human erythrocytes by the malarial parasite *Plasmodium vivax*[28] and the related monkey parasite *Plasmodium knowlesi.*[29] The C-X-C chemokines IL-8 and MGSA have

[25] M. Mandel and A. Higa, *J. Mol. Biol.* **53,** 159 (1970).

[26] J. Lee, R. Horuk, G. C. Rice, G. L. Bennett, T. Camerato, and W. I. Wood, *J. Biol. Chem.* **267,** 16283 (1992).

[27] R. Horuk, C. E. Chitnis, W. C. Darbonne, T. J. Colby, A. Rybicki, T. J. Hadley, and L. H. Miller, *Science* **261,** 1182 (1993).

[28] L. H. Miller, S. J. Mason, D. F. Clyde, and M. H. McGinniss, *N. Engl. J. Med.* **295,** 302 (1986).

[29] L. H. Miller, S. J. Mason, J. A. Dvorak, M. H. McGinniss, and I. K. Rothman, *Science* **189,** 561 (1975).

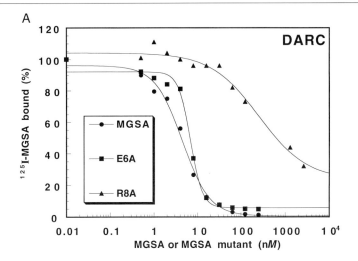

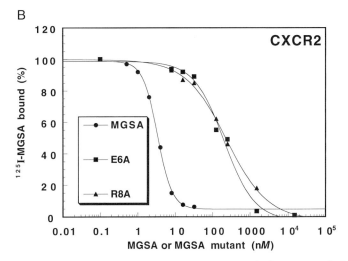

FIG. 2. MGSA competition binding studies. Competition binding was studied between ^{125}I-MGSA and increasing concentrations of unlabeled MGSA mutants E6A and R8A for (A) human erythrocytes and (B) human kidney cells transfected with CXCR2. Cells were incubated 1 hr at 4° with ^{125}I-MGSA in the presence of increasing amounts of the unlabeled MGSA mutants. (Adapted from Ref. 11 with permission.)

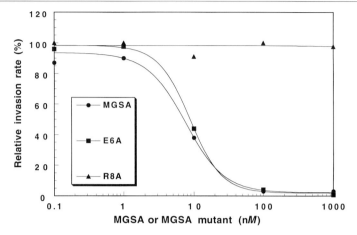

FIG. 3. Inhibition of erythrocyte invasion by *P. knowlesi* by MGSA and MGSA mutants. The invasion rates are expressed as a percentage of the rate of invasion in the absence of chemokines. Inhibition of invasion EC_{50} values are as follows: MGSA, 7 nM; E6A, 8.6 nM; R8A, >1 μM. (Adapted from Ref. 11 with permission.)

been shown to dose-responsively inhibit both parasite binding and invasion.[27] Thus, the ability of MGSA and the MGSA mutants E6A and R8A to
block *P. knowlesi* invasion of human erythrocytes is assessed as previously
described.[11] As shown in Fig. 3 the E6A MGSA mutant is able to block

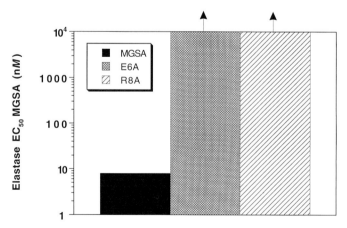

FIG. 4. MGSA and MGSA mutant stimulation of elastase release from human neutrophils. The EC_{50} value is defined as the concentration of MGSA or mutant required for half-maximal release of elastase from neutrophils. (Adapted from Ref. 11 with permission.)

malarial parasite invasion in a dose-responsive manner similar to wild-type MGSA (for MGSA $EC_{50} = 7$ nM, for E6A $EC_{50} = 8.6$ nM). However, in line with the change in binding affinity of R8A to DARC, the R8A mutant is unable to block malarial invasion at concentrations up to 1 μM.

The mutants are then tested for biological activity on neutrophils, which express CXCR2, using elastase release assays. The data from the elastase release assay (Fig. 4) clearly show that the MGSA mutants E6A and R8A are biologically inactive (EC_{50} 10 μM) compared to wild-type MGSA (EC_{50} 8 nM). Thus, the use of alanine scan mutagenesis of MGSA has allowed us to identify an MGSA mutant, E6A, that can block malarial invasion of human erythrocytes but will not activate neutrophils, properties that may be useful therapeutically in the design of small molecules that inhibit erythrocyte invasion by *P. vivax* but have no effect on neutrophils.

Conclusion

The example of alanine scan mutagenesis described for MGSA clearly illustrates the powerful nature of this technique not only for identifying specific residues involved in the binding of ligands to their native receptors but also for generating potentially useful drugs. Because chemokines and chemokine receptors play an important role in a variety of acute and chronic diseases, the use of alanine scanning mutagenesis to accurately define receptor binding domains could aid in the design of molecular antagonists that would be useful therapeutically.

[6] Biological Assays for C-X-C Chemokines

By Deborah Baly, Ursula Gibson, David Allison, and Laura DeForge

Introduction

Chemokines play a major role in mobilizing leukocytes to ward off attacks by invading pathogens.[1] Each of the two major classes of chemokines is selective for a particular group of immune cells. In this chapter, we discuss the assays available to measure the biological activities of one of these

---

[1] T. Schall, *in* "The Cytokine Handbook" (A. Thompson, ed.). p. 419. Academic Press, San Diego, 1994.

groups of chemoattractants, the C-X-C chemokines. The C-X-C chemokines, which include interleukin-8 (IL-8) and melanoma growth stimulating activity (MGSA), preferentially attract neutrophils and induce their activation by producing changes in neutrophil shape, transient increases in cellular calcium concentration, and the upregulation of surface adhesion proteins.[2] These chemokines play a major role in acute inflammatory responses. In contrast, the C-C chemokines, which include monocyte chemoattractant proteins (MCPs) and RANTES, have no effects on neutrophils but are chemoattractant for monocytes and induce their degranulation.[1] The C-C chemokines are involved in chronic inflammatory responses. Their biological properties are presented elsewhere,[2a] and they are not discussed further here.

Acute inflammatory responses are associated with increased adherence of neutrophils to vascular endothelium, which involves the upregulation of adhesion molecules, such as macrophage differentiation antigen-1 (MAC-1), also known as CD11B/CD18.[3] This is often followed by extravasation of neutrophils into tissues with subsequent release of tissue-damaging proteases (elastase and β-glucuronidase, among others) and toxic oxygen metabolites.[2] The various activities of neutrophils can be measured in a variety of assays, some of which are detailed in this chapter. We describe a procedure for isolating neutrophils from human blood and then discuss the biological effect of C-X-C chemokines on these cells as measured in a number of assays, including upregulation of CD18 adhesion proteins, elastase and β-glucuronidase release assays, and the oxidative burst.

Isolation of Neutrophils from Human Blood

Reagents and Equipment

Dextran T500 (Pharmacia, Piscataway, NJ)
Lymphocyte separation medium, 100 ml, sterile (Organon Teknika Cappel, Durham, NC)
Vacutainer brand blood collection sets (Baxter, Edison, NJ)
60-ml Syringes (Becton Dickinson, Lincoln Park, NJ)
Hemocytometer (Reichert Scientific Instruments)

---

[2] M. Baggiolini and I. Clark-Lewis, *FEBS Lett.* **307,** 97 (1992).
[2a] J. Van Damme, S. Struyf, A. Wuyts, G. Opdenakker, P. Proost, P. Allavena, S. Sozzani, and A. Mantovani, *Methods Enzymol.* **287** [8], 1997 (this volume); A.-M. Buckle, S. Craig, and L. G. Czaplewski, *Methods Enzymol.* **287** [9], 1997 (this volume); R. C. Newton and K. Vaddi, *Methods Enzymol.* **287** [12], 1997 (this volume).
[3] T. A. Springer, *Nature (London)* **346,** 425 (1991).

Stock Solutions

3% (w/v) Dextran solution: Make up dextran (30 g to 1 liter) in 0.15 M NaCl. It can be stored at 2–8° for 1 month from date of preparation.

Low-salt solution (34 mM NaCl), which can be stored at 2–8° for up to 2 months

High-salt solution (274 mM NaCl), which can be stored at 2–8° for 2 months from date of preparation

Buffer A: 0.1% (w/v) bovine serum albumin (BSA) in phosphate-buffered saline (PBS)

Isolation

1. Draw approximately 45 ml of blood into a 60-ml heparinized syringe. Keep at ambient temperature until neutrophil isolation. For elastase release assays also collect blood into one 10-ml serum separation tube without anticoagulants for opsonization with zymosan A and store in an ice bucket until ready for assay.

2. Transfer 15 ml of the blood into fresh syringes and add 30 ml of 3% dextran solution into each syringe. Gently mix the blood and dextran solution by inverting the syringe several times and placing it upright in a test tube rack. Repeat this process for the remaining blood.

3. Incubate the syringes for 30 min at ambient temperature without disturbing them. Two layers form in the syringes; the bottom, dark red layer consists of erythrocytes, whereas the clear, top layer contains the neutrophils.

4. Add 14 ml of lymphocyte separation medium (LSM) to three 50-ml centrifuge tubes. Carefully layer the neutrophil containing layer from step 3 on top of the LSM. This is done using sterile 12-inch tubing with a female Luer to connect it to the syringe. Stop adding material to the top of the LSM when the lower layer containing erythrocytes becomes visible in the tubing. Repeat this step for the remaining syringes.

5. Centrifuge the tubes at 550 g for 40 min at ambient temperature with the brake in the "off" position. During the centrifugation several layers form in the tubes; an upper clear yellowish layer contains plasma, and a narrow opaque band in the middle contains mononuclear cells. The pellet contains the neutrophils and some erythrocytes.

6. After centrifugation aspirate the supernatant, leaving the pellet containing neutrophils behind. Lyse any contaminating erythrocytes by adding 10 ml of ice-cold low-salt solution, gently vortexing the neutrophil pellet, and immediately adding 10 ml of the cold high-salt solution. It is important that these salt solutions are used at 4° because they are less effective at room temperature.

7. Centrifuge the neutrophils at 550 *g* for 10 min and aspirate the supernatant. Repeat steps 6 and 7 two more times for a total of three washes. It is extremely important that any contaminating erythrocytes are removed because these cells express the DARC receptor that can bind several chemokines and may complicate experimental data if they contaminate the neutrophil preparation.

8. After the last wash step, resuspend the neutrophils in a small volume of buffer for counting. Calculate the number of neutrophils isolated and add the appropriate assay buffer (for subsequent assays) to adjust the concentration to 1.0×10^7 cells/ml. The isolated neutrophils can be stored for up to 1 hr at 2–8° in a refrigerator. Avoid prolonged exposure to ice.

CD11b/CD18 Assay

MAC-1 on neutrophils consists of an 180-kDa α subunit, CD11b, and a 95-kDa β subunit, CD18, that are noncovalently held together by ionic interactions.[3] The MAC-1 complex is membrane bound and glycosylated and binds to intercellular adhesion molecules (ICAMs) that belong to the immunoglobulin superfamily. MAC-1 is expressed on all leukocytes and is involved in the adhesion of leukocytes to endothelial cells that promotes their extravasation across the tissue space. Chemokines have been shown to upregulate the CD11b/CD18 complex,[4] and with the advent of specific monoclonal antibodies that recognize these proteins we have constructed an assay based on fluorescence-activated cell sorting (FACS), using the FACScan from Becton Dickinson, that measures the ability of C-X-C chemokines to upregulate the expression of the MAC-1 adhesion proteins in neutrophils. This assay is extremely sensitive and can thus be used to measure the biological activity of C-X-C chemokines like IL-8.

Reagents and Equipment

Cytochalasin B (Sigma, St. Louis, MO)
Paraformaldehyde (Sigma)
IL-8 (Genentech)
Anti-CD18 antibody (Genentech)
Anti-IL-8 antibody (Genentech)
Anti-HER-2 antibody (4D5) (Genentech)
Goat anti-human F(ab')2 FITC conjugate (Cappel)
12 × 75-mm tubes (Falcon, Lindoln Park, NJ)

[4] P. A. Detmers, S. K. Lo, E. E. Olsen, A. Walz, M. Baggiolini, and Z. A. Cohn, *J. Exp. Med.* **171,** 1155 (1990).

Stock Solutions

Assay buffer: 0.1% (w/v) BSA in PBS + 1 mM CaCl$_2$ and MgCl$_2$
1% (v/v) Paraformaldehyde in PBS

Assay

1. Neutrophils are isolated as described above and resuspended in Buffer A to a final concentration of 10^7 cells/ml.

2. The neutrophils are activated by incubation with cytochalasin B (5 μg/ml) at 37° for 30 to 45 min.

3. Antibodies to IL-8 (A5.12.14 and 6G425), or a nonspecific control antibody (4D5), are added to IL-8 at various antibody to IL-8 molar ratios (8 : 1 to 0.125 : 1) in assay buffer containing Ca^{++} and Mg^{++} ions. The final IL-8 concentration in each case is 12.5 nM. This assay format allows processing a large number of samples. The incubations are done in 96-well format polypropylene micronic tubes, enabling dilution, mixing, and washing steps to be done with the aid of a multichannel pipettor.

4. The anti-IL-8/IL-8 micronic tubes are preincubated at 37° for 30 min prior to adding the activated neutrophils. In addition, an IL-8 standard curve (12.5 to 0.034 nM IL-8), including a baseline control (no added IL-8), is prepared in assay buffer by 1 : 3 serial dilutions of the stock 25 nM IL-8.

5. Next, 100 μl of the activated neutrophils are added to the 100-μl aliquots of the IL-8 standard curve, including the baseline control, and the preincubated anti-IL-8/IL-8 mixtures. The cells, at a final concentration of 10^6 cells/tube, are incubated for 2 hr at 37°.

6. After incubation, the neutrophils are fixed by the addition of 200 μl of the 1% paraformaldehyde solution (2–6°) to each tube. The cells are fixed for 20 min at 2–6°. The tubes are then centrifuged (400 g, 5 min, 4°) in a 96-well plate centrifuge carrier modified for minitubes. After the supernatant is aspirated, the pellet is washed in 0.5 ml assay buffer (2–6°) and centrifuged a second time. We have found fixing the cells to be critical to obtaining good pellets on centrifugation of the sample.

7. The pellet from step 6 is resuspended in 200 μl of assay buffer containing the antibody of interest. Antibodies to CD18 are prepared at 50 μg/ml in assay buffer and are added to all tubes excluding the baseline controls. The cells are mixed and incubated at 2–6° for 60 min. After incubation, the cells are centrifuged (400 g, 5 min, 4°) twice, with the supernatant being aspirated and the pellet washed in 0.5 ml assay buffer (2–6°).

8. The pellet from step 7 is resuspended in 200 μl of a 1 : 200 dilution of the FITC conjugate in assay buffer. The FITC conjugate is also added to the baseline controls. The cells are mixed and incubated at 4° for 30 min.

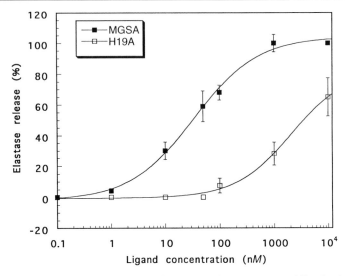

FIG. 1. Elastase release from cytochalasin B-treated human neutrophils stimulated with MGSA and mutant H19A. Elastase activity was measured as the amount of substrate (methoxysuccinyl-alanyl-alanyl-prolyl-valyl-p-nitroanilide) converted to p-nitroanilide per minute per 10^6 cells and has been normalized as percentage biological activity. Data are mean values from four separate experiments $\pm$ SEM.

9. At the end of the incubation the cells are centrifuged (400 g, 5 min, 4°) twice, discarding the supernatant by aspiration and washing the pellet in 0.5 ml of assay buffer (2–6°). Then 400 μl of assay buffer is added to each tube and the cell pellet is mixed.

10. The mean fluorescence intensity (MFC) of the CD18 or other markers for the IL-8 stimulated neutrophils and the anti-IL-8/IL-8 treated neutrophils are measured by flow cytometry. The MFC between the baseline cells, the IL-8 stimulated cells, and the anti-IL-8/IL-8 cells are compared and evaluated for changes in the MFC.

In our hands, the FACScan CD18 assay is the most sensitive assay for neutrophil function. We have used it to assess the biological activity of MGSA and MGSA alanine scan mutants.[5] MGSA, like IL-8,[4] can upregulate the expression of CD18 in neutrophils, which is important for extravasation. Thus, we have compared the ability of MGSA to stimulate neutrophil function by both the elastase release assay and the FACScan CD18 assay. The relative potencies of an H19A mutant of MGSA were compared to that of the wild-type MGSA in both assays (Figs. 1 and 2). As shown in

[5] J. Hesselgesser, C. Chitnis, L. Miller, D. J. Yansura, L. Simmons, W. Fairbrother, C. Kotts, C. Wirth, B. Gillece-Castro, and R. Horuk, *J. Biol. Chem.* **270,** 11472 (1995).

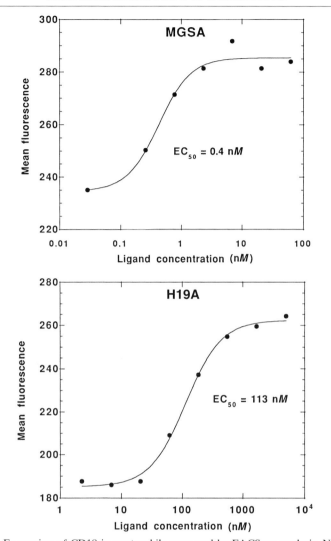

FIG. 2. Expression of CD18 in neutrophils measured by FACScan analysis. Neutrophils were incubated in the presence of increasing concentrations (0 to 1 μM) of MGSA (top) and H19A (bottom) for 2 hr at 37°. To determine CD18 upregulation the cells were incubated with an anti-CD18 antibody as described in the assay protocol. CD18 up-regulation was detected using an anti-human F(ab')2 FITC-conjugated antibody and analyzed using a FACScan (Becton Dickinson) flow cytometer.

Fig. 1, the concentration of the H19A mutant required to induce a half-maximal biological effect in the elastase release assay was 3595 nM compared to 27 nM for wild-type MGSA, a 133-fold reduction in potency. The FACScan analysis of the expression of CD18 was more sensitive than the elastase release assay of neutrophil function, and the half-maximal effect was observed with an MGSA concentration of 0.4 nM compared to 115 nM for the H19A mutant, a 282-fold reduction in potency (Fig. 2).

Elastase Release Bioassay

The elastase assay measures the ability of the C-X-C chemokines to stimulate the release of elastase from activated neutrophils. Opsonized zymosan A, which is coated with both complement and IgG, serves as a positive control in the assay. Binding of the complement fraction C3bi on the opsonized zymosan A to CD11b/CD18 receptors stimulates the neutrophils to release elastase from intracellular azurophil storage granules. Elastase concentrations are determined by measuring the protease's activity on MeOSuc-Ala-Ala-Pro-Val-PNA, a synthetic nitroanilide substrate.

Reagents

Zymosan A (Sigma)
MeOSuc-Ala-Ala-Pro-Val-PNA (Calbiochem, La Jolla, CA)
BSA, fraction V (Sigma)

Stock Solutions

Assay buffer
  10 ml of 1 M HEPES, pH 7.2
  2 g α-D-Glucose, anhydrous
  0.353 g Sodium bicarbonate (NaHCO$_3$)
  10 g BSA, fraction V (Sigma or equivalent)
  Prepare the reagents in Hanks' balanced salt solution (HBSS) with Ca^{2+}/Mg^{2+} 1× Liquid without phenol red (GIBCO, Grand Island, NY, or equivalent) and bring the volume to 1 liter. Store at 2–8°. Expiration is 1 month from date of preparation. Note: BSA, fraction V, should be used in the assay buffer. Other types of BSA, such as RIA grade, do not work as well.
Zymosan A stock solution: Add 100 mg of zymosan A to 10 ml of sterile water and boil for 1 hr. After washing the zymosan twice, resuspend at 50 mg/ml in saline. This stock solution can be stored at 2–8° for 1 month.

Elastase substrate stock solution: Dissolve 50 mg of MeOSuc-Ala-Ala-Pro-Val-PNA in 8 ml dimethyl sulfoxide (DMSO). This reagent can be stored at 2–8° for 2 weeks from date of preparation.

Elastase substrate buffer
8 ml of 5 M NaCl
4 ml 1 M HEPES, pH 7.2
4 ml Elastase substrate stock solution
18 ml Sterile water
Prepare the elastase substrate buffer immediately before use.

Assay Procedure

1. Neutrophils are isolated as described above and resuspended in Buffer A to a final concentration of 10^7 cells/ml.

2. The neutrophils are activated by incubation with cytochalasin-B (5 μg/ml) at 37° for 15 to 30 min.

3. Activated neutrophils (100 μl) are aliquoted into eight-tube strips of clean polypropylene minitubes. Polypropylene micronic tubes must be used in this assay because elastase sticks to polystyrene. It is important that the neutrophils are thoroughly mixed before each addition since they settle out quickly. However, they should be mixed with care since too vigorous manipulation can activate the cells.

4. Incubate the neutrophils with chemokines (IL-8) at room temperature for 2 hr.

5. Preparation of opsonized zymosan A: Add 12.5 mg/ml of zymosan A stock solution to fresh autologous serum and incubate for 1 hr at 37°. The opsonized zymosan A is washed twice in saline and resuspended at 3.33 mg/ml in elastase assay buffer.

6. Add zymosan A to the positive control samples: Add 75 μl of the 3.33 mg/ml opsonized zymosan A solution to each positive control minitube containing neutrophils. Mix the opsonized zymosan A carefully before each addition and incubate at room temperature for 2 hr.

7. Isolation of elastase: At the end of the 2-hr incubation place the strips of minitubes into a 96-well plate centrifuge carrier modified for minitubes. Cover the minitubes with Parafilm to contain any spills or aerosols. Centrifuge the tubes at 400 g for 10 min at 4–10° in a bench-top centrifuge.

8. Carefully transfer the top 100 μl of the elastase-containing supernatant to clean polypropylene micronic tubes using a multichannel pipettor. Do not disturb the neutrophil-containing pellet. The elastase-containing supernatant can be stored at 4° overnight.

9. For the chromogenic assay, add 30 μl of elastase-rich supernatant to each well of a 96-well flat-bottomed tissue culture plate. Add 170 μl of

elastase substrate solution to each well. Incubate the elastase-rich supernatant and elastase substrate solution for 0.5 hr at 37°. The reaction can be
stopped by adding 20 μl of glacial acetic acid per well. Read the optical
densities at 405 nm. The amount of elastase released varies significantly
with each blood donor. If the optical density (OD) of the zero standard is
less than 1.0 OD unit after a 1-hr incubation at 37°, the incubation time
can be increased to 1.5 or 2.0 hr.

Data Analysis. A standard curve is generated by plotting absorbance
at 405 nm versus IL-8 concentration (nM). Sample concentrations and
ED_{50} values are computed from a four-parameter data reduction program.
Sample concentrations should be read in the range of 10 to 100 nM. Standard curves in which the difference between the maximum and minimum
OD is less than 0.9 should be discarded. Data from typical assays are shown
in Figs. 1 and 3 (lower panel). In general the assay is not as sensitive as
the FACScan assay for CD18 up-regulation, with which it is compared in
Fig. 2.

β-Glucuronidase Assay

Reagents

Hanks' balanced salt solution (GIBCO)
Cytochalasin B (Sigma)
4-Methylumbelliferyl-β-D-glucuronide (Calbiochem)

Stock Solutions

Assay buffer: Hanks' balanced salt solution (HBSS, GIBCO) supplemented with 1% (w/v) BSA (fraction V, Sigma), 2 mg/ml glucose,
4.2 mM NaHCO$_3$, and 10 mM HEPES, pH 7.2
Substrate solution: 10 mM 4-Methylumbelliferyl-β-D-glucuronide,
made up in 0.1 M sodium acetate, pH 4.0, 0.1% (v/v) Triton X-100
Stop solution: 50 mM Glycine, 5 mM EDTA, pH 10.4

Assay Procedure

1. Neutrophils are isolated as previously described and suspended in
assay buffer at a concentration of 1×10^7 cells/ml.
2. Chemokines are serially diluted in assay buffer, and 50-μl aliquots
are added to sterile polypropylene 96-well plates (Costar, Cambridge, MA)
such that the final concentrations after addition of neutrophils would range
between 0.01 and 1000 nM.

3. After stimulation for 15 min at 37° with 5 μg/ml cytochalasin B, 100 μl of the neutrophil suspension is added to each sample well, and the plates are incubated at room temperature for up to 3 hr. The neutrophils are pelleted by centrifuging the plates at 500 g for 7 min, and the cell-free supernatants are harvested.

4. Neutrophil lysates are prepared by mixing equal volumes of the neutrophil suspension with 0.4% Triton X-100 in HBSS and incubating on ice for 20 min or longer. The lysates are diluted to the same buffer composition as the cell free lysates and cleared of particulate material by centrifugation as described for the samples. Samples and lysates are stored at 4° until they are analyzed for β-glucuronidase activity.

β-Glucuronidase Activity

1. For measuring β-glucuronidase activity, we have adapted the method of Dewald and Baggiolini[6] to a 96-well format. Samples (25 μl) are mixed with equal volumes of HBSS plus 0.1% Triton X-100 in 96-well polystyrene tissue culture plates. Fifty microliters of substrate solution is added to each sample, and the plates incubated at 37° in a tissue culture incubator for 90 min.

2. The reaction is stopped by diluting 20 μl of each reaction mixture with 290 μl/well of stop solution. The plates are then read in a microplate fluorometer (Cambridge Technology, Watertown, MA) using an excitation wavelength of 360 nm and an emission wavelength of 460 nm.

Elastase and β-glucuronidase are both contained in the azurophilic granules of neutrophils and typically show very similar patterns of release in experiments where neutrophils are stimulated with various agonists.[7] The enzymes differ substantially in the time course of their release, their assay stability, and their susceptibility to interfering substances in the samples. Analysis of supernatants from neutrophils stimulated with IL-8 for varying periods of time demonstrated that the extent of release of elastase continues to increase for 3 hr or more. Release of β-glucuronidase, in contrast, is essentially complete after a 15-min incubation (Fig. 3). The reason for this difference is unclear but can possibly be attributed to different granule subtypes. Another difference in the measurement of the activity of these two enzymes is that the β-glucuronidase reaction can be readily stopped by diluting samples in a high pH glycine solution, whereas the

[6] B. Dewald and M. Baggiolini, *Methods Enzymol.* **132,** 267 (1986).

[7] H. B. Lowman, P. H. Slagl, L. E. DeForge, C. M. Wirth, B. L. Gillece-Castro, J. H. Bourell, and W. J. Fairbrother, *J. Biol. Chem.* **271,** 14344 (1996).

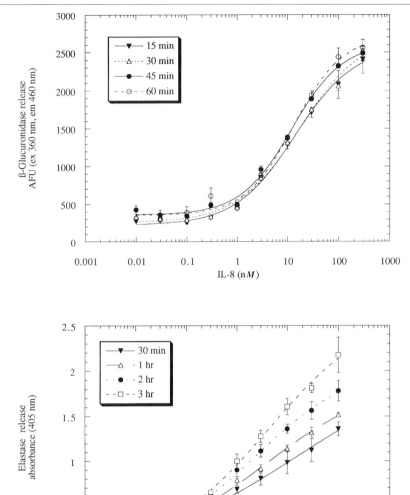

FIG. 3. Time course of release of β-glucuronidase (top) and elastase (bottom) from human neutrophils. The two graphs show differing time frames of incubation; longer incubation times do not result in a greater release of β-glucuronidase.

elastase assay is not typically run using a stopped end point. The rapid time course of β-glucuronidase release combined with a stable assay end point reduces the risk of artifacts caused by assay drift in experiments involving large numbers of samples on multiple plates; assay drift is potentially a greater problem when analyzing elastase activity.

Another key issue in assaying for β-glucuronidase and elastase activity is interference of sample components with enzyme activity. As shown in Fig. 4, elastase activity is extremely sensitive to inhibition by plasma, with concentrations as low as 0.05% (v/v) causing a precipitous drop in elastase activity. β-Glucuronidase activity, in contrast, is unaffected by plasma concentrations up to approximately 5%, although plasma components interfere with the fluorescence measurements at higher concentrations. In addition, it is often useful to express the extent of enzyme release as a percentage of the total cellular enzyme activity. Whereas β-glucuronidase activity is readily assessed in neutrophil lysates prepared using Triton X-100, substances in these lysates interfere with the determination of total cellular elastase activity.

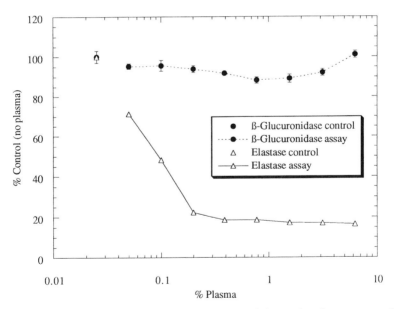

FIG. 4. Effect of plasma on release of β-glucuronidase and elastase from human neutrophils. Increasing concentrations of plasma were added to neutrophils stimulated with 100 nM IL-8. The resulting supernatants were analyzed in both assays, and the results expressed as a percentage of the enzyme activity observed in the absence of plasma.

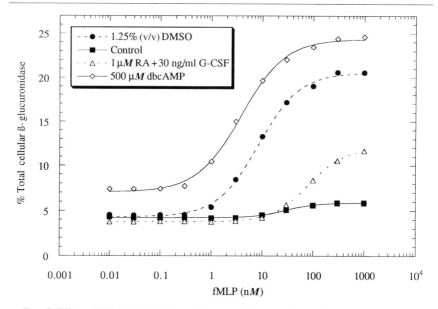

FIG. 5. Effect of HL-60 cell differentiation on β-glucuronidase release. HL-60 cells were differentiated for a period of 3 days in the presence of the indicated concentrations of DMSO, RA + G-CSF, and dbcAMP. The cells were then washed, resuspended at the same density in fresh medium, dosed with the same concentration of differentiating agent(s), and incubated for an additional 3 days. The results show that whereas undifferentiated cells (control) show little or no response to fMLP, the differentiated cells release β-glucuronidase in response to fMLP in a dose-dependent manner.

Treatment of the promyelocytic cell line HL-60 with various agents causes the cells to differentiate into a more mature form that bears similarity to primary neutrophils. Although not uniformly suitable as a substitute, differentiated HL-60 cells can be used as a reasonable approximation of neutrophils in certain types of experiments and can circumvent some disadvantages associated with the use of primary neutrophils, such as the time required to isolate the cells and interdonor variation in response. In the experiment shown in Fig. 5, HL-60 cells were induced to differentiate by treatment for 6 days with either 1.25% (v/v) DMSO,[8] 1 μM retinoic acid (RA) combined with 30 ng/ml granulocyte colony-stimulating factor (G-CSF),[9] or 500 μM dibutyryl cyclic AMP (dbcAMP).[10] Although these

[8] P. E. Newburger, M. E. Chovaniec, J. S. Greenberger, and H. J. Cohen, *J. Cell Biol.* **82,** 315 (1979).

[9] A. Sakashita, T. Nakamaki, N. Tsuruoka, and Y. Honma, *Leukemia* **5,** 26 (1991).

[10] M. Hozumi, T. J. Chaplinski, and J. E. Niedel, *J. Clin. Invest.* **70,** 953 (1982).

cells contained little or no elastase activity, differentiation with dbcAMP, DMSO, and, to a lesser extent, RA + G-CSF resulted in a cell population that released β-glucuronidase activity in response to the chemotactic peptide formylmethionylleucylphenylalanine (fMLP) in a dose-dependent manner, similar to the response seen in freshly isolated neutrophils (Fig. 5). The utility of differentiated HL-60 cells may be limited to studies involving only certain agonists, however, as a much lower level of β-glucuronidase release is seen on stimulation of the cells with IL-8 (data not shown).

Oxidative Burst Assay

Activation of human neutrophils occurs in response to a variety of particulate as well as soluble membrane stimulants. When cells are activated by invading pathogens, cellular migration to the site of infection is followed by the binding and phagocytosis of the pathogens, activation of the oxidative burst, and finally the release of degradative granular enzymes. The respiratory burst response involves enhanced oxygen consumption, glucose oxidation via the hexose monophosphate shunt, and production of reactive oxygen intermediates initiated by the NADPH oxidase-catalyzed reduction of O_2, using NADPH as the electron donor.[11] One method of monitoring the oxidative burst of the stimulated neutrophils is by measuring the generation of hydrogen peroxide (H_2O_2). When neutrophils are incubated with dichlorofluorescein diacetate (DCFH-DA), a stable, nonfluorescent, nonpolar compound, the DCFH-DA diffuses into the cells where it is deacetylated by intracellular esterases to nonfluorescent DCFH. This DCFH, which is polar and trapped within the cell, then serves as a substrate for the hydrogen peroxide generated during the oxidative burst, forming the polar and highly fluorescent 2′,7′-dichlorofluorescein (DCF).[11–13] The fluorescence generated within the neutrophils can then be measured by flow cytometry.

The assay described uses zymosan opsonized with both complement and immunoglobulin (opsonized zymosan) and zymosan opsonized with heat-inactivated serum (IgG–zymosan), as well as phorbol myristate acetate (PMA), to activate the neutrophils. Opsonized zymosan binds to CR3 (CD11b/CD18 receptor) via its complement fraction C3bi and to the Fc receptor (FcR) via the IgG coated on its surface, whereas the IgG–zymosan binds only to the FcR on the neutrophil surface. The oxidative burst trig-

[11] D. A. Bass, J. W. Parce, L. R. Dechatelet, P. Szejda, M. C. Seeds, and M. Thomas, *J. Immunol.* **130**, 1910 (1983).
[12] A. R. Rosenkranz, S. Schmaldienst, K. M. Stuhlmeier, L. Chen, W. Knapp, and G. J. Zlabinger, *J. Immunol. Methods* **156**, 39 (1992).
[13] J. M. Zeller, L. Rothberg, and A. L. Landay, *Clin. Exp. Immunol.* **78**, 91 (1989).

gered in neutrophils by PMA, however, is receptor independent, with PMA diffusing into the cell. The assay can be adapted to measure chemokine-induced oxidative burst simply by including chemokines such as IL-8 in the protocol. IL-8 stimulates oxidative burst at similar concentrations to the previous assays.[14]

Reagents

2',7'-Dichlorodihydrofluorescein diacetate (DCFH-DA) (Molecular Probes, Eugene, OR)
Phorbol 12-myristate 13-acetate (PMA) (Sigma)
N-Ethylmaleimide (NEM) (Sigma)

Stock Solutions

Opsonized zymosan and IgG–zymosan: This reagent is prepared using pooled human serum as follows. (1) Forty milliliters of human serum is collected; 20 ml of the serum is heated for 30 min at 56° to inactivate complement, and the remaining 20 ml serum is left with the complement intact. (2) A 50 mg/ml zymosan A stock solution is prepared as described for the elastase release bioassay. (3) Add 5 ml of 50 mg/ml zymosan A stock solution to each of the two serum-containing tubes and incubate for 1 hr at 37°, mixing every 10 min. The opsonized zymosan and IgG–zymosan are then washed twice in 0.15 M NaCl and resuspended in assay buffer at 13.8 mg/ml. Each zymosan preparation is divided into Eppendorf tubes and stored at −20° until needed.
Assay buffer: PBS containing 0.1% gelatin, 5 mM glucose, 0.5 mM MgCl$_2$. Sterile filter through a 0.2-μm filter and store at 2–8°. Expiration is 2 weeks from date of preparation.
10× lysing solution
1.68 M NH$_4$Cl
0.1 M KHCO$_3$
1.27 M EDTA, tetrasodium salt
Dissolve reagents in distilled water. Adjust pH to 7.3 and bring to 1 liter. Store in a tightly closed container at 2–8°. Expiration is 6 months from date of preparation.
DMSO:H$_2$O: Mix 1:1 (v/v) (stable for 1 month at 2–8°)

---

[14] M. P. Wymann, V. von Tscharner, D. A. Deranleau, and M. Baggiolini, *Anal. Biochem.* **165,** 371 (1987).

PMA stock solution (3.24 mM): Dissolve 10 mg PMA in 5.0 ml DMSO. Divide into aliquots and store at $-20°$ in the dark. Expiration is 3 months from date of preparation.

PMA working solution (79 mM)

1.0 ml DMSO/H_2O

25 ml PMA stock solution

Store in dark until use. Prepare fresh daily.

DCFH-DA stock solution: 5 mM in ethanol. Store at $-20°$ in the dark.

NEM stock solution: 100 mM NEM in distilled water. Store at 2–8°. Expiration is 3 months from date of preparation.

NEM working solution: 23 mM NEM in assay buffer. Prepare fresh daily.

Assay Procedure

Preparation of Lysed Whole Blood

1. Using blood collected on heparin, obtain the white blood cell count (WBC/ml) and percent lymphocytes by Coulter counter. Assume 10% monocytes and calculate the percent neutrophils/2 ml. Calculate the volume of assay buffer to adjust the neutrophil concentration to 7×10^5 cells/ml.

2. Add 2.0 ml heparinized whole blood to 28 ml of $1\times$ lysing solution in a 50-ml polypropylene centrifuge tube. Mix by inversion and incubate 5 min at ambient temperature. Do not extend this incubation beyond 5 min.

3. Centrifuge the tubes (300 g) for 5 min at ambient temperature and aspirate the supernatant, leaving approximately 1–1.5 ml of the lysing solution. Quickly add 30 ml of PBS to the cell pellet and mix by inversion. Avoid forming bubbles.

4. Centrifuge the tubes as in step 3 and aspirate the supernatant, leaving approximately 1–1.5 ml of the PBS solution. Transfer cell pellets to a 15-ml polypropylene centrifuge tube and add PBS to 14 ml. Centrifuge as above, aspirate as much of the supernatant as possible, and resuspend pellet in 0.5 ml of assay buffer. Add the remaining assay buffer as calculated in step 1 above and mix.

Loading Cells with DCFH-DA. DCFH-DA is added to the neutrophils to a final concentration of 5 mM, and the cells are incubated for 15 min, gently mixing twice, in the dark at 37° for dye loading.

Stimulation of Oxidative Burst and Measurement of Fluorescence

1. Thaw the opsonized zymosan A and IgG–zymosan A. Break up aggregates with a tuberculin syringe by pulling up and expelling zymosan A approximately 20 times. Avoid foaming.

2. Arrange 16 clean 12 × 75 mm polypropylene tubes in a test tube rack. Label the tubes in duplicate as shown below. Add the following to each pair of tubes:

PMA baseline control and PMA time control: 1 μl of 50/50 (v/v) DMSO and 60 μl of assay buffer

PMA: 1 μl of 79 mM PMA and 60 μl of assay buffer

Zymosan baseline control and zymosan time control: 60 μl of assay buffer

Opsonized zymosan: 50 μl of 13.8 mg/μl opsonized zymosan and 10 μl of assay buffer

Opsonized zymosan and NEM (negative control): 50 μl of 13.8 mg/μl opsonized zymosan, 10 μl of 23 mM NEM, and 10 μl of assay buffer

IgG–zymosan: 50 μl of 13.8 mg/μl IgG–zymosan and 10 μl of assay buffer

3. Cover tubes with Parafilm and store at ambient temperature for up to 2 hr. To start the reaction, add 0.4 ml of the DCFH-DA loaded cells to each tube and incubate in a 37° water bath as follows:

a. PMA baseline control and zymosan baseline control: 0 min
b. PMA time control and PMA: 10 min
c. Zymosan time control, opsonized zymosan, opsonized zymosan + NEM, and IgG–zymosan: 30 min

4. Terminate the incubation by transferring the tubes to an ice bath and store in the dark for 30 min before reading. The fluorescence signal is stable for 2.5 hr after the initial 30 min on ice.

5. Measure the median fluorescence intensity (MFC) of the cells in each tube by flow cytometry.

For optimal sensitivity in the oxidative burst assay, a number of assay parameters need to be considered. These include the choice of anticoagulant, the stability of the activated neutrophils and of the DCFH-DA stock solution in the assay, the effect of magnesium and calcium ions, and the opsonized zymosan A concentration in the assay.

Of the anticoagulants that we have evaluated, we find that the oxidative burst in response to PMA and opsonized zymosan is greater in neutrophils obtained from heparinized blood than in neutrophils obtained from ACD (acid–citrate–dextran) anticoagulated blood. In neutrophils from EDTA anticoagulated blood, the oxidative burst was similar to that in neutrophils from heparinized blood; however, the use of EDTA anticoagulated blood is not recommended. EDTA is a chelator of divalent cations such as Ca^{2+}

and Mg^{2+}, and CD11b has been shown to be a calcium-dependent epitope in human neutrophils.[14]

The stability of neutrophils in whole blood was found to be variable and donor dependent, and we recommend that blood be stored no longer than 2 hr before being assayed. If that should not be feasible, all samples should be stored at ambient temperature for the same length of time before being assayed to avoid variation due to neutrophil deterioration. We have noted that PMA activated neutrophils show a gradual increase in corrected median channel fluorescence (CF) from 104 at 5 min to 142 at 3 hr, with no further increase at 4 hr. The opsonized zymosan activated neutrophils show no significant change in corrected median CF for up to 3 hr of storage on ice and a decrease of less than 10% at 4 hr. We therefore recommended that the assay is carried out as rapidly as possible and within at least 0.5–3 hr after storage on ice.

We have compared the stability of DCFH-DA stock solutions in DMSO in side-by-side assays, using neutrophils from a single donor. Cells loaded with 6-day or 27-day-old DCFH-DA stock solution showed no significant difference in activation. Thus we make up DCFH-DA stock solutions and dispense aliquots into Eppendorf tubes and store them at $-20°$ for up to 3 months. This avoids assay variations due to variable loading of cells with DCFH-DA.

Addition of Ca^{2+} ions to the assay medium is not required and does not result in an increase of the oxidative burst, because the metal-binding sites of CR3 are of relatively high affinity and are usually occupied by Ca^{2+} ions, even in Ca^{2+}-free medium.[15] Addition of Ca^{2+} ions to the assay medium only leads to homotypic aggregation of neutrophils and will require the subsequent addition of EDTA to reverse this process. Therefore, Ca^{2+} ions should not be added to the assay medium.

Finally, zymosan forms aggregates of various sizes during opsonization and freezing, and these aggregates can interfere with the gating of neutrophils in the FACS assay. Therefore, immediately prior to use, the opsonized zymosan and IgG–zymosan solutions should be dispersed by means of a tuberculin syringe.

Data Analysis

The data obtained from oxidative burst experiments can be reported in a variety of ways, using the statistics supplied with each histogram by the Becton Dickinson Flow Cytometer software.

---

[15] L. Leino and K. Sorvjarvi, *Biochem. Biophys. Res. Commun.* **187,** 195 (1992).

The FACS FL1 histogram (gated on neutrophils) is divided into two marker areas defined as follows:

Marker 1 (M1): Area of background or negative fluorescence

Marker 2 (M2): Area of positive fluorescence

The positive and negative areas of the histogram are determined by the user. In this assay, the zero-minute cell control was used as the cutoff for negative cell fluorescence. However, fluorescence in the M2 region is not always a good indicator of oxidative burst, especially when only a few highly fluorescent cells are present.

The statistics given for each of the above markers include as follows:

1. *Mean Channel Fluorescence.* Mean channel fluorescence can be used when cells show normal distribution (as for PMA stimulated neutrophils).
2. *Median Channel Fluorescence.* Median channel fluorescence should be used when the curve is skewed or biphasic, as for IgG-zymosan stimulated cells.
3. Percentage of cells present in each marker region.

We routinely use a corrected median channel fluorescence, which is calculated by multiplying the median channel fluorescence of positive cells (M2) by the percentage of positive cells in M2. This method provides a measure of both the number of positive cells, as well as their degree of fluorescence and eliminates artificially high values due to false staining of dead cells or contamination. It is, therefore, a truer measure of oxidative burst induction, or suppression, in neutrophils.

Acknowledgments

The authors thank Cindy Wirth, Claire Kotts, Susan McCabe, Paula Jardieu, Coreen Booth, Chris Jung, and Elena Lee for efforts in the development and characterization of these assays.

[7] Characterization of Quaternary Structure of Interleukin-8 and Functional Implications

By Krishnakumar Rajarathnam, Cyril M. Kay, Ian Clark-Lewis, and Brian D. Sykes

Introduction

Chemokines are a family of low molecular mass proteins (~8 to 10 kDa) characterized by four conserved cysteines and are categorized as either C-X-C or C-C chemokines. The C-X-C chemokines predominantly attract neutrophils, whereas C-C chemokines attract monocytes, lymphocytes, and eosinophils.[1,2] Interleukin-8 (IL-8), a member of the C-X-C family, is the best characterized chemokine, and the structure–function relationships of IL-8 have been the focus of a number of studies.[3–9]

The structure of IL-8 determined by nuclear magnetic resonance (NMR) and X-ray methods showed it to be a noncovalent homodimer (Fig. 1).[10,11] The monomer unit in the dimer is well structured and consists of a series of turns in the N terminus followed by three antiparallel β strands and a C-terminal α helix. The dimer interface consists of residues in the first β strand, and part of the helix and is stabilized by a number of hydrogen bonds and hydrophobic interactions. The four cysteines are involved in disulfide bond formation (Cys-7 and Cys-34; Cys-9 and Cys-50) and have been shown to be essential for tertiary structure and function. On the

[1] M. D. Miller and M. S. Krangel, *Crit. Rev. Immunol.* **12,** 17 (1992).

[2] M. Baggiolini, B. Dewald, and B. Moser, *Adv. Immunol.* **55,** 97 (1994).

[3] I. Clark-Lewis, C. Schumacher, M. Baggiolini, and B. Moser, *J. Biol. Chem.* **266,** 23128 (1991).

[4] I. Clark-Lewis, B. Dewald, M. Loetscher, B. Moser, and M. Baggiolini, *J. Biol. Chem.* **269,** 16075 (1994).

[5] C. A. Hébert, R. V. Vitangcol, and J. B. Baker, *J. Biol. Chem.* **266,** 18989 (1991).

[6] K. Rajarathnam, I. Clark-Lewis, and B. D. Sykes, *Biochemistry* **33,** 6623 (1994).

[7] K. Rajarathnam, B. D. Sykes, C. M. Kay, B. Dewald, T. Geiser, M. Baggiolini, and I. Clark-Lewis, *Science* **264,** 90 (1994).

[8] I. U. Schraufstätter, M. Ma, G. O. Zenaida, D. S. Barritt, and C. G. Cochrane, *J. Biol. Chem.* **270,** 10428 (1995).

[9] B. Moser, B. Dewald, L. Barella, C. Schumacher, M. Baggiolini, and I. Clark-Lewis, *J. Biol. Chem.* **268,** 7125 (1993).

[10] G. M. Clore, E. Appella, M. Yamada, K. Matsushima, and A. M. Gronenborn, *Biochemistry* **29,** 1689 (1990).

[11] E. T. Baldwin, I. T. Weber, R. Charles, J.-C. Xuan, E. Appella, M. Yamada, K. Matsushima, B. F. P. Edwards, G. M. Clore, A. M. Gronenborn, and A. Wlodawer, *Proc. Natl. Acad. Sci. U.S.A.* **88,** 502 (1991).

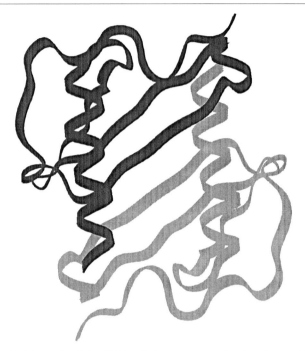

FIG. 1. Schematic of the NMR solution structure of native dimeric IL-8.

basis of the dimeric structure, which is strikingly similar to that of major histocompatibility complex HLA-1, it was proposed that the groove between the dimer interface was essential for receptor binding. However, by necessity, structural determination of proteins are done at millimolar (mM) concentrations, whereas chemokines show maximal activity at nanomolar (nM) concentrations. Naturally, this leads to the question: what is the state of association at functional concentrations? Here, we discuss the various methods that have been used to address this issue. Initially, we discuss the association state of IL-8 and related chemokines as probed by different techniques and then discuss the various studies that directly probe the importance of the dimer interface at the amino acid level.

Methods to Address Association State of Proteins

The monomer (M)–dimer (D) equilibrium and the equilibrium constant (K_a) are related by the following equations:

$$M + M \rightleftharpoons D \qquad (1)$$
$$K_a = [D]/[M]^2 \qquad (2)$$

The dissociation constant (K_d), which is more commonly used in the discussion of monomer–dimer equilibrium, is related to the equilibrium constant (K_a) by

$$K_d = 1/K_a \tag{3}$$

K_a and the thermodynamic parameters Gibbs free energy ($\Delta G°$), enthalpy ($\Delta H°$), and entropy ($\Delta S°$) are related by Eqs. (4) and (5):

$$\Delta G° = -RT \ln K_a \tag{4}$$
$$\Delta G° = \Delta H° - T\Delta S° \tag{5}$$

where R is the gas constant and T is the temperature (kelvin).

The relative amounts of the monomer and dimer present at different total concentrations of the protein for three different K_d values are given in Fig. 2. A K_d value of approximately 1 nM can be considered as strong, a K_d of approximately 1 μM as moderate, and a K_d of approximately 1 mM as weak association. The chosen experimental technique should be sensitive in the concentration range defined by the K_d. For example, techniques such as gel filtration and fluorescence anisotropy are useful from 10 nM to 10 μM and can be used to detect the association state of the protein in this range. At the lowest concentration (10 nM), these techniques will detect monomers for proteins with K_d values in the range of 1 μM to 1 mM and predominantly dimers for a protein with a K_d of 1 nM. Techniques

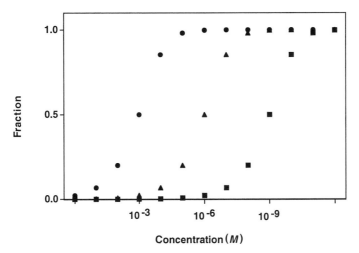

FIG. 2. Distribution profile of monomers of a protein showing monomer–dimer equilibrium as a function of total concentration for K_d values of 1 nM (■), 1 μM (▲), and 1 mM (●). The fraction of monomer (f) as a function of total protein concentration expressed in monomeric units (P_t) and dissociation constant (K_d) is calculated using the following equation: $f = [-K_d + (K_d^2 + 8P_t K_d)^{1/2}]/4P_t$.

such as NMR are effective in the 100 μM to 1 mM range. At a 1 mM concentration, NMR will detect dimers for a protein with a K_d value of 1 nM to 1 μM and equal amounts of monomer and dimer for a protein with a K_d of 1 mM.

For calculating K_d, we need to detect both monomers and dimers. In the ideal case, the technique used will measure the relative concentrations of monomers and dimers at a given single concentration. In reality, this is seldom observed as both monomer and dimer contribute to the observed property. Hence, data are collected over a large concentration range, and the amplitude of the measured property over this range is curve fitted to calculate the distribution of monomer and dimer, and hence K_d. Only techniques such as isothermal titration calorimetry provide reliable data over the entire concentration range required for calculating K_d values, the major limiting factor with other techniques being the lack of sensitivity at lower concentrations. In contrast, techniques such as sedimentation equilibrium and reversible denaturation equilibrium are not dependent on any particular property of the protein and can provide K_d from data at a single concentration (see later discussion). The latter techniques are relatively robust and are widely used for measuring K_d values.

An extended and critical review of all of the techniques that can be used to address monomer–dimer equilibrium of proteins is beyond the scope of this article, and we confine our discussion to techniques that have been used directly to address the association state and/or to measure the dissociation constant of IL-8 and related chemokines.

Monomer–Dimer Equilibrium of Chemokines

A summary of the association states and calculated dissociation constants of IL-8 and related neutrophil-activating chemokines by various techniques are shown in Table I.

Sedimentation Equilibrium

Of the different methods available for characterizing protein association, sedimentation equilibrium ultracentrifugation is the most widely used. The method requires less than 0.5 mg of protein, all of which can be recovered at the end of experiment. Proteins, with molecular masses ranging from a few kilodaltons to 10^7 Da, can be studied under a variety of conditions. The availability of the Beckman XL-A and XL-I ultracentrifuges has considerably automated all aspects of data acquisition and processing, and it is possible to quantitate weak and moderate strength interactions (K_d values

TABLE I
ASSOCIATION STATES OF IL-8, NAP-2, AND MGSA[a]

Chemokine	Technique	State[b]	K_d (μM)	pH	Temperature (°)	Buffer	Ref.
IL-8	NMR	D	—	5.1	40	20 mM Sodium acetate	10
	X-Ray	D	—	8.5	—	40% (NH$_4$)$_2$SO$_4$	11
	Sedi. eqbm.	M/D	0.6 ± 0.3	7.4	8	PBS	18
	Sedi. eqbm.	M/D	14 ± 4	7.4	25	20 mM NaP$_i$, 150 mM NaCl	17
	Calorimetry	M/D	18 ± 6	7.4	37	20 mM NaP$_i$, 150 mM NaCl	17
	Fluorescence	M	—	7.4	25	20 mM NaP$_i$, 150 mM NaCl	17
	Size exclusion	M	—	6.4	25	50 mM Tris, 150 mM NaCl	28
	Cross-linking	M/D	0.77	7.4	25	PBS	29
IL-8(1–66)	Sedi. eqbm.	M	—	5.0	20	50 mM NaP$_i$, 100 mM NaCl	This article
NAP-2	X-Ray	T	—	4.6	—	100 mM sodium acetate, 200 mM ammonium acetate	21
	NMR	M/D/T	300, 900	7.0	30	250 mM NaCl	26
	Sedi. eqbm.	M/D	53	7.0	20	50 mM NaP$_i$, 100 mM NaCl	20
	Sedi. eqbm.	M/D	102	5.0	20	50 mM NaP$_i$, 100 mM NaCl	20
	Cross-linking	M/D	0.32	7.4	25	PBS	29
MGSA	NMR	D	—	5.1	30	—	22
	NMR	D	—	5.6	30	50 mM P$_i$	23
	Sedi. Eqbm.	M/D	4	7.0	20	50 mM NaP$_i$, 100 mM NaCl	20
	Sedi. eqbm.	M/D	43	5.0	20	50 mM NaP$_i$, 100 mM NaCl	20

[a] Sedi. eqbm., Sedimentation equilibrium; PBS, phosphate-buffered saline.
[b] M, Monomer; D, dimer; T, tetramer.

from 0.1 μM to 1 mM) using this ultracentrifuge.[12,13] Although lacking the sophistication of the Beckman XL-A and XL-I, the Beckman E ultracentrifuge can provide K_d values from approximately 10 μM and upward in a reliable fashion. The equations for calculating the K_d are derived from first principles and are mathematically stringent. Refer elsewhere for excellent reviews on theory and experimental procedures.[12–15]

Sedimentation equilibrium experiments measure the equilibrium concentration profile that results when the sedimentation due to centrifugal force is balanced by diffusional transport. From the sedimentation profile at equilibrium, the molecular weight of the protein in solution can be determined. The average molecular weight (M_{av}) is calculated using the following equation:

[12] T. M. Laue, *Methods Enzymol.* **259**, 427 (1995).
[13] G. Ralston, "Analytical Ultracentrifugation," Vol. 1. Beckman Instruments, Palo Alto, California, 1993.
[14] D. K. McRorie and P. J. Voelker, "Analytical Ultracentrifugation," Vol. 2. Beckman Instruments, Palo Alto, California, 1994.
[15] J. C. Hansen, J. Lebowitz, and B. Demeler, *Biochemistry* **33**, 13155 (1990).

$$M_{av} = \frac{2RT}{(1 - \nu\rho)\omega^2} \frac{d(\ln C)}{d(r^2)} \tag{6}$$

where R is the universal gas constant, T is the temperature in kelvin, ρ is the solvent density, ω is the angular velocity (rpm $\times$ $2\pi/60$) in radians per second, ν is the partial specific volume of the protein, C is the concentration of the protein, and r is the distance from the axis of rotation.

A linear slope indicates that the protein exists as a single species, and the calculated molecular weight should correspond to that expected of a monomer or dimer (assuming a simple monomer–dimer equilibrium). A nonlinear slope indicates that multiple species exist in solution, and the calculated molecular weight will reflect the same. This is shown to be the case for neutrophil-activating peptide-2 (NAP-2, see below) as illustrated in Fig. 3. In this case, the dissociation constant for the monomer–dimer equilibrium can be calculated from Eq. (7)[16]:

$$C_r = C_0 \exp(HM\delta) + C_0^2 K_a \exp(2HM\delta) + E \tag{7}$$

where δ is $(r^2 - r_0^2)$, C_r and C_0 are the concentrations at positions r and r_0 respectively, H is equal to $(1 - \nu\rho)\omega^2/2RT$, M is the molecular weight of the monomer, K_a is the association constant for the monomer–dimer equilibrium, and E is the offset.

Two independent groups have determined the K_d for IL-8 using sedimentation equilibrium on a Beckman XL-A ultracentrifuge.[17,18] The relevant experimental details and the calculated values are given in Table I. The 20-fold difference (14 μM at 25° and 0.6 μM at 8°) between the two studies may be directly attributed to the difference in temperature, as thermodynamic calculations show that, in general, lowering the temperature favors dimer formation.[19]

NAP-2 and melanoma growth stimulating activity (MGSA) are two neutrophil-activating members of the C-X-C family (Fig. 4). We have calculated K_d values (Table I) for NAP-2 and MGSA using a Beckman E ultracentrifuge.[20] Although the structure of NAP-2 determined by X-ray

[16] M. L. Johnson, J. J. Correia, D. A. Yphantis, and H. R. Halvorson, *Biophys. J.* **36,** 575 (1981).
[17] S. D. Burrows, M. L. Doyle, K. P. Murphy, S. G. Franklin, J. R. White, I. Brooks, D. E. McNulty, M. O. Scott, J. R. Knutson, D. Porter, P. R. Young, and P. Hensley, *Biochemistry* **33,** 12741 (1994).
[18] M. S. Krangel, personal communication. The K_d of IL-8 was initially reported erroneously as 21 $\pm$ 10 μM by Paolini *et al.*[28]
[19] J. Janin, *Proteins: Struct. Funct. Genet.* **21,** 30 (1995).
[20] K. Rajarathnam, C. M. Kay, B. Dewald, M. Wolf, M. Baggiolini, I. Clark-Lewis, and B. D. Sykes, *J. Biol. Chem.* **272,** 1725 (1997).

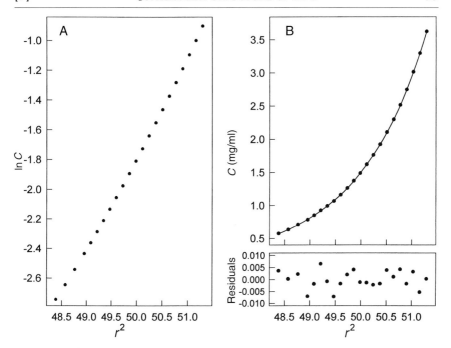

FIG. 3. (A) Average molecular weight determination by sedimentation equilibrium of native NAP-2. The sedimentation run was performed on a Beckman Spinco Model E analytical ultracentrifuge using interference optics. A plot of ln C versus r^2 is shown, and the molecular weight was calculated from the slope of the plot [Eq. (6)]. The native NAP-2 was run at an initial concentration of 1 mg/ml at 26,000 rpm in 50 mM sodium phosphate, 100 mM sodium chloride at pH 5. (B) Calculation of the dissociation constant of native NAP-2 by sedimentation equilibrium. The concentrations calculated in fringe displacement units (in mg/ml) are plotted versus r^2. The dissociation constant was calculated from curve-fitting the data using Eq. (7). Residuals of the corresponding least squares fit are also shown. There was no need to fit the data to a more complex monomer–dimer–tetramer equilibrium, and when tried there was no significant improvement of the residuals, indicating that the monomer–dimer equilibrium adequately fits the data.

```
                  5     10    15    20    25    30    35    40    45    50    55    60    65    70
IL-8              SAKELRCQCIKTYSKPFHPKFIKELRVIESGPHCANTEIIVKLSD-GRELCLDPKENWVQRVVEKFLKRAENS
Groα/MGSA       ASVATELRCQCLQTLQ-GIHPKNIQSVNVKSPGPHCAQTEVIATLKN-GRKACLNPASPIVKKIIEKMLNSDKSN
Groβ            APLATELRCQCLQTLQ-GIHLKNIQSVKVKSPGPHCAQTEVIATLKN-GQKACLNPASPMVKKIIEKMLKNGKSN
Groγ            ASVVTELRCQCLQTLQ-GIHLKNIQSVNVRSPGPHCAQTEVIATLKN-GKKACLNPASPMVQKIIEKILNKGSTN
NAP-2               AELRCMCIKTTS-GIHPKNIQSLEVIGKGTHCNQVEVIATLKD-GRKICLDPDAPRIKKIVQKKLAGDESAD
ENA-78     AGPAAAVLRELRCVCLQTTQ-GVHPKMISNLQVFAIGPQCSKVEVVASLKN-GKEICLDPEAPFLKKVIQKILDGGNKEN
PF4           EAEEDGDLQCLCVKTTS-QVRPRHITSLEVIKAGPHCPTAQLIATLKN-GRKICLDLQAPLYKKIIKKLLES
MIG             TPVVRKGRCSCISTNQGTIHLQSLKDLKQFAPSPSCEKIEIIATLKN-GVQTCLNPDSADVKELIKKWEKQVSQ
IP-10             VPLSRTVRCTCISISNQPVNPRSLEKLEIIPASQFCPRVEIIATMKKKGEKRCLNPESKAIKNLLKAVSKEMSKRSP
SDF-1           GKPVSLSYRCPCRFFES-HVARANVKHLKILN-TPNCALQIVARLKNNN-RQVCIDPKLKWIQEYLEKALNK
```

FIG. 4. Alignment of the amino acid sequences of human C-X-C chemokines.

crystallography[21] showed it to be a tetramer, sedimentation studies showed only a monomer–dimer equilibrium, and the calculated K_d values of 50 μM at pH 7.0 and 100 μM at pH 5.0 indicate a weaker dimer interface than that observed for IL-8. The NMR solution structure of MGSA[22,23] showed it to be a dimer. Sedimentation data at pH 7.0 indicated a monomer–dimer equilibrium with a K_d of approximately 4 μM and, at pH 5.0, a monomer–dimer equilibrium with a K_d of 43 μM (Table I).

Titration Calorimetry

Isothermal titration calorimetry (ITC) has been increasingly used due to improvements in instrumentation,[24] and dissociation constants up to the nanomolar region can be calculated. In ITC, the heat absorbed or released, on dimerization or dissociation, is directly measured at a constant temperature at different concentrations of the protein, which provides the dimerization enthalpy change ($\Delta H°$) and dissociation constant (K_d). In addition, dimerization heat capacity ($\Delta C°$) can be obtained by measuring $\Delta H°$ as a function of temperature, and this parameter has been shown to be useful in quantitating the relative amounts of polar and nonpolar residues that are buried on dimerization. The ITC technique does require an *a priori* knowledge of the nature of the transition as it cannot discriminate whether the transition is between a monomer and dimer or between a dimer and tetramer.

For IL-8, the heat of dissociation was measured by diluting the protein inside the calorimeter from a concentration where it is predominantly dimeric to that where it is monomeric.[17] On the basis of the enthalpy data, an equilibrium constant of 18 μM was calculated (Table I). If the structure of a protein is available, for example, for native IL-8,[10,11] it is possible to calculate the structure-based dimerization enthalpy change ($\Delta H°$) and the surface area of the amino acids that become solvent-exposed on dimer dissociation, to determine whether significant structural changes occur on dimer dissociation. For IL-8, it was observed that experimental and the calculated values were similar, indicating that the monomeric structure is similar to the monomeric unit of the dimeric structure. We have determined the tertiary structure of a monomeric IL-8[25] that cannot dimerize (see later

[21] M. G. Malkowski, J. Y. Wu, J. B. Lazar, P. H. Johnson, and B. F. P. Edwards, *J. Biol. Chem.* **270**, 7077 (1995).

[22] K.-S. Kim, I. Clark-Lewis, and B. D. Sykes, *J. Biol. Chem.* **269**, 32909 (1994).

[23] W. J. Fairbrother, D. Reilly, T. J. Colby, J. Hesselgesser, and R. Horuk, *J. Mol. Biol.* **242**, 252 (1994).

[24] E. Friere, O. L. Mayorga, and M. Straume, *Anal. Biochem.* **62**, 950 (1990).

[25] K. Rajarathnam, I. Clark-Lewis, and B. D. Sykes, *Biochemistry* **34**, 12983 (1995).

discussion), and the structure is similar but not identical to the monomeric unit in the dimer. Titration calorimetry offers this additional advantage over other methods in its ability to provide insights into structural changes that may take place on dissociation into monomers.

Nuclear Magnetic Resonance

In NMR spectroscopy, a variety of parameters such as chemical shift, line width, and relaxation rates can be measured. The chemical shift of a nucleus is extremely sensitive to its environment, and is the easiest and the first parameter that is measured. If the chemical shift of a proton is different between a monomer and dimer, it should be possible to determine a dissociation constant from the relative intensities of the resolved signals of the two species, providing they are in slow exchange. However, unlike other techniques such as fluorescence spectroscopy, NMR requires a high concentration of the protein and is a method of choice only for proteins with weak dissociation constants ($>100 \ \mu M$).

A dissociation constant of 10 μM for IL-8 rules out using NMR as a tool to address the monomer–dimer issue, as even at the lowest concentrations it is unlikely that the monomeric species will be detected. However, Yang et al. have studied the association properties of NAP-2 by NMR[26] by exploiting the resonances of tyrosine (Tyr) ring protons, which are, in general, well resolved from the resonances of the rest of the protein and are therefore useful structural probes. As NAP-2 lacks Tyr, they engineered a Tyr at the C-terminal end. On the basis of curve-fitting the spectra of the multiple peaks for Tyr, the NMR data were interpreted as due to presence of monomers, dimers, and tetramers, and the relative distribution was shown to be sensitive to changes in pH and ionic strength. At 250 mM NaCl and pH 7.0, the authors calculated a dissociation constant for the monomer–dimer equilibrium as approximately 300 μM and that for a dimer–tetramer equilibrium as approximately 900 μM (Table I). The observation of a tetramer in these experiments is consistent with the structure of NAP-2 solved by X-ray crystallography, which showed it to be a tetramer.[21] As discussed earlier, sedimentation experiments showed only a weak monomer–dimer equilibrium. This can be attributed to the fact that sedimentation experiments are performed at lower concentrations (~1 mg/ml), whereas both NMR and X-ray studies are carried out at much higher concentrations. These studies showed no differences in the K_d between pH 5.0 and pH 7.0, which is also consistent with the sedimentation data (Table I).

[26] Y. Yang, K. H. Mayo, T. J. Daly, J. K. Barry, and G. J. La Rosa, *J. Biol. Chem.* **269,** 20110 (1994).

Chemical Cross-Linking

A bifunctional cross-linker, which consists of a spacer with similar reactive groups at both ends, is commonly used for detecting noncovalent homodimers in solution. A variety of cross-linkers, with different functional groups and with spacers of different length, are commercially available.[27] Most widely used cross-linkers have esters as functional groups and hence can cross-link lysines in proteins. The proteins are radiolabeled so that the cross-linking experiments can be carried out at lower concentrations than afforded by detection using conventional methods such as Coomassie Blue or silver staining. For the cross-linking to take place, reactive groups on the protein, in this case lysines, should be optimally oriented. In an ideal case, at concentrations in which the protein is a monomer, there should be no cross-linking; at concentrations where there are dimers, cross-linking should be evident.

Two studies have addressed the association state of IL-8 by chemical cross-linking. Paolini *et al.*[28] used ethylene glycol bis(sulfosuccinimidyl succinate) (sulfo-EGS, Pierce, Rockford, IL) a cross-linker with a spacer 16 Å long and succinimide ester as the leaving group, to covalently cross-link ^{125}I-labeled IL-8. The relative ratios of monomers and dimers at different concentrations of IL-8 (from 8.3 nM to 10 μM) were visualized by sodium dodecyl sulfate–polyacrylamide gel electrophoresis (SDS–PAGE). At 10-μM concentration substantial amounts of dimers were observed, whereas at 8.3 nM no dimers were seen.

Schnitzel *et al.*[29] have addressed the association state of IL-8 in a similar fashion but using a different cross-linker, bis(sulfosuccinimidyl) suberate (BS, Pierce). The BS cross-linker is similar to sulfo-EGS except that the length of the spacer is shorter, 11 versus 16 Å. The cross-linking experiments were carried out with IL-8 concentrations ranging from 0.8 nM to 74 μM. Theoretically, it is possible to calculate a K_d by quantitating the intensities of dimers and monomers if one assumes that the intensities reflect the actual ratio of the dimers and monomers in solution. On the basis of their experiments, Schnitzel *et al.*[29] calculated a K_d of 0.77 μM. However, they observed formation of oligomers at IL-8 concentrations exceeding 20 μM. Paolini *et al.*[28] did not observe oligomer formation up to 10 μM IL-8 and it is unclear whether oligomers will be seen at higher concentration, similar to what was observed when cross-linked with BS. Formation of higher

[27] "Pierce Catalog & Handbook." Pierce, Rockford, Illinois.
[28] J. F. Paolini, D. Willard, T. Consler, M. Luther, and M. S. Krangel, *J. Immunol.* **153,** 2704 (1994).
[29] W. Schnitzel, U. Monschein, and J. Besemer, *J. Leukocyte Biol.* **55,** 763 (1994).

order oligomers is inconsistent with the structural studies by both NMR and X-ray crystallography, which show that IL-8 is a dimer up to 2 mM concentrations. In a similar fashion, Schnitzel *et al.* have also calculated the K_d for NAP-2 to be 0.32 μM, which is inconsistent with the results from both the sedimentation and the NMR data (Table I). It is noteworthy that cross-linking at concentrations up to 20 μM did not demonstrate tetramers for NAP-2 even though the X-ray structure of NAP-2 showed it to be a tetramer. In the case of IL-8, tetramers and higher order oligomers were found by cross-linking at 20 μM, and the structures showed it to be a dimer.

These two studies reveal the intrinsic limitation of cross-linking experiments, as the nature and extent of cross-linking is not known at the molecular level. The choice of the cross-linker is empirical, and accordingly the outcome of the results and hence the interpretation can vary. Although chemical cross-linking is a useful method, we suggest that it be used as a complementary method to techniques such as sedimentation equilibrium.

Thermal/Chemical Denaturation

Denaturation or unfolding is widely used as a method for calculating dissociation constants. In a denaturation experiment, the protein is unfolded by increasing the temperature or by increasing the concentration of the denaturant such as urea or guanidinium hydrochloride (GnHCl). The process is monitored by either circular dichroism (CD) or fluorescence spectroscopy. Using CD, the unfolding can be followed by monitoring the ellipticity at 220 or 280 nm. Alternately, the intensity and the wavelength of the fluorescence emission of tryptophan (Trp) are used as probes for monitoring unfolding. The denaturation process can be represented by Eq. (8):

$$D \rightleftharpoons 2M \rightleftharpoons 2U \tag{8}$$

where D is the dimer, M is the folded monomer, and U is the unfolded monomer.

If the monomer exists as a discrete species during unfolding, it should be possible to follow the two-step unfolding process and hence calculate the K_d. A number of dimeric proteins do not unfold as in Eq. (8), but unfold as indicated in Eq. (9):

$$D \rightleftharpoons 2U \tag{9}$$

In the case of native IL-8, thermal and urea denaturation were ineffective, because the protein retained substantial secondary structure up to 80° and in 8 M urea; however, it could be unfolded by GnHCl.[30] This suggests that the protein is extremely stable and that the stability arises from the hydrophobic core and the two disulfide bonds. The unfolding profile showed a two-state transition when followed by both circular dichroism and fluorescence, but the profiles were nonsuperimposable, which is indicative of the two-step unfolding mechanism via the monomeric intermediate as shown in Eq. (8).[7] In addition, a similar $[GnHCl]_{1/2}$, or GnHCl concentration at which 50% of the protein is unfolded, was obtained when denaturation was carried out at protein concentrations differing by 10-fold. This is also indicative that the protein is unfolding in a similar fashion [Eq. (8)]. These observations indicate that both CD and fluorescence are insensitive in differentiating the two steps of the unfolding process.[7] This is not surprising in the context of the solution structure of the monomer being similar to the monomeric unit in the dimer structure.[25]

The observations made for IL-8 would apply to other chemokines, implying that denaturation is not a method of choice for calculating K_d values. However, it is possible that denaturation methods may provide a dissociation constant for MIG, a C-X-C chemokine. MIG has a single Trp (Fig. 4), and if we assume that MIG adopts a structure similar to IL-8, this Trp is present at the dimer interface. In this case, unfolding followed by fluorescence will show a biphasic transition if the tryptophan is in a different environment in the dimer, monomer, and the unfolded state.

Gel Filtration/Size-Exclusion Chromatography

Gel filtration is widely used to determine the oligomerization state, but not for calculating the K_d, of proteins. It is a standard tool in any biochemistry laboratory, is easy to use, and is a good technique to obtain an idea of the association state of a protein. Initially, the elution times of a set of standard proteins of known molecular weights are obtained, and a standard curve of molecular weight versus retention time is plotted. Then, on the basis of the retention time of the unknown protein, the molecular weight and hence the oligomerization state are determined. The sensitivity of the gel filtration experiments, in terms of the lowest concentration that can be used, is dictated by the method of detection. Usually the eluants are detected by measuring the absorbance at 280 or 222 nm.

In the case of IL-8, even at the lowest concentrations the only species detected is a dimer. To look at the oligomerization state at lower concentrations than afforded by conventional detection by absorbance, Paolini *et al.*[28] metabolically labeled cysteines with ^{35}S in IL-8 and used radioactivity

---

[30] K. Rajarathnam, C. M. Kay, and B. D. Sykes, unpublished results.

to detect the elution time. They observed IL-8 to be a monomer by this method.

Methods to Disrupt Dimer Interface

The above discussion clearly indicates that at functional nanomolar concentrations, independent of the actual technique used to measure dissociation constants, chemokines exist as monomers. However, this does not rule out the possibility that dimerization may occur during or after binding at the surface of the receptor. Dimerization, if essential, may play a role by orienting critical residues, which are otherwise ineffective in the monomer, for interaction with the receptor. This can be studied by mutating residues at the dimer interface or by designing analogs with decreased propensity to form dimers. Both approaches have been used in the case of IL-8 and related chemokines.

Mutational Studies

The relationships between the structure and function of IL-8 have been extensively studied. A number of studies have shown that only residues 4 to 22 and 31 to 34 are essential for activity.[3–5] This suggests that the rest of the protein, namely, the β strands and α helix, provide a scaffold for the orientation of the functionally important residues for their interaction with the receptor. The sequences of the C-X-C chemokines are shown in Fig. 4. It is seen that residues which constitute the dimer interface are not highly conserved except for some of the hydrophobic residues. Leu-25, Val-27, and Leu-66 are residues at the dimer interface which are conserved or similar among IL-8, NAP-2, and MGSA, indicating that hydrophobic interactions stabilize the tertiary and quaternary interactions in these proteins.

Substitution of the charged residues of the first β strand at the dimer interface with residues of opposite charge (K23E, R26E, E29K, and a double mutant K23E and E24K) in IL-8 had no effect on either binding or activity.[4] Substitution of the complete β strand of IL-8 with that of IP-10, a C-X-C chemokine which is not a neutrophil activator, also had no effect on activity.[4] Similarly, substitution of Leu-25, a conserved hydrophobic residue, with Tyr had no effect on activity.[31] These substitutions, however, may or may not have an effect on dimer stability. Sedimentation equilibrium experiments showed that K23E and L25Y analogs were still largely dimeric.[30]

---

[31] I. Clark-Lewis, B. Dewald, and M. Baggiolini, unpublished results.

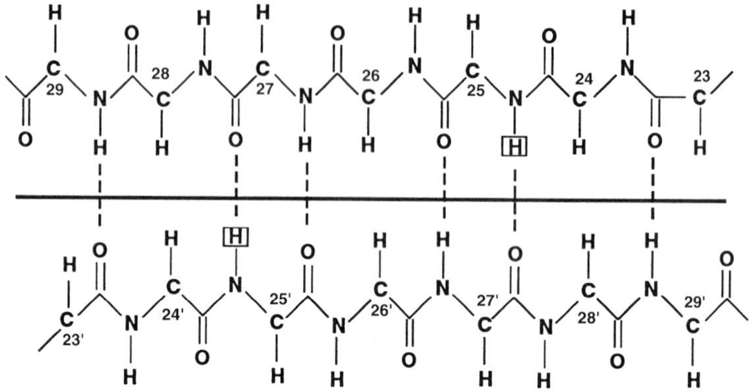

FIG. 5. Schematic presentation of the dimer interface in interleukin-8 showing the hydrogen bonding by dashed lines. The NH of Leu-25, which is boxed, was modified to NCH_3 ($NH \rightarrow NCH_3$) in the L25NMe analog.

Lowman et al.[32] used an approach of multiple mutations of the residues at the dimer interface to design a monomeric analog. The strategy was to weaken the dimer interface and at the same time retain the activity. An analog with seven substitutions (E24R, I28R, T37E, F65H, L66E, A69E, and S72E) was shown to be monomeric, with a K_d of 2 mM. The analog displayed about five-fold lower binding than the native protein.

Using Nonnatural Amino Acids at Dimer Interface

We were interested in selectively disrupting the dimer interface with minimal modification of the protein. Examination of the IL-8 structure showed that the residues of the first β strand are involved in hydrogen-bond formation across the dimer interface (Fig. 5). The use of nonnatural amino acids such as N-methyl amino acids at the interfacial site provides a novel and convenient method to disrupt the dimer interface. This can be accomplished by conventional molecular biology methods or by peptide synthesis. Clark-Lewis et al.[33] have shown that IL-8 is amenable to chemical synthesis, and approximately 20–50 mg of protein can be obtained in a routine synthesis. We adopted the chemical synthesis strategy and introduced N-methylleucine (NMe-Leu) at position 25 (NH $\rightarrow$ NCH_3), which

[32] H. B. Lowman, P. H. Slagle, L. E. DeForge, C. M. Wirth, B. L. Gillece-Castro, J. H. Bourell, and W. J. Fairbrother, *J. Biol. Chem.* **271,** 14344 (1996).

[33] I. Clark-Lewis, B. Moser, A. Walz, M. Baggiolini, G. J. Scott, and R. Aebersold, *Biochemistry* **30,** 3128 (1991).

disrupts two hydrogen bonds about the two-fold symmetry in the putative dimer (Fig. 5). In addition, the introduction of the bulky methyl group will prevent two monomers from approaching each other.

The L25NMe analog could be folded and purified in a fashion similar to that adopted for the native protein. The analog had functional characteristics that were indistinguishable from that of the native protein.[7] The L25NMe analog was shown to be a monomer by sedimentation equilibrium and NMR studies.[7] The observation that this analog is a monomer up to a concentration of 20 mg/ml indicates that it has no propensity to dimerize, and the same factors that prevent its dimerization in solution will also prevent dimerization on the receptor. The solution structure of L25NMe IL-8 was solved by NMR spectroscopy[25] and was shown to be similar but not identical to that of the monomeric unit in the dimer. The major differences are in the turn residues 31–35 and the N-terminal residues 4 to 6, which are disulfide linked via Cys-7 and Cys-34. This part of the structure is more similar to the X-ray than the NMR structure, indicating that this region exhibits significant motion. In addition, the last six residues of the C-terminal helix were unstructured in the monomer, whereas they are structured in the dimer. As the monomer is the receptor binding species, the structural features of the monomer are critical in our understanding of the structure–function relationships and for serving as a more realistic template for structure-aided design of low-molecular-weight antagonists.

Using a similar strategy, we have studied the importance of the dimer interface in NAP-2 and MGSA. Analogs of NAP-2 and MGSA were synthesized in which the backbone amide proton to Leu-22 in NAP-2 and Val-26 in MGSA were substituted with the bulky methyl group.[20] These analogs were shown to be monomeric by sedimentation equilibrium and were found to have the same activity as the native protein in neutrophil elastase release assays.

Deletion Studies

Another approach to alter the dimerization affinity would be to minimize the contribution of the C-terminal helix residues. Functional studies had indicated that deleting residues of the C-terminal helix had no effect on activity; in fact, an IL-8 analog missing residues 67–72 had a slightly higher activity than the native protein.[3] Examination of the IL-8 structure indicates that the helix stabilizes the dimer interface, and residues Leu-25, Val-27, Phe-65, and Leu-66 form a hydrophobic core. However, deletion of the terminal residues 67–72 will minimize this interaction due to entropic factors. Sedimentation studies indicate that the 1–66 analog is a monomer with no propensity to dimerize (Fig. 6).

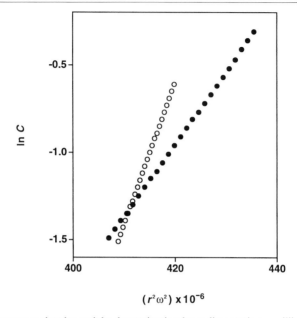

FIG. 6. Average molecular weight determination by sedimentation equilibrium of native IL-8 (○) and the IL-8(1–66) truncation analog (●). Sedimentation equilibrium experiments were performed on a Beckman Spinco Model E analytical ultracentrifuge using absorbance optics. Plots of ln C versus $r^2\omega^2$ are shown, where C is the concentration, r is the distance from the axis of rotation, and ω is the angular velocity. As the sedimentation runs of the two proteins were at different rotor speeds, the data are plotted as ln C versus $r^2\omega^2$ rather than ln C versus r^2. For clarity of presentation, a constant value of -0.5 and 240 were added to ln C and $r^2\omega^2$, respectively, for native IL-8. Molecular weight is proportional to the slope, and the calculated molecular weights of native IL-8 and of IL-8(1–66) correspond to those of a dimer and monomer, respectively. Samples were in 50 mM sodium phosphate, 100 mM sodium chloride at pH 7.

Conclusion

A variety of studies have addressed the association state of IL-8. Similarly, a number of studies have addressed the importance of the dimer interface by mutating residues at the dimer interface and designing monomers that show no propensity to dimerize. All of them consistently indicate that a monomeric unit is sufficient to bind to a neutrophil receptor for functional activation. Similarly, calculation of K_d values and functional studies of monomeric analogs of NAP-2 and MGSA that cannot dimerize indicate that the monomer is the functionally relevant species. Of the different techniques available to study the significance of the dimer interface, calculating dissociation constants by sedimentation equilibrium and designing nonassociating monomers address the issue in an unequivocal manner.

Acknowledgments

We thank Drs. Narayanaswami, Lavigne, Chao, and Crump for valuable discussions and critical reading of the manuscript and Les Hicks for performing the ultracentrifugation studies.

Section II

C-C and C Chemokines

[8] Isolation of Human Monocyte Chemotactic Proteins and Study of Their Producer and Responder Cells by Immunotests and Bioassays

By Jo Van Damme, Sofie Struyf, Anja Wuyts,
Ghislain Opdenakker, Paul Proost, Paola Allavena,
Silvano Sozzani, and Alberto Mantovani

Introduction

Historical Background

Before the term "chemokines" was assigned to a novel group of small proinflammatory proteins with four conserved cysteines, the C-X-C chemokine platelet factor 4 (PF4, originally discovered as a platelet granule protein with high affinity for heparin) had already been described as a chemoattractant of both neutrophils and monocytes. Monocyte chemotactic protein-1 was the first C-C chemokine that was selectively isolated on the basis of its capacity to attract monocytes in an *in vitro* migration assay. In the meantime, other C-C chemokines, originally characterized as macrophage-derived proinflammatory proteins (MIP-1), were found to possess other unrelated properties including stimulation or inhibition of hematopoiesis. Regulated on activation normal T cell expressed and secreted (RANTES), another C-C chemokine, was molecularly identified by subtractive hybridization, and the recombinant protein was chemotactic for both monocytes and T lymphocytes. Subsequently, MCP-1 and the C-X-C chemokine interleukin-8 (IL-8) were reported to attract lymphocytes. In this chapter our strategy to isolate and identify novel monocyte chemotactic proteins (MCPs) from natural cellular sources is discussed. Methods to measure the chemotactic potential and selectivity of the related chemokines MCP-1, -2, and -3[1] toward various leukocyte populations are described. In addition, specific immunological assays to study chemokine expression and to detect MCPs in body fluids during disease are discussed. Finally, methods to investigate differences in receptor usage and signal transduction pathways are summarized.

---

[1] P. Proost, A. Wuyts, and J. Van Damme, *J. Leukocyte Biol.* **59,** 67 (1996).

Migration Assays as Tools to Isolate Novel Chemokines: Conditions and Limitations

Many novel cytokines and chemokines have been discovered by the classic approach of purifying natural ligands from crude conditioned media of *in vitro* cultured and stimulated cells. Essential for this strategy is the use of a bioassay, which should be as sensitive and specific as possible. As an example, the development of a hybridoma cell line, which is dependent for its growth on low concentrations (pg/ml) of a factor from cell culture conditioned medium, allowed for the isolation of IL-6. In addition, such a bioassay needs to be reproducible and applicable to test large series of column fractions obtained during the purification of the novel factor through classic chromatographic procedures.

The *in vitro* migration of phagocytes, using the Boyden chamber as a test system, has been studied for decades. However, testing many samples was not feasible until a microchamber for chemotaxis became available.[2] In addition, the visual counting of cells that migrated through the micropore filters remains labor intensive even today. This probably explains in part why most chemokines have been identified only rather recently. Furthermore, the chemotaxis assay is not sufficiently sensitive (minimal effective chemoattractant concentration of 1 nM) nor specific (several chemokines are active on the same cell type). Nevertheless, about half of the known chemokines were identified by protein purification with the help of this type of bioassay. Other related molecules have been discovered using molecular cloning technology. They were classified as putative chemokines merely on the basis of their primary protein structure characterized by the conservation of four cysteine residues. After subsequent expression and testing of recombinant protein in the chemotaxis assay, authentic chemokine activity was later confirmed for some but not all molecules. As an example, human IL-8 was molecularly identified by cDNA cloning (without functional characterization), but the protein was independently isolated from cell cultures by virtue of its chemotactic activity and has subsequently been identified by amino acid sequence analysis.

In vivo migration tests have served as a useful tool to identify novel chemokines such as IL-8 and eotaxin.[3,4] For *in vivo* tests, the factors need to be injected intradermally in laboratory animals (e.g., guinea pigs or rabbits), and the chemotactic effect is evaluated by measuring the inflam-

[2] W. Falk, R. H. Goodwin, Jr., and E. J. Leonard, *J. Immunol. Methods* **33,** 239 (1980).
[3] J. Van Damme, J. Van Beeumen, G. Opdenakker, and A. Billiau, *J. Exp. Med.* **167,** 1364 (1988).
[4] P. J. Jose, D. A. Griffiths-Johnson, P. D. Collins, D. T. Walsh, R. Moqbel, N. F. Totty, O. Truong, J. J. Hsuan, and T. J. Williams, *J. Exp. Med.* **179,** 881 (1994).

matory response, including the local cell infiltration in skin biopsies. To increase the reproducibility and sensitivity of such *in vivo* tests, leukocytes can be radiolabeled. The radioactivity measured in biopsies functions as a parameter for cell accumulation.[5] Because the number of samples that can be tested in a single animal is limited, such a complex *in vivo* assay is better tuned to study the physiological significance of a pure chemokine, rather than to assay large series of column fractions.

Purification and Identification of Natural Human Monocyte Chemotactic Proteins

Cellular Sources of Natural Chemokines

In our laboratory, a number of C-X-C chemokines [IL-8, neutrophil-activating peptide-2 (NAP-2), and granulocyte chemotactic protein-2 (GCP-2)] and C-C chemokines (MCP-1, -2, and -3), as well as natural truncated derivatives thereof, have been biochemically and biologically characterized by isolation of the proteins from natural cellular sources using the microchamber chemotaxis assay.[6,7] The developed strategy consists of the production and purification of sufficient amounts of natural chemotactic protein to identify completely its primary structure by amino acid sequence analysis. For that purpose, two major cellular sources have been used, namely, connective tissue cells grown *in vitro* on a large scale and leukocytes freshly isolated from buffy coats derived from blood donations. Although cultured diploid fibroblasts are good producers of chemokines, growing these cells *in vitro* is hampered by their finite life span, and their growth rate is strongly influenced by the quality of the tissue culture conditions (growth medium and physiological parameters). For this reason, we have developed in our laboratory tumor cell lines, for example, MG-63 osteosarcoma cells,[8] selected on the basis of their *in vitro* growth characteristics and their capacity to produce cytokines. As an alternative, the use of unfractionated leukocytes has allowed for the isolation of more cell specific chemokines, including the platelet-derived chemokines PF4 and β-thromboglobulin (βTG)/NAP-2.[7]

Proper stimulation of the cells is essential to obtain maximal chemokine secretion in culture media containing a minimal protein content (possibly

[5] M. Rampart, J. Van Damme, L. Zonnekeyn, and A. G. Herman, *Am. J. Pathol.* **135,** 21 (1989).

[6] P. Proost, C. De Wolf-Peeters, R. Conings, G. Opdenakker, A. Billiau, and J. Van Damme, *J. Immunol.* **150,** 1000 (1993).

[7] J. Van Damme, B. Decock, J.-P. Lenaerts, R. Conings, R. Bertini, A. Mantovani, and A. Billiau, *Eur. J. Immunol.* **19,** 2367 (1989).

[8] J. Van Damme and A. Billiau, *Methods Enzymol.* **78,** 101 (1981).

serum-free). For MG-63 osteosarcoma cells, we first studied which inducers are most appropriate. Good production of interferon-β (IFN-β), IL-6, granulocyte–macrophage colony-stimulating factor (GM-CSF), and IL-8 is obtained after induction with IL-1. However, to obtain synergistic induction, a mixture of cytokines (e.g., IL-1 and IFN-γ) has often been used as a stimulus. Mononuclear leukocytes, which form an alternative cellular source for cytokines, are stimulated with a mixture of endotoxin (1 μg/ml) and mitogen (e.g., concanavalin A at 1 μg/ml) to induce chemokine production in monocytes and lymphocytes, respectively. As a consequence, a chemokine cocktail is generated to allow simultaneous isolation of different natural proteins from a single batch. In principle, an identified cytokine (e.g., IL-6) or chemokine (e.g., IL-8) is used as a marker to test the induction level and the recovery of biological activity during the purification procedure. This is done with a specific bioassay (IL-6)[9] or with a selective immunotest (IL-8).

Procedure for Chemokine Isolation

The procedure to purify and identify monocyte chemotactic proteins produced by human MG-63 osteosarcoma cells is similar to that for granulocyte chemotactic proteins described in detail elsewhere in this volume.[9a] Because no specific reagents (e.g., monoclonal antibodies) are available for the isolation of novel molecules, the proteins are purified by consecutive steps using conventional chromatographic methods. During the initial purification step it is essential to reduce the large volume of crude conditioned medium (1 to 10 liters) containing the chemokine activities. Batch adsorption to controlled pore glass or to silicic acid beads results in a 10- to 20-fold reduction of the initial volume. Importantly, this method also allows for a partial purification of chemokines (20- to 40-fold) and a good recovery (up to 75%) of biological activity. Because the activity can be eluted with a pH 2.0 buffer, this indicates that most chemokines are acid-stable. The second purification step on a heparin-Sepharose column is based on the affinity of chemokines for heparin (MCPs have a high isoelectric point). The human, tumor cell-derived monocyte chemotactic proteins MCP-1, -2, and -3 elute at 0.7, 1.1, and 1.0 M in the NaCl gradient, respectively

[9] J. Van Damme, J. Van Beeumen, B. Decock, J. Van Snick, M. De Ley, and A. Billiau, *J. Immunol.* **140,** 1534 (1988).

[9a] A. Wuyts, P. Proost, G. Froyan, A. Haelens, A. Billiau, G. Opdenakker, and J. Van Damme, *Methods Enzymol.* **287** [2], 1997 (this volume).

TABLE I

PURIFICATION PROCEDURE AND BIOCHEMICAL CHARACTERISTICS OF NATURAL MONOCYTE CHEMOTACTIC PROTEINS

Chromatographic purification step[a]	Elution condition	Chemokine behavior/elution			
		MCP-1	MCP-2	MCP-3	PF4
CPG adsorption	pH 2.0	Stable	Stable	Stable	Stable
Heparin-Sepharose	0–2 M NaCl	0.7 M	1.1 M	1.0 M	2.0 M
Cation-exchange FPLC	0–1 M NaCl	0.45 M	0.7 M	0.6 M	1.0 M
RP-HPLC	0–80% CH$_3$CN	26%	30%	27.5%	ND

[a]For further details, see Wuyts et al.[9a]

(Table I). Owing to these minor but relevant differences in the affinity for heparin, the MCPs can be partially separated by this step. In contrast, the bulk of monocyte chemotactic activity from thrombin-stimulated platelets elutes at 2 M NaCl.[7] This activity corresponds to the chemokine PF4, and this is in agreement with the reported high affinity of this chemokine for heparin. Thus, if unfractionated buffy coats are used as a cell substrate to produce chemokines, PF4 (but not βTG) can be separated from MCPs by heparin-Sepharose chromatography.

Further purification and fractionation of MCPs is achieved by cation-exchange fast protein liquid chromatography (FPLC) at pH 4.0. This yields three distinct peaks of monocyte chemotactic activity separated in the NaCl gradient (Fig. 1). These peak fractions contain proteins of 10 kDa (0.45 M), 7 kDa (0.7 M), and 11 kDa (0.6 M) corresponding to MCP-1, MCP-2, and MCP-3, respectively (Table I). The identity of chemokines can normally be obtained by sequence analysis after final purification to homogeneity by reversed-phase high-performance liquid chromatography (RP-HPLC). Because the three MCPs show a distinct elution pattern on HPLC (26, 30, and 27.5% acetonitrile for MCP-1, -2, and -3, respectively, see Table I), complete separation is obtained at this stage. This excludes the possibility that the activity measured with the small amounts of recovered MCP-2 and MCP-3 is due to cross-contamination with the larger quantities of MCP-1 (Fig. 1).

The identity of the MCPs could not be disclosed by NH$_2$-terminal sequence analysis because the pure proteins are blocked for Edman degradation with a pyroglutamate residue. After proteolytic cleavage with trypsin and endoproteinases Asp-N, Lys-C, and Asn-C and subsequent purification of the fragments by RP-HPLC, the near complete amino acid sequence of human MCPs was determined (for technical details, see Wuyts et al.[9a]).

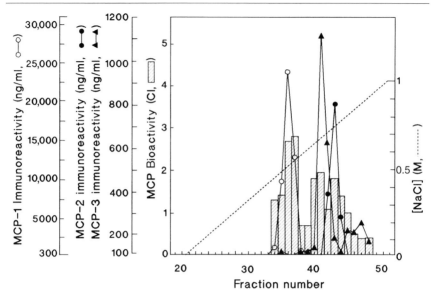

Fɪɢ. 1. Fractionation of natural MCPs by Mono S cation-exchange chromatography (FPLC). After heparin-Sepharose affinity chromatography, fractions containing monocyte chemotactic activity purified from osteosarcoma cell-derived conditioned medium were dialyzed against 50 mM formate, pH 4.0, and loaded on a Mono S (Pharmacia) column. Proteins were eluted in a linear NaCl gradient, and MCP-1, -2, or -3 immunoreactivity was measured by radioimmunoassay.[20] Monocyte chemotactic activity was determined in the microchamber assay. The chemotactic index (CI) was calculated by dividing the number of cells that migrated toward the samples by the number of cells that migrated to the control medium.

Determination of Monocyte Chemotactic Protein-Responsive Cell Types

Synthetic Chemokine as Alternative for Natural Monocyte Chemotactic Proteins

Before starting a detailed investigation of all possible target cells for a novel chemokine, it is important to obtain a standard preparation of high purity and potency. Because only small amounts (Fig. 1) of natural MCP-2 and MCP-3 (in contrast to IL-8 and MCP-1) can be obtained, a more efficient production method is necessary. Because chemokines are rather small proteins (70–80 amino acids), automated chemical solid-phase peptide synthesis can serve as a fast alternative to molecular cloning. Different chemical approaches [9-fluorenylmethoxycarbonyl (Fmoc) and *tert*-butoxycarbonyl (tBoc) protecting groups] are available to synthesize chemokines at a 0.1- to 1-mmol scale. The method of Fmoc chemistry is described elsewhere in this volume for the synthesis of human GCP-2.[9a] Briefly, after chemokine chain assembly, the peptide is cleaved from the resin, and the side-chain-protecting groups are removed. Intact synthetic chemokine is subsequently purified from incomplete peptides by RP-HPLC and the two disulfide bridges (Cys[1]-Cys[3] and Cys[2]-Cys[4]) essential for biological activity of chemokines are formed in a mixture of oxidized and reduced glutathione. Folded chemokines are finally purified to homogeneity by RP-HPLC and biochemically analyzed by sodium dodecyl sulfate–polyacrylamide gel electrophoresis (SDS–PAGE) under reducing and nonreducing conditions, by amino acid sequence analysis, and by molecular mass determination. MCP-2 (7.5 kDa) and MCP-3 (11 kDa) were synthesized in our laboratory using this procedure. They show biological and biochemical properties identical to those observed with the natural chemokines.[10] To prepare a standard chemokine solution, the MCPs are divided into aliquots at 10 μg/ ml in physiological saline (9 g/liter NaCl) supplemented with 0.01% (v/v) Tween 20 and stored at −70°.

Isolation of Leukocyte Cell Types as Targets for Monocyte Chemotactic Proteins

Because chemokines are active on several types of leukocytes, isolation of these cells from fresh blood is an important step in the biological characterization of novel molecules. In general, the isolation procedure used is a compromise between several parameters including viability, purity, and number of isolated cells. To obtain high cell viability, the speed of isolation is more crucial for granulocytes than for mononuclear cells, although the latter should not be activated during the isolation procedure. The lower percentage of basophils, natural killer (NK) cells, and dendritic cells in the peripheral blood from healthy volunteers necessitates a multistep isolation procedure based on the removal of irrelevant cells. Depending on the cell type and the experience of the laboratory, a combination of different techniques is used (Table II).

During the initial isolation phase, it is essential to keep the blood cells in buffer containing anticoagulant. If large volumes of blood need to be processed (e.g., >1000 ml), leukocyte-enriched buffy coats are prepared by centrifugation. Further, remaining red blood cells need to be removed by aggregation or lysis (Table II). Erythrocytes can be lysed by treatment with NH₄Cl (8.3 g/liter) in phosphate-buffered saline (PBS) for 10 min or by hypotonic shock in double-distilled water for 30 sec. Alternatively, red blood cells can be removed by aggregation and sedimentation with hydroxy-

---

[10] P. Proost, P. Van Leuven, A. Wuyts, R. Ebberink, G. Opdenakker, and J. Van Damme, *Cytokine* **7,** 97 (1995).

TABLE II
METHODS FOR ISOLATION OF VARIOUS CELL TYPES FROM BLOOD

Isolation method	Cell type[a]						
	Mo	Ly	NK	DC	Eo	Ba	Neu
Erythrocyte removal							
Aggregation (Plasmasteril)	+[b]	+			+	+	+
Lysis (hypotonic shock)					+		+
Density gradient centrifugation							
Lymphoprep	+				+		+
Ficoll-Paque		+	+	+			
Percoll			+	+		+	
Magnetic sorting							
(Dynabeads, MACS)							
CD1a				>80%			
CD2				−[c]		−	
CD3		>94%[d]	<2%	<2%			
CD6			−				
CD14	+	<1%	<2%	−			
CD16		−	>80%			−	
CD19		<1%		−		−	
CD20				<4%			
CD56		<1%	>80%				
Purity of cells	>90%	>94%	>80%	>80%	90%	65–80%	>95%
Ref.	11	12	13	14	17	15	11

[a] Mo, Monocytes; Ly, lymphocytes; NK, natural killer cells, DC, dendritic cells; Eo, eosinophils; Ba, basophils; Neu, neutrophils.

[b] Plus symbol (+) means enriched by positive selection with indicated method or marker.

[c] Minus symbol (−) means depleted by negative selection with indicated method or marker.

[d] Percentage of cells positive for CD marker.

ethyl-starch (Plasmasteril, Fresenius, Bad Homburg, Germany) for 30 min.[11] To separate different leukocyte types, density gradient centrifugation is an essential tool (Table II). Centrifugation of leukocytes on Lymphoprep (sodium metrizoate–Ficoll solution, Nycomed Pharma, Oslo, Norway) or Ficoll-Paque (Pharmacia Biotech, Uppsala, Sweden) with a density of 1.077 (g/cm^3) permits the separation of mononuclear cells from granulocytes.[12–14] If granulocytes are still contaminated with erythrocytes, further removal

[11] J. Van Damme and R. Conings, in "Cytokines, a Practical Approach" (F. R. Balkwill, ed.), p. 215. IRL Press, Oxford, 1995.

[12] D. D. Taub, P. Proost, W. J. Murphy, M. Anver, D. L. Longo, J. Van Damme, and J. J. Oppenheim, J. Clin. Invest. 95, 1370 (1995).

[13] P. Allavena, G. Bianchi, D. Zhou, J. Van Damme, P. Jilek, S. Sozzani, and A. Mantovani, Eur. J. Immunol. 24, 3233 (1994).

[14] S. Sozzani, F. Sallusto, W. Luini, D. Zhou, L. Piemonti, P. Allavena, J. Van Damme, S. Valitutti, A. Lanzavecchia, and A. Mantovani, J. Immunol. 155, 3292 (1995).

of the latter (e.g., by hypotonic shock) is mandatory. If only a small number of cells needs to be isolated, whole blood can be directly fractionated by density gradient centrifugation. Discontinuous or continuous (linear or nonlinear) gradients generated with Percoll silica particles (Pharmacia) are used to separate leukocyte populations in layers with different densities.[11,13–16] However, to obtain higher cell purity or to isolate cell types that are present only at low percentages in peripheral blood (e.g., basophils and dendritic cells), magnetic cell sorting is a powerful tool (Table II). With respect to preparing cells for chemotaxis assays, any reaction that activates the cells needs to be avoided. Positive selection through labeling of cells with magnetic microbeads via monoclonal antibodies against cell-specific membrane markers (MACS, Miltenyi Biotec, Bergisch Gladbach, Germany) does not affect the migration capacity of the cells. Alternatively, cell populations can also be enriched by negative selection by retaining the labeled cells on the column with a strong magnetic field.[17] Taken together, by a combination of the different isolation techniques, the purity of individual cell populations is usually higher than 80% and may be higher than 95% (Table II).

Parameters Affecting Migration of Different Leukocyte Populations In Vitro

Isolated leukocytes are washed with Hanks' balanced salt solution (HBSS) supplemented with 1 mg/ml of human serum albumin. After determination of the total cell number, as well as their purity and viability, leukocytes are suspended at a concentration appropriate for the chemotaxis assay using the Boyden microchamber (Table III). The upper compartment (containing the cells) is separated by a micropore filter from the lower compartment (containing the chemoattractant). Depending on the cell type, different filter materials (polycarbonate, nitrocellulose) and pore sizes need to be used. Furthermore, for monocytes, polycarbonate filters are treated with a wetting agent (polyvinylpyrrolidone, PVP), whereas for lymphocyte chemotaxis coating of the membranes with fibronectin (20 μg/ml) or collagen (type I or IV) is important for adherence and to prevent migrating lymphocytes from dropping through the filter.[18] Finally, the optimal incubation period for chemotaxis through the filter is variable (45 to 240 min) depending on the test cell type (Table III). After incubation, migrated cells (at the bottom side of the filter) are fixed and counted microscopically.

[15] R. Alam, P. Forsythe, S. Stafford, J. Heinrich, R. Bravo, P. Proost, and J. Van Damme, *J. Immunol.* **153**, 3155 (1994).

[16] E. Morita, J.-M. Schröder, and E. Christophers, *J. Immunol.* **144**, 1893 (1990).

[17] A. Wuyts, N. Van Osselaer, A. Haelens, I. Samson, P. Herdewijn, P. Proost, and J. Van Damme, *Biochemistry* **36**, 2716 (1997).

[18] A. M. Pilaro, T. J. Sayers, K. L. McCormick, C. W. Reynolds, and R. H. Wiltrout, *J. Immunol. Methods* **135**, 213 (1990).

TABLE III
PARAMETERS AFFECTING MIGRATION OF VARIOUS CELL TYPES IN THE MICROCHAMBER

Parameter	Cell population						
	Mo	Ly	NK	DC	Eo	Ba	Neu
Cell density (cells/ml)	2×10^6	10^7	10^6	$0.7–1 \times 10^6$	10^6	2×10^6	10^6
Filter							
Polycarbonate							
Pore size	5 μm	5 μm	$+^a$	5 μm	3–5 μm	5 μm	5 μm
PVP treatment	+	−	−		−		−
Coating	−	$+^b$	−	−	−	−	−
Nitrocellulose			8 μm				
Incubation time (hr)	2	4	2	1.5	1	1	0.75

[a] Combination of nitrocellulose filter covered by polycarbonate membrane.
[b] Fibronectin or collagen type I or IV.

The chemotactic activity is usually expressed as the number of migrated cells counted in 10 microscopic fields, or as the percentage of the maximal number of cells migrated to a reference chemoattractant. Alternatively, a chemotactic index can be calculated by dividing the number of cells migrated to the chemoattractant by the number of cells migrated to the buffer medium (Fig. 1). In view of the bell-shaped dose–response curve obtained in this test (inhibition of migration at high chemokine concentration), a dilution series (e.g., from 1 μg/ml to 1 ng/ml) should be tested for each sample. This allows one to compare the maximal chemotactic response (at concentrations yielding the highest number of migrated cells or migration index) among different chemokines on the various cell types. Table IV compares the chemotactic potencies of MCP-1, -2, and -3 using their minimal effective concentration as a parameter. It can be seen that the three chemokines are equipotent on both monocytes and T lymphocytes, but their chemotactic activity varies considerably on dendritic cells and eosinophils. By comparing the various cell types it is clear that mononuclear cells are more responsive to MCPs than are granulocytes. MCP-3 is the most pluripotent chemokine in that most leukocyte cell types respond quite well.

Immunoassays to Study Monocyte Chemotactic Protein Gene Regulation and Production During Disease

Preparation of Antibodies

Migration of cells *in vitro* can be studied with the help of various biological assays, but none of these is specific for a particular chemokine.

TABLE IV
POTENCY OF MCP-1, -2, AND -3 AS CHEMOTACTIC
FACTORS FOR VARIOUS LEUKOCYTE CELL TYPES

Cell type	Minimal effective concentration (nM)		
	MCP-1	MCP-2	MCP-3
Monocytes	0.1	0.1–0.3	0.1–0.3
T lymphocytes	0.1	0.1	0.1
Dendritic cells	>10	>10	1
NK cells	3–6	3	3
Basophils	3	10	3
Eosinophils	>300	10	3
Neutrophils	>100	>100	>100

The agarose assay,[19] which allows testing of a large series of samples by measuring the migration distance of leukocytes under agarose, is not very sensitive nor applicable for all leukocyte cell types and chemokines. Nevertheless, this method is useful to test potent chemoattractants (e.g., IL-8) for neutrophils. The Boyden chamber test[2] is more sensitive and can be applied to various leukocyte types (with minor adaptations of cell density, microfilter, and incubation time, see Table III) but remains nonspecific if chemokine mixtures or impure chemoattractants are tested. Finally, because this chemotaxis assay is based on microscopic counting it is rather labor-intensive, and automation is not routinely available.

A general strategy in immunochemical protein research is to develop polyclonal and/or monoclonal antibodies that are helpful for the purification and the detection by immunoassay of the factor studied. No specific antibody is available at the start of isolating novel chemokines from natural cellular sources. Once the natural chemokine is obtained in its pure form, often insufficient protein (e.g., MCP-2 and MCP-3) is available to raise antibodies. For these reasons, chemically synthesized chemokines serve as a good protein source (*vide supra*) to develop antibodies. Alternatively, polyclonal antibodies can be raised against partially purified natural chemokines. Polyclonal antibodies are routinely purified by protein A or G affinity chromatography or by antigen affinity (e.g., synthetic MCP-2) chromatography.

Radioimmunoassays for Monocyte Chemotactic Proteins

To study the regulation of chemokines, the availability of relatively small amounts of pure protein (controlled by amino acid sequence analysis)

[19] R. D. Nelson, P. G. Quie, and R. L. Simmons, *J. Immunol.* **115**, 1650 (1975).

and a nonspecific antibody (against semipure factor) are sufficient to develop a specific immunotest. Indeed, the pure antigen (e.g., 10 μg of chemically synthesized MCP-2 or natural MCP-1) can be radioactively (^{125}I) labeled and purified from free label by gel filtration.[20] It is subsequently used in a standard radioimmunoassay (RIA), based on competition for antibody binding between labeled and unlabeled chemokine (standard or test sample). Immunoreactivity is measured by precipitation of antibody–antigen complexes with protein A bacterial adsorbent, and detection of radioactivity is achieved using a gamma-counter. Table V shows the results obtained for MCP-1 and MCP-2 in such RIAs using polyclonal antibodies. With these RIAs no cross-reactivity is observed with other chemokines, despite the high structural relationship of MCP-1 with MCP-2 and MCP-3 (65 and 71% identity at the amino acid sequence level, respectively) (Fig. 1). The detection limit for these RIAs is about 0.5 ng/ml, irrespective of whether a nonspecific (against semipure natural MCP-2) or specific polyclonal antibody (e.g., against pure synthetic MCP-2) is used.[20] These RIAs allow one to investigate the cellular sources (fibroblasts and leukocytes) and inducers (cytokines, viral and bacterial products) of MCP-1 and MCP-2 *in vitro* (Table V). It is concluded that in fibroblasts MCP-2 is often coproduced with MCP-1, although lower absolute amounts of MCP-2 are secreted by the cells tested. In addition, the regulation of expression is not identical for the two chemokines because IL-1β is a good inducer of MCP-1, whereas for MCP-2 induction by IFN-γ is superior.

Competitive and Sandwich ELISAs for Monocyte Chemotactic Proteins

With the use of the same reagents (pure antigen, nonspecific antibodies), nonradioactive enzyme-linked immunosorbent assays (ELISAs) can also be worked out. In this case, pure antigen is coated on plastic microtiter plates (Maxisorp Immunoplates, Nunc, Roskilde, Denmark) and binding of antibody is competed for by chemokine from the standard or the test samples. Samples containing chemokines will reduce antibody binding to the solid phase resulting in diminished absorbance. In these assays, the antibody is usually biotinylated and detected by streptavidin–peroxidase conjugates. The effect can also be measured with an enzyme-labeled antiserum directed against the chemokine antibody (indirect ELISA). The advantage of a competitive ELISA over a RIA is that radioactive isotope (short half-life) handling and radiation damage can be avoided. However, in the competitive ELISA the coating requires a relatively higher amount of pure antigen.

[20] J. Van Damme, P. Proost, W. Put, S. Arens, J.-P. Lenaerts, R. Conings, G. Opdenakker, H. Heremans, and A. Billiau, *J. Immunol.* **152**, 5495 (1994).

TABLE V

INDUCTION OF MCP-1 AND MCP-2 IN HUMAN DIPLOID FIBROBLASTS AND OSTEOSARCOMA CELLS[a]

| | | Chemokine (ng/ml) production by | | | |
| | | Diploid fibroblasts | | MG-63 osteosarcoma cells | |
Inducer	Concentration	MCP-1[b]	MCP-2[b]	MCP-1	MCP-2
Poly(rI):poly(rC)	100 μg/ml	75	6.8	24	<0.4
	10 μg/ml	63	2.7	20	0.6
	1 μg/ml	4	<0.5	—	—
Measles	10^5 TCID$_{50}$/ml	42	1.8	109	6.2
	10^4 TCID$_{50}$/ml	20	0.6	31	2.8
	10^3 TCID$_{50}$/ml	—	—	6	0.4
PMA	100 ng/ml	11	0.8	5	<0.4
	10 ng/ml	23	0.7	8	<0.4
IL-1β	100 U/ml	91	3.0	173	0.9
	10 U/ml	34	1.2	83	1.2
	1 U/ml	16	<0.6	24	0.6
IFN-β	3000 U/ml	9	1.2	10	<0.4
	300 U/ml	9	0.7	10	<0.4
IFN-γ	200 ng/ml	25	7.2	23	2.8
	20 ng/ml	23	2.4	19	3.6
	2 ng/ml	14	1.3	14	0.4
Control	—	8	<0.6	16	<0.4

[a] Cultures of human embryonic skin muscle fibroblasts and MG-63 osteosarcoma cells are grown to confluency in 25-cm^2 flasks (Nunc) and stimulated for 48 hr with different inducers: the double-stranded RNA poly(rI):poly(rC) (P-L Biochemicals, Milwaukee, WI), measles virus [Attenuvax strain, $10^{6.4}$ 50% tissue culture infectious dose (TCID$_{50}$)/ml], phorbol myristate acetate (PMA, Sigma, St. Louis, MO), pure natural human interleukin-1β (IL-1β), pure natural interferon-β (IFN-β), purified recombinant human interferon-γ (Boehringer Mannheim, Mannheim, Germany), lipopolysaccharide (LPS) from *Escherichia coli* (0.111:B4, Difco, Detroit, MI), or concanavalin A (ConA, Calbiochem, San Diego, CA).

[b] Detection of immunoreactivity by RIA, using ^{125}I-labeled pure recombinant MCP-1 and ^{125}I-labeled pure synthetic MCP-2 and polyclonal antibodies against recombinant MCP-1 and against semipurified natural MCP-2, respectively.

At a later stage, such as when specific polyclonal antibodies and/or monoclonal antibodies are available, a sensitive and specific sandwich ELISA can be developed. As an example, a specific polyclonal antibody against purified recombinant MCP-1 or synthetic MCP-2 simultaneously served as capture and detector antibody. Such ELISAs also allow detection of MCP-1 and MCP-2 in plasma of sepsis patients, the detection limit being

TABLE VI
Comparison of MCP-1 RIA and MCP-1 ELISA for in vitro Detection of MCP-1
Induction in Human Mononuclear Leukocytes[a]

| Inducer | Concentration | Chemokine production (ng/ml) | |
		MCP-1 ELISA[b]	MCP-1 RIA[c]
Poly(rI):poly(rC)	100 μg/ml	28[d]	19[d]
	10 μg/ml	19	16
Measles	10^5 TCID$_{50}$/ml	55	79
	10^4 TCID$_{50}$/ml	9	4
LPS	50 μg/ml	18	15
	5 μg/ml	28	32
ConA	10 μg/ml	59	68
	1 μg/ml	35	39
PMA	100 ng/ml	3	<4
	10 ng/ml	4	<4
IL-1	100 U/ml	17	10
	10 U/ml	9	5
	1 U/ml	2	<4
IFN-β	3000 U/ml	30	26
	300 U/ml	17	9
IFN-γ	200 ng/ml	16	8
	20 ng/ml	3	<4
Control	—	3	<4

[a] Human mononuclear leukocytes, isolated from fresh blood by density gradient centrifugation (Lymphoprep, see text), are stimulated for 48 hr with different inducers (See footnote a in Table V).

[b] Sandwich ELISA based on coating with a polyclonal antibody against recombinant MCP-1 and detection with the same biotin-labeled antibody.

[c] RIA based on ^{125}I-labeled purified recombinant MCP-1 and polyclonal antibody against recombinant MCP-1.

[d] Values are averages of four independent induction experiments.

0.1 ng/ml.[21] Table VI compares the results obtained with the MCP-1 RIA (see Table V) and the MCP-1 ELISA in a study of the induction of MCP-1 in human peripheral blood mononuclear leukocytes (isolated by density gradient centrifugation). It can be seen that the data of both immunotests correlate well. Measles virus, the mitogen concanavalin A, and cytokines such as IFN-β are good inducers of MCP-1 in mononuclear leukocytes. Introduction of a monoclonal antibody against MCP-1 as detector antibody improved the detection limit by reducing the background levels.

[21] A. W. J. Bossink, L. Paemen, P. M. Jansen, C. E. Hack, L. G. Thijs, and J. Van Damme, *Blood* **86**, 3841 (1995).

Differences between Monocyte Chemotactic Proteins in Receptor
  Usage and Signal Transduction Mechanisms

Induction of Intracellular Calcium Increase in Monocytes

The different spectrum of target cells activated by MCP-1, -2, and -3
can in part be explained by their use of different receptors for signaling.
Indeed, MCP-1 and MCP-3 can activate monocytes via the C-C chemokine
receptor CCR-2, whereas MCP-2 does not.[22,23] In addition, the signal transduction pathways of MCP-1/MCP-3 and MCP-2 are not identical, for example, with respect to their effect on intracellular calcium concentration
($[Ca^{2+}]_i$). Changes in $[Ca^{2+}]_i$ are measured with the fluorescent indicator
Fura-2 as described by Grynkiewicz *et al.*[24] Lymphoprep-purified monocytes
(*vide supra*) (10^7 cells/ml) are incubated for 30 min at 37° with 2.5 μM
Fura-2/AM (Molecular Probes Europe BV, Leiden, The Netherlands) and
0.01% Pluronic F-127 (Sigma, St. Louis, MO) in Eagle's minimal essential
medium (EMEM) supplemented with 0.5% (v/v) fetal calf serum (FCS).
After incubation, cells are washed twice and resuspended (10^6 cells/ml) in
HBSS (1 mM Ca^{2+}) supplemented with 0.1% (v/v) FCS and buffered at
pH 7.4 with 10 mM HEPES. Cells are allowed to equilibrate at 37° for
10 min before Fura-2 fluorescence is measured in a Perkin-Elmer LS50B
luminescence spectrophotometer fitted with a water-thermostattable, stirra-
ble four-position cuvette holder (Perkin-Elmer, Norwalk, CT). Excitation
wavelengths used are 340 and 380 nm; the emission is measured at 510 nm.
$[Ca^{2+}]_i$ is calculated using the Grynkiewicz equation:

$$[Ca^{2+}]_i = K_d \left(\frac{R - R_{min}}{R_{max} - R}\right) \left(\frac{Sf_2}{Sb_2}\right)$$

where R is the ratio of the fluorescence intensity at 340 nM to the fluorescence intensity at 380 nM; Sf_2 and Sb_2 are the measured fluorescence intensities at 380 nM when all indicator is Ca^{2+} free and Ca^{2+}-bound, respectively. R_{max} is obtained after lysis of the cells with 50 μM digitonin; R_{min}
is determined by addition of 10 mM EGTA to the lysed cells after adjusting
the pH to 8.5 with 20 mM Tris. The K_d used for calibration is 224 nM.

Figure 2 illustrates that at low concentration (1 nM) MCP-1 and
MCP-3 and at high concentration (10 nM) MCP-2 induce a significant
increase of $[Ca^{2+}]_i$, indicating their difference in signal transduction in

[22] C. Franci, L. M. Wong, J. Van Damme, P. Proost, and I. F. Charo, *J. Immunol.* **154,**
6511 (1995).
[23] C. Combadiere, S. K. Ahuja, J. Van Damme, H. L. Tiffany, J.-L. Gao, and P. M. Murphy,
J. Biol. Chem. **270,** 29671 (1995).
[24] G. Grynkiewicz, M. Poenie, and R. Y. Tsien, *J. Biol. Chem.* **260,** 3440 (1985).

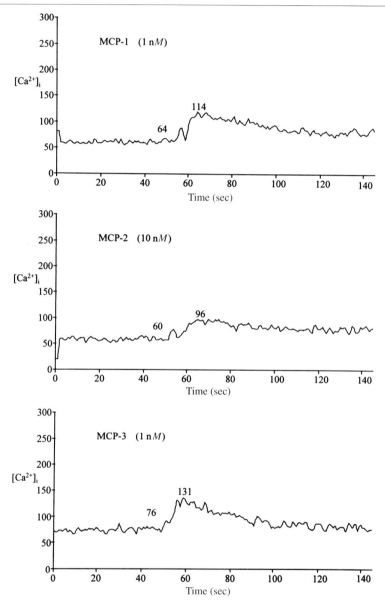

FIG. 2. Induction of $[Ca^{2+}]_i$ influx in mononuclear cells by MCPs. Mononuclear cells (loaded with Fura-2/AM) were stimulated with natural MCP-1, synthetic MCP-2, or synthetic MCP-3 at the indicated concentration. $[Ca^{2+}]_i$ (nM) was measured by the method of Grynkiewicz.[24]

monocytes. Increases in the $[Ca^{2+}]_i$ of mononuclear cells in response to MCP-1 and MCP-3 were inhibited with pertussis toxin (PT), whereas cholera toxin (CT) had little or no effect on the increase in the $[Ca^{2+}]_i$ or chemotaxis.[25] PT catalyzes the ADP-ribosylation of G_α proteins at a Cys residue resulting in G protein uncoupling from the receptor. Cholera toxin catalyzes the ADP-ribosylation of arginine on the G_α protein and suppresses GTPase activity locking the G_α protein in the GTP-bound active conformation. On binding of MCP-2 to monocytes, a CT-sensitive G protein transduced the signal.

Activation of Phospholipase A_2 in Monocytes by Monocyte Chemotactic Proteins

Cells of the monocytic lineage possess at least three different types of phospholipase A_2 (PLA$_2$): a low molecular mass ($\sim$14 kDa) secreted form, a cytosolic 85-kDa PLA$_2$ (cPLA$_2$), and a Ca^{2+}-independent cytosolic PLA$_2$.[25] C-C chemokines induce activation of cPLA$_2$ and release of arachidonic acid (AA) in human monocytes.[26] Release of AA was shown to be closely associated with the chemotactic response. Inhibitors of PLA$_2$ decreased in a concentration-dependent manner monocyte migration to chemokines.[27] In parallel, some agonists (i.e., platelet activating factor and 5-oxoETE [eicoratetraenoic acid]) were able to act in a synergistic fashion on both AA release and monocyte migration with C-C chemokines.[27–29] Release of [^{3}H]AA by MCP-1 and related chemokines (e.g., MCP-3, RANTES, MIP-1α) is rapid ($<$15 sec) and transient ($\sim$15 min) and can be efficiently used to investigate the early signaling mechanisms following chemokine receptor activation. However, it must be considered that the great majority of free [^{3}H]AA is retained intracellularly and less than 20% of the [^{3}H]AA released is found in cell-free supernatants from MCP-stimulated monocytes (Fig. 3). Thus, total (released and cell associated) [^{3}H]AA needs to be evaluated. A second concern is that a great number of platelets are usually associated with monocytes, and precautions must be taken to lower platelet contamination.[30]

[25] S. Sozzani, D. Zhou, M. Locati, M. Rieppi, P. Proost, M. Magazin, N. Vita, J. Van Damme, and A. Mantovani, *J. Immunol.* **152**, 3615 (1995).

[26] E. A. Dennis, *J. Biol. Chem.* **269**, 13057 (1994).

[27] M. Locati, G. Lamorte, W. Luini, M. Introna, S. Bernasconi, A. Mantovani, and S. Sozzani, *J. Biol. Chem.* **271**, 6010 (1996).

[28] M. Locati, D. Zhou, W. Luini, V. Evangelista, A. Mantovani, and S. Sozzani, *J. Biol. Chem.* **269**, 4746.

[29] S. Sozzani, M. Rieppi, M. Locati, D. Zhou, F. Bussolino, J. Van Damme, and A. Mantovani, *Biochem. Biophys. Res. Commun.* **199**, 761 (1994).

[30] S. Sozzani, D. Zhou, M. Locati, S. Bernasconi, W. Luini, A. Mantovani, and J. T. O'Flaherty, *J. Immunol.* **157**, 4664 (1996).

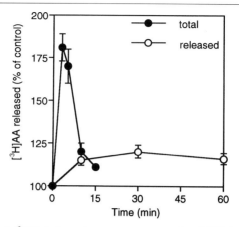

FIG. 3. Release of [³H]AA from human monocytes after MCP-1 stimulation. Human monocytes were diluted in RPMI 1640 containing 0.2% BSA and stimulated with 100 ng/ml MCP-1. [³H]AA release is expressed as the percentage of the respective control at different incubation time points.

Preparation and Labeling of Human Monocytes. Human monocytes can be obtained from buffy coats of normal blood donors. To reduce platelet contamination, anticoagulated whole blood is diluted 1:4 with cold phosphate-buffered isotonic saline without Ca^{2+} and Mg^{2+} (PBS) and centrifuged at 150 g at 4° for 20 min. The supernatant is discarded and the cell pellet is washed with PBS in the same conditions. Cells are resuspended in PBS containing 0.3 mM EDTA, layered on top of Ficoll (Biochrom, Berlin, Germany), and centrifuged at 800 g at room temperature for 25 min. Mononuclear cells are recovered, diluted, and washed twice in PBS at 4°. To remove platelets specifically adherent to monocytes, mononuclear cells are resuspended in FCS (Hyclone, Logan, UT) containing 5 mM EDTA and subjected to two sequential incubations (15 min) at 37°. Platelet-free mononuclear cells are recovered by centrifugation at 400 g at room temperature for 15 min. Monocytes need to be further purified (>90% pure) by centrifugation at 600 g on a 46% isosmotic Percoll (Pharmacia) gradient.[27] The monocyte preparation obtained normally does not release [³H]AA when challenged with 10 U/ml thrombin. Monocytes are then diluted in RPMI 1640 with 10% FCS (10^6/ml) and labeled with 1 μCi/ml [³H]AA (200 Ci/ mmol; Amersham International, Amersham, United Kingdom) in Petriperm dishes (Aereus, Vienna, Austria) for 16–18 hr. Incubation does not affect cell viability (>95% (v/v), by trypan blue dye exclusion) nor the ability of monocytes to migrate in response to MCP-1.

Evaluation of [³H]Arachidonic Acid Release. At the end of the incubation, cells are washed twice and resuspended in RPMI 1640 supplemented with 0.2% (v/v) fatty acid free serum albumin (BSA; Sigma). Monocytes $(5 \times 10^6$ cells/ml) are prewarmed at 37° for 5 min and then stimulated. The reaction is terminated by the addition of 2 ml of chloroform/methanol/ formic acid $(1:2:0.2, v/v)$ followed by agitation. Cell extracts are transferred to centrifuge tubes, and 1 ml of water and 2 ml of chloroform are added. Tubes are mixed and spun at 2000 rpm for 10 min, and the organic phase (lower phase) is collected. Chromatographic separation of lipids is performed by evaporating the organic phase under a stream of nitrogen, redissolving the residue in chloroform, and loading the extract on silica gel G plates (Merck, Darmstadt, Germany). Fatty acids can be separated by thin-layer chromatography (TLC) using hexane/ethyl ether/formic acid $(15:10:1, v/v)$ as solvent system, for 30 min. The position of free fatty acids on TLC plates is determined by comigration with commercially available standards after exposure to iodine vapors. Quantitative determinations are obtained by scraping portions of the silica gel into scintillation vials followed by liquid scintillation spectrometry. Results are expressed as percentage of radioactivity in the fatty acid band of the total radioactivity recovered from each lane (Fig. 3).

Acknowledgments

A. W. is a research assistant of the Belgian National Fund for Scientific Research (NFWO). This work has been supported by the Concerted Research Actions of the Regional Government of Flanders, the Interuniversity Attraction Pole of the Belgian Federal Government (IUAP), the Belgian NFWO (including "Levenslijn"), the General Savings and Retirement Fund (ASLK), and the European Community.

[9] Assays for Macrophage Inflammatory Proteins

By Anne-Marie Buckle, Stewart Craig, and Lloyd G. Czaplewski

Introduction

A major focus in the ever-expanding field of chemokine research is to understand the involvement of these molecules in the pathobiology of disease states and to, thereby, define how they may be applied therapeutically. This chapter specifically describes methodology intended to provide the reader with a basic experimental grounding in two clinically relevant areas of chemokine research, namely, the regulation of hematopoietic progenitor cell proliferation and the mechanism of human immunodeficiency virus (HIV) infection. Preclinical studies using the *in vivo* and *in vitro* assays described have supported progression of macrophage inflammatory protein-1α (MIP-1α) as a candidate biopharmaceutical with potential application in cancer and HIV therapy. Although these assays are described for MIP-1α, they have of course been used for analysis of other members of the macrophage inflammatory protein (MIP) subfamily and chemokines in general.

In terms of cancer therapy, the clinical potential of the stem cell inhibitor (SCI) activity first observed by Lord *et al.* in 1976[1] was immediately recognized for protection of bone marrow stem cells from the effects of cycle specific cytotoxic chemotherapy agents. The factor responsible for the murine (mu) SCI activity was eventually identified as muMIP-1α,[2] a protein originally purified as an inflammatory mediator.[3] *In vivo* and *in vitro* proof of principle for the theory of stem cell protection by muMIP-1α[4,5] and its predicted[6] human (hu) homolog,[7] LD78,[8] soon followed. These results provided a strong rationale for clinical evaluation of huMIP-1α as an adjunct to alleviate the side effects of cytotoxic cancer chemotherapy, and a genetically modified variant of huMIP-1α[9] (called BB-10010) with improved formulation properties is now in five separate phase II trials.[10]

Publications have highlighted a possible application of MIP-1α for the inhibition of HIV infection of macrophages.[11,12] These reports have stimulated significant interest in the clinical potential of MIP-1α for control of the progression of HIV infection. As described in this chapter, a number of assay systems have been developed and implemented in this exciting new area of chemokine research.

[1] B. I. Lord, K. J. Mori, E. G. Wright, and L. G. Lajtha, *Br. J. Haematol.* **34**, 441 (1976).
[2] G. Graham, E. G. Wright, R. Hewick, S. D. Wolpe, N. Wilkie, D. Donaldson, S. A. Lorimore, and I. B. Pragnell, *Nature (London)* **344**, 442 (1990).
[3] S. D. Wolpe, G. Davatelis, B. Sherry, B. Beutler, D. G. Hesse, H. T. Nguyen, L. L. Moldawer, C. F. Nathan, S. F. Lowry, and A. Cerami, *J. Exp. Med.* **167**, 570 (1988).
[4] B. I. Lord, T. M. Dexter, J. M. Clements, M. G. Hunter, and A. J. H. Gearing, *Blood* **79**, 2605 (1992).
[5] R. Maze, B. Sherry, B. S. Kwon, A. Cerami, and H. E. Broxmeyer, *J. Immunol.* **149**, 1004 (1992).
[6] S. Blum, R. E. Forsdyke, and D. R. Forsdyke, *DNA Cell. Biol.* **9**, 589 (1990).
[7] K. Obaru, M. Fukuda, M. Maeda, and K. A. Shimada, *J. Biochem. (Tokyo)* **99**, 885 (1986).
[8] D. J. Dunlop, E. G. Wright, S. Lorimore, G. J. Graham, T. Holyoake, D. J. Kerr, S. D. Wolpe, and I. B. Pragnell, *Blood* **79**, 2221 (1992).
[9] M. G. Hunter, L. Bawden, D. Brotherton, S. Craig, S. Cribbes, L. G. Czaplewski, T. M. Dexter, A. H. Drummond, C. M. Heyworth, B. I. Lord, M. McCourt, P. G. Varley, L. M. Wood, and R. M. Edwards, *Blood* **86**, 4400 (1995).
[10] British Biotech plc., Annual Report, p. 18. British Biotech, Oxford, U.K., 1996.
[11] B. Canque and I. C. Gluckman, *Blood* **84** (Suppl. 1), 480a (1994).
[12] F. Cocchi, A. L. deVico, A. Garzino-Demo, S. K. Ayra, R. C. Gallo, and P. Lusso, *Science* **270**, 1811 (1995).

General Considerations When Assaying Macrophage Inflammatory Proteins

General considerations to be taken into account when studying the biological activity of macrophage inflammatory proteins are as follows. (1) *In vitro* progenitor cell assays are notoriously variable with respect to batches of serum; therefore, we advise that several samples of serum be tested for plating efficiency under standard conditions before purchasing a large quantity. In this respect, companies such as Stem Cell Technologies, Inc., Vancouver, B.C., now sell ready-tested sera and reagents. (2) Chemokines such as MIP-1α and MIP-1β self-associate to form multimeric complexes ranging in mass from 8 kDa (monomer) to >1000 kDa. Researchers should be aware of the effect of multimerization on calculation of active concentrations and the effects of different buffer concentrations on self-association (see Ref. 13 for an overview). (3) Alternative N termini for mature huMIP-1α exist. Theoretical peptide cleavage sites have variously predicted the mature peptide to be 66, 69, and 70 amino acids in length.[14–17] Manufacturers of recombinant huMIP-1α produce molecules of different length depending on which prediction is used, as noted in the tabulation below.

Source	Location	Recombinant huMIP-1α product
R&D Systems	Minneapolis, MN	66 amino acids
British Biotech	Oxford, UK	69 amino acids
Peprotech	Rocky Hill, NJ	70 amino acids

Because the N-terminal regions of monocyte chemotactic protein-1 (MCP-1),[18,19] interleukin-8 (IL-8),[20] MIP-1α,[21] and MIP-2[22] have been impli-

[13] S. Craig and R. Hoffman, *in* "Chemoattractant Ligands and Their Receptors" (R. Horuk, ed.), p. 193. CRC Press, Boca Raton, Florida, 1996.

[14] K. D. Brown, S. M. Zurawski, T. R. Mosmann, and G. Zurawski, *J. Immunol.* **142,** 679 (1989).

[15] P. F. Zipfel, J. Balke, S. G. Irving, K. Kelly, and U. Siebenlist, *J. Immunol.* **142,** 1582 (1989).

[16] I. B. Pragnell, D. D. Donaldson, G. J. Graham, and G. G. Wong, International Patent Application WO91/04274 (1991).

[17] M. Nakao, H. Nomiyama, and K. Shimada, *Mol. Cell. Biol.* **10,** 3646 (1990).

[18] J.-H. Gong and I. Clark-Lewis, *Cytokine* **6,** 546 (1994).

[19] I. Clark-Lewis, J.-H. Gong, B. Dewald, M. Baggiolini, K. Rajarathnam, K. S. Kim, and B. D. Sykes, *Cytokine* **6,** 539 (1994).

[20] I. Clark-Lewis, C. Schumacher, M. Baggiolini, and B. Moser, *J. Biol. Chem.* **266,** 23128 (1991).

[21] S. Craig, M. G. Hunter, R. M. Edwards, L. G. Czaplewski, and R. J. Gilbert, International Patent Application WO 93/13206 (1993).

[22] L. M. Pelus, P. K. Bhatnagar, A. J. King, and J. M. Balcarek, International Patent Application WO 94/29341 (1994).

cated in receptor binding, it is possible that a range of biological activities and specific activities could be observed with different forms of huMIP-1α. It is also important that the source and sequence of huMIP-1α be considered when designing experiments or comparing literature studies. Protein from different suppliers cannot easily be substituted without adequate comparison.

Assays to Determine Effect of MIP-1α on Primitive Hematopoietic
  Progenitor Cell Proliferation

*Tissue Sources of Hematopoietic Progenitor and Hematopoietic
Stem Cells*

Bone marrow is probably the most common source of hematopoietic progenitor cells (HPCs) and hematopoietic stem cells (HSCs) for use in these assays, and detailed procedures for isolation of hematopoietic cells from both murine and human bone marrow tissue are provided in this section. Blood represents another source of cells; however, the number of circulating HPCs and HSCs is usually very low in both mice and humans. To use blood as a tissue source, therefore, the content of HPCs and HSCs must normally be increased by mobilizing cells from the bone marrow into the circulation using chemotherapy agents[23] and/or cytokines. A procedure for collecting blood from mice is described below.

Clinically, harvesting of mobilized blood cells for use in hematopoietic reconstitution following myeloablative chemotherapy or as targets for gene therapy is very attractive because it is less invasive than bone marrow aspiration, and current opinions of mobilization regimens that use agents such as granulocyte colony-stimulating factor (G-CSF), granulocyte–macrophage colony-stimulating factor (GM-CSF), or G-CSF + Cytoxan (a cytotoxic chemotherapeutic) have been presented.[24] Although the mechanism of blood mobilization is poorly understood at this time, combinations of treatments are being clinically tested to optimize the quantity and quality of HPCs and HSCs released into the circulation. In this context, it is interesting to note that in mice MIP-1α has itself been shown to rapidly mobilize both mature and primitive hematopoietic cells into the circulation.[25,26] If this finding can be reproduced in humans, then it may offer an

---

[23] R. Pettengel, N. G. Testa, R. Swindell, D. Crowther, and T. M. Dexter, *Blood* **82,** 2239 (1993).
[24] *Drug & Market Development* **7,** 194 (1996).
[25] M. McCourt, D. Brotherton, M. Comer, S. Cribbes, L. G. Czaplewski, I. Hemingway, C. M. Heyworth, B. I. Lord, L. M. Wood, and M. G. Hunter, *Bone Marrow Transplant.* **14** (Suppl. 2), S34 (1994).
[26] B. I. Lord, L. B. Woolford, L. M. Wood, L. G. Czaplewski, M. McCourt, M. G. Hunter, and R. M. Edwards, *Blood* **85,** 3412 (1995).

alternative cytokine for use in clinical mobilization regimens. Scientifically, the latter finding provides a classic illustration of the multifunctional action of chemokines because MIP-1α is also reported to induce the adhesion of HPCs to bone marrow-derived stromal layers.[27,28]

Preparation of Murine Bone Marrow Cells. Donor mice (e.g., from standard mouse strains such as C57/Bl or BDF1) ages 8–12 weeks are sacrificed by cervical dislocation, and the fur is sprayed with alcohol for sterilization purposes. The legs from the hip to the ankle are removed and placed in a petri dish, and muscular tissue scraped away using a scalpel. After removal from the animal, it is important that the tissue be kept on ice to maximize cell viability. Conventionally the bones are flushed by cutting open the ends of the femurs with a scalpel, then forcing ice-cold Hanks' buffered saline solution with 5% heat-inactivated fetal calf serum (HBSS + 5% HIFCS) through them with a syringe and 21-gauge needle to remove the bone marrow plug. The plug is then resuspended using vigorous agitation with a Pasteur pipette. After flushing, the cells are washed by centrifugation at 1200 rpm for 5 min at 4°, resuspended in HBSS + 5% (v/v) HIFCS, and filtered through 100-μm nylon mesh (Fisons, Loughborough, UK) to remove clumps prior to counting on a hemocytometer using trypan blue to exclude dead cells. With this method approximately 10^8 bone marrow leukocytes can be recovered from a single mouse.

Very primitive hematopoietic cells are believed to be more intimately associated with the bone stromal tissue. An alternative method designed to isolate such primitive cells is to grind the bones using a precooled mortar and pestle, with ice-cold HBSS + 5% HIFCS added. The cells are then centrifuged, washed, and counted as above.

Preparation of Human Bone Marrow Cells. Human bone marrow aspirates are separated by density gradient centrifugation on 1.077 g/ml Ficoll-Hypaque (Pharmacia, Piscataway, NJ). This process depletes the sample of red blood cells and enriches for colony-forming mononuclear cells. This procedure typically yields around $1–5 \times 10^6$ bone marrow leukocytes per milliliter of bone marrow.

Mobilized Blood. In humans, blood is harvested from peripheral veins (e.g., in the arm), hence the term peripheral blood stem/progenitor cells. In mice, blood is harvested by cardiac puncture, which is central rather than peripheral. We, therefore, generically refer to cells from this tissue as mobilized blood cells (MBCs). In some studies, the tissue is used directly and in others the primitive hematopoietic cells are further enriched by immunoselection. For human tissue, MBCs are immunoselected using

[27] J. Migas, R. Hurley, and C. M. Verfaillie, *Exp. Hematol.* **22**, 736 (1994).
[28] R. Bhatia, P. B. McGlave, and C. M. Verfaillie, *Exp. Hematol.* **22**, 797 (1994).

monoclonal antibodies raised against the CD34 glycoprotein[29,30] that is present on the surface of HPCs and HSCs. Various simple, commercial immunomagnetic human CD34 selection kits are available for research use (e.g., Baxter, Irvine, CA; CellPro, Bothwell, WA; Miltenyi Biotec, Inc., Auburn, CA).

Mice provide a model system in which to investigate the effects of cytokines on mobilization with blood collection by terminal cardiac puncture under Ethrane anesthesia. Blood from groups of donor mice is pooled with heparin (25 U/ml) as an anticoagulant, and the white blood cell count in the pooled sample is estimated. In such investigations, the concentration of cytokine and timing of harvest must often be empirically determined by dose and mobilization kinetics studies. As a specific example, murine hematopoietic cells can be mobilized using 100 μg/kg of MIP-1α injected subcutaneously (s.c.) into the mouse (e.g., 25–30 g B6D2F$_1$ animals, 10 per group) 30 min prior to harvesting of the blood.[25,26]

In vivo Assays for Hematopoietic Progenitor Cells

The bone marrow is severely compromised when animals are subjected to high doses of irradiation, because proliferating cells such as hematopoietic progenitors and cycling stem cells are killed. This leads to a deficit of the mature blood cells being formed in the marrow, resulting in severe anemia and immunosuppression. The animals can, however, be rescued by administration of an intravenous inoculation of bone marrow from a donor animal, which contains progenitor cells capable of repopulating the bone marrow and blood.

Principle of CFU-S Assay. The colony-forming unit—spleen (CFU-S) assay was first described by Till and McCulloch in 1961,[31] when they observed macroscopic nodules forming in the spleens of irradiated mice injected with bone marrow cells. Arising from infused bone marrow cells lodged in the spleen, the nodular colonies proliferate in two waves: (1) the first wave at day 8–9 (CFU-S d8) derive from more mature HPCs and (2) the second wave at day 11–13 (CFU-S d11) are believed to derive from more primitive HPCs that take longer to differentiate to the point at which they form colonies in the spleen. The colonies then gradually disappear over the course of 3–4 days as they differentiate into mature circulating blood cells. Using genetic markers, it has been shown that each colony is derived from a single cell, and that the number of colonies formed is linearly related to the number of bone marrow cells injected into the irradiated

[29] D. S. Krause, M. J. Fackler, C. I. Civin, and W. S. May, *Blood* **87**, 1 (1996).
[30] C. I. Civin, L. C. Strauss, C. Brovall, M. J. Fackler, J. F. Schwartz, and J. H. Shaper, *J. Immunol.* **133**, 157 (1984).
[31] J. E. Till and E. A. McCulloch, *Radiat. Res.* **14**, 213 (1961).

mouse.[32] Some spleen colonies are capable of differentiation into multiple lineages and of producing further spleen colonies. Indeed a precursor activity (pre-CFU-S) has been identified[33] that is believed to detect an even more primitive HPC, which migrates to the bone marrow before seeding to the spleen to form a colony. The CFU-S assay is, therefore, considered to be a classic indicator of very early HPCs, most likely a primitive multipotential myeloerythroid precursor.

In vivo, CFU-S can be recruited into the cell cycle by the use of cytotoxic drugs such as hydroxyurea (HU), cytosine arabinoside, and 5-fluorouracil that ablate those cells in the marrow which are actively synthesizing DNA. Hydroxyurea, for example, is believed to block further entry of cells into DNA synthesis for about 4 hr, but after this time noncycling primitive CFU-S cells are released into the synchronous cell cycle to repopulate the ablated marrow.[34] Around 7 hr after the first treatment with hydroxyurea, therefore, the actively cycling CFU-S are prime targets for a second dose of hydroxyurea. Hence, the propensity of MIP-1α to inhibit CFU-S proliferation can be estimated by its ability to protect CFU-S from a second dose of hydroxyurea.

Experimental Protocol. Recipient mice ages 8–12 weeks are exposed to a sublethal dose of radiation. The dose used varies depending on the radiation source, exposure rate, and tolerance of the mouse strain. We have found that two doses of 4.5 Gy at 0.6 Gy/min from a cesium source given approximately 4 hr apart works well; however, others[4] successfully use a single dose of 15.25 Gy at 0.95 Gy/hr. The dose will initially need to be titrated such that sufficient radiation is given to ablate host colony formation yet is sublethal. Because a sublethal dose can allow subclinical infections to take hold and result in high mortality, the assay is best performed on specific pathogen-free housed animals. If animals are to be kept in conventional conditions, then they may benefit from the use of antibiotics such as a mixture of polymixin B and neomycin sulfate.

After irradiation, mice are allowed to recover for 2 hr before injecting bone marrow cells intravenously. Cells should be injected using a volume of 100–200 μl, which can be accurately measured with an insulin syringe. Because the assay is subject to substantial biological variation, groups of at least 3–5 animals are normally used per data point. However, when assaying the effect of chemokines, it is recommended that a larger group of about 10 animals per data point be used to obtain statistically significant

[32] E. A. McCulloch, *in* "Regulation of Haematopoiesis" (A. S. Gordon, ed.), p. 133. Appleton Century and Crofts, New York, 1970.

[33] G. S. Hodgson and T. R. Bradley, *Nature (London)* **281,** 381 (1979).

[34] G. S. Hodgson, T. R. Bradley, R. F. Martin, M. Sumner, and P. Fry, *Cell Tissue Kinet.* **8,** 51 (1975).

results. Mice are housed for 8–11 days, then sacrificed and the spleen removed into fixative for enumeration of colonies.

Generally, the colonies are very large nodules (Fig. 1); hence, a maximum number of around 10 can be accurately counted before the spleen is confluent. To quantitate inhibitory effects, therefore, it is important to inject sufficient bone marrow cells or MBCs that would generally yield around 10 colonies, thereby allowing a range of 1–10 colonies per spleen. As a guideline, 10^5 bone marrow leukocytes normally give about 5–10 colonies, and researchers describe the dose either in terms of absolute number of whole bone marrow cells (or MBCs) given or as a fraction (e.g., 0.01) of a femur harvest. At day 11, there are occasionally signs of very small colonies that may be the remains of day 8 CFU-S. The results of the assay can be expressed as CFU-S per 10^5 bone marrow cells or per femur, or per milliliter of mobilized blood.

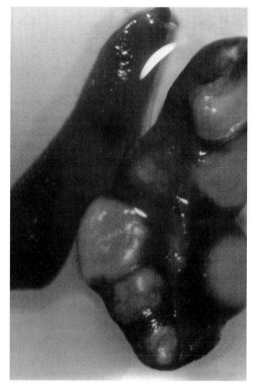

Fig. 1. Day 11 spleen colonies. Colonies formed in mouse spleen 11 days after irradiation and injection of bone marrow cells (right) compared to a control spleen from an irradiated mouse with no bone marrow inoculation (left).

Pre-CFU-S Assay. The sample containing HPCs is injected into irradiated hosts as described for the CFU-S assay, but on day 11 when the spleens are harvested, the bone marrow is also harvested and intravenously inoculated at varying doses into a secondary irradiated host. After 13 days, the spleens are harvested and colonies counted as described above.

In vitro Clonogenic Assays for Hematopoietic Progenitor Cells

Principle. Several *in vitro* clonal assays for multipotential and monopotential hematopoietic progenitor cells have been developed. In these assays hematopoietic cells (murine or human) are suspended in a semisolid medium of agar or methylcellulose and cultured in the presence of stimulatory hematopoietic growth factors. During the culture period single or mixed colonies of macrophages, granulocytes, erythroid cells, and megakaryocytes can develop from colony-forming HPCs. The use of a soft gel localizes the progeny of the HPC as a colony that can subsequently be counted, the number of colonies being proportional to the number of bone marrow cells inoculated. One key advantage of this type of assay is that different combinations of growth factors can be added to control and manipulate the differentiation of specific lineages. As with the CFU-S assay, there are waves of colony formation, and plates are typically scored at two time points to monitor both early and late progenitor cells. Agar or methylcellulose can be used for this type of assay, although the latter is most suitable for growing erythroid colonies.

A variation of this type of assay is the colony-forming unit—type A (CFU-A) as defined by Pragnell *et al.*[35] in which a more primitive cell, with similar characteristics to the CFU-S, is quantitated. The exact combination of hematopoietic growth factors that promote CFU-S *in vivo* is not known; however, in the CFU-A assay, colony formation can be supported by a combination of supernatants containing the hematopoietic growth factors macrophage colony-stimulating factor (M-CSF) and GM-CSF.

Considerations. The *in vitro* assays described below use sterile reagents and should be performed using sterile techniques in a tissue culture cabinet to maintain contamination-free cultures. Three major considerations must be taken into account when establishing soft gel assays.

First, because the growth requirements of different bone marrow cells are complex, the choice of growth factors added to stimulate colony formation will vary depending on the types of HPC populations under study. For example, the influence of MIP-1α on erythroid colony formation requires the addition of erythropoietin (Epo) to the cultures, whereas myeloid differ-

---

[35] I. B. Pragnell, E. G. Wright, S. A. Lorimore, J. Adam, M. Rosendaal, J. F. De Lamarter, M. Freshney, L. Eckman, A. Sproul, and N. Wilkie, *Blood* **72,** 196 (1988).

entiation requires factors such as M-CSF and GM-CSF but can progress in the absence of erythropoietin. As a guideline, Broxmeyer et al.[36] have reported potent inhibitory effects of MIP-1α and MIP-2α in human soft gel cultures stimulated with various combinations of steel factor [also knowns as stem cell factor, c-Kit ligand, or mast cell growth factor (50 ng/ml)], GM-CSF (100 U/ml), and Epo (1–2 U/ml). Chasty et al.[37] have also successfully used steel factor (100 ng/ml) and GM-CSF (10 ng/ml) in human cultures. Lord et al.[4] observed inhibition in murine cultures stimulated with interleukin-3. The range of inhibitory activity of MIP-1α in this type of assay will depend on the stimulatory source, but 150 ng/ml has been shown to be an appropriate dose in some assays.[4,37]

Second, the number of cells plated per assay will depend on the plating efficiency achieved using the optimal growth factor and serum conditions. As with the CFU-S assay, it is important to inoculate sufficient cells to enable detection of both increases and decreases in colony numbers without the culture becoming too confluent to see individual colonies, or too sparse to be able to derive statistically useful data. As a rough guide, conditions that generate around 50 colonies per 1-ml culture dish are advisable, and this number of colonies can normally be generated from about 2×10^5 bone marrow leukocytes. The assays are routinely performed in triplicate, using 1 ml per 35-mm petri dish.

Finally, scoring procedures vary from laboratory to laboratory, and even from individual to individual. To ensure reproducible scoring, therefore, it is important not only to use a specific set of scoring criteria, but also for individuals to perform cross-comparisons of their counts.

Experimental Protocols for Clonogenic Assays

Agar Assay. A stock solution of 3.3% (w/v) agar is warmed in a boiling water bath until molten, then added to a plating mix prepared from the following: 20% (v/v) fetal calf serum (FCS), 10% (w/v) bovine serum albumin (BSA), 10% (v/v) growth factors, 10% (v/v) cells (at 10× final concentration), and 40% (v/v) Iscove's modification of Dulbecco's medium (IMDM). The molten agar is then added at 10% (v/v) to give a final concentration of 0.33%, followed by rapid mixing using a 16-gauge needle. The agar mix is immediately plated out into 35-mm petri dishes using a 16-gauge needle. The petri dishes are swirled to distribute the agar evenly and left for a few minutes to set. The cultures are then incubated at 37° in a humidified atmosphere with 5% (v/v) CO_2. Because the petri dishes may

[36] H. E. Broxmeyer, B. Sherry, S. Cooper, L. Lu, R. Maze, M. P. Beckmann, A. Cerami, and P. Ralph, *J. Immunol.* **150,** 3448 (1993).

[37] R. C. Chasty, G. S. Lucas, P. J. Owen-Lynch, P. J. Pierce, and A. D. Whetton, *Blood* **86,** 4270 (1995).

dry out somewhat during the culture period, they can be placed inside a larger petri dish along with a dish containing sterile water, which serves to provide a humidified microenvironment for the cultures. The dishes are then scored at days 7 and 11 for murine colonies and at days 10 and 20 for human cultures, using an inverted light microscope. A grid scored on the bottom of the plate can help when counting dishes with high numbers of colonies.

The most frequent problem that we have encountered with this type of agar-based assay has been nonuniform solidification of the agar. Therefore, it is important to mix and plate out the agar rapidly to minimize this happening. Likewise temperature is also important, and the plating mix may need some warming to 37° before adding the agar to ensure that there is no premature solidification.

CFU-A Assay. The assay according to Pragnell *et al.*[35] is a variation of the assay described above. A stock solution of 6% agar is heated in a boiling water bath until molten. Meanwhile a plating mix is prepared that consists of 25% (by volume) HIFCS, 20% growth factors (10% L929 + 10% AF1-19T conditioned medium or recombinant M-CSF + GM-CSF), and 45% α-modified Eagle's medium (α-MEM). Then 10% agar is added to this mixture, giving a final concentration of 0.6% agar. The tube is mixed well, and 1-ml triplicates are immediately plated out into 35-mm petri dishes using a syringe and 16-gauge needle, after which the agar is allowed to set.

Following this, a stock solution of 3% agar is warmed until molten while a second plating mix is prepared consisting of 25% HIFCS, 10% cells, and 55% α-MEM. Then 10% agar is added as above to give a final concentration of 0.3% agar, which is then used to overlay the 0.6% agar. The cultures are incubated at 37° in a humidified atmosphere with 5% CO_2 as described above. At the end of the 11-day culture period the colonies are evaluated on an inverted light microscope as described below.

Methylcellulose Assay. Forty milliliters methylcellulose (Stem Cell Technologies) is mixed with 30 ml FCS, 10 ml of 10% (w/v) BSA, 0.1 ml 2-mercaptoethanol ($10-1$ M), and 1 ml L-glutamine (200 mM) and stored at −80°. This methylcellulose combination can be purchased ready made from Stem Cell Technologies. After warming to 37° the methylcellulose is mixed with cells as follows: 10% cells, 10% growth factors, and 80% methylcellulose (0.9% final concentration). As with the agar assays, the cells are immediately mixed and plated using a 16-gauge needle and syringe, then cultured at 37° in a humidified atmosphere with 5% CO_2.

Scoring Soft Gel Assays. Agar and methylcellulose colonies are usually scored in a similar fashion, whereby colony-forming cells are defined as those HPCs capable of producing at least 50 progeny over 7 days in murine assays and 10–12 days in human assays. Cells that generate less than 50 progeny are termed cluster-forming cells and represent a more mature HPC population. Colony-forming cells (CFU-C) are essentially a combination of myeloid progenitors that can be subdivided into progenitors restricted to the granulocyte and/or macrophage pathway, termed CFU-G (granulocyte), CFU-M (macrophage), and CFU-GM (granulocyte–macrophage). Other specific colony types include the following. (1) Megakaryocyte colonies fall into a separate category, as they can undergo DNA synthesis in the absence of cell division. In this case, a colony consists of at least three megakaryocytes and no cells from other lineages. (2) Colony-forming unit—granulocyte, erythroid, monocyte, macrophage, megakaryocyte (CFU-GEMM)—is a colony that contains cells of multiple lineages. Believed to be primitive progenitors, these are evaluated in methylcellulose at day 11 in murine cultures or at days 18–20 in human cultures. (3) Erythroid colonies are readily distinguishable by virtue of their red tinge, but they are more difficult to score because they are composed of multiple clusters. Colony-forming unit—erythroid (CFU-E) consist of one or two clusters that are believed to lyse when fully mature and are scored at day 7 for murine or day 10–12 for human. (4) Burst-forming unit—erythroid (BFU-E) denotes an earlier erythroid progenitor with greater proliferative capacity than CFU-E that forms 3–8 clusters per colony, which are scored at day 7 (day 10–12 human). A further, more primitive BFU-E can be scored on day 11 (day 18–20 human) as colonies comprising greater than 9 clusters. With experience these CFU types can be distinguished directly in the culture; initially, however, it is advisable to remove the colonies onto slides and stain them using Wright–Giemsa stain.

The CFU-A assay is analyzed using the parameters described by Pragnell *et al.*,[35] where after 11 days of culture of murine cells macroscopic colonies greater than 2 mm in diameter are scored.

In Vitro Stromal Assays for Hematopoietic Progenitor Cells

Principle. The soft gel assays described above have the disadvantage that they do not support growth of the very primitive HPCs responsible for long-term engraftment of ablated bone marrow. To study such primitive HPCs, assays based on the long-term bone marrow culture (LTBMC) system first described by Dexter *et al.*[38] have been developed. Unlike the soft gel assays, which rely on the addition of exogenous growth factors, coculture systems mimic the crucial *in vivo* relationship between hematopoietic cells and their microenvironment, where close association with the stroma regulates and supports maintenance of HPCs.

The original LTBMC assay[38] was not clonal and could not readily be quantitated. However, the assay has been set up for quantitation using

---

[38] T. M. Dexter, T. D. Allen, and L. G. Lajtha, *J. Cell Physiol.* **91,** 335 (1977).

limiting dilution analysis, and subsequently the frequency of an LTBMC initiating cell (LTC-IC) has been established by scoring cultures of human bone marrow growing on irradiated human primary stromal cells for replatable CFU-C.[39] A second variation is the cobblestone area forming cell assay (CAFC), a miniaturized LTBMC in which bone marrow cells are overlaid at limiting dilution on stromal cells and the percentage of wells with at least one phase-dark cobblestone area (a primitive hematopoietic clone under the stroma) is determined over several weeks.[40] In the murine system the CAFC day 7–10 measure is reported to be a correlate to the CFU-S d11, whereas CAFC results at week 4 and above are related to the long-term engrafting cells. This assay has now been performed with both mouse and human[41] bone marrow cells, and it has the advantage that it can be performed using murine cloned cell lines as stroma rather than having to prepare primary mouse or human stroma.

Considerations. Coculture methods can use cell lines or primary cells[39] to provide the basal stromal layer. The method described below uses a murine stromal cell line. Not all stromal cell lines support HPC cultures, and therefore selection of the correct line is important. As a guideline, FBMD-1 cells,[41] MS-5 cells,[42] and genetically engineered cells[43] have all successfully been used for coculture assays. Because each stromal cell line has its own growth medium and culture specifications, it is important to ensure that highly detailed instructions on how to culture the cells are provided by the supplier, once the line has successfully been set up, aliquots of frozen stock should be established.

Because cultures are maintained for up to 8 weeks in the same tissue culture plate with weekly feeds, a major problem encountered with CAFC assays is contamination of the cocultures. A very high standard of cell culture technique is required to maintain contamination-free cultures.

Experimental Protocol for Stromal Assays

Maintenance of Stroma. Stromal cells are adherent and can be cultured in 75-cm^2 flasks at 37° in a humidified incubator with 5% CO_2. Cells are normally passaged once a week using the following trypsin/EDTA procedure. Culture medium is removed gently with a 10-ml pipette, then 5 ml of HBSS is added to the flask, which is gently swirled to wash the cells free

[39] H. J. Sutherland, P. M. Lansdorp, D. H. Henkelman, A. C. Eaves, and C. J. Eaves, *Proc. Natl. Acad. Sci. U.S.A.* **87,** 3584 (1990).

[40] R. E. Ploemacher, J. P. Van der Sluijs, J. S. A. Voerman, and N. H. C. Brons, *Blood* **74,** 2755 (1989).

[41] D. A. Breems, E. A. W. Blokland, S. Neben, and R. E. Ploemacher, *Leukemia* **8,** 1095 (1994).

[42] I. Auffrey, A. Dubart, B. Izac, W. Vainchenker, and L. Coulombel, *Exp. Haematol.* **22,** 417 (1994).

[43] C. J. Eaves and A. C. Eaves, *Blood Cells* **20,** 83 (1994).

of any remaining medium. Five milliliters of trypsin/EDTA (Sigma, St. Louis, MO), prewarmed to 37°, is added, and the flask is incubated for 5 min at 37°. During this time the cells will lift off the flask, and gentle tapping of the flask against the palm of the hand will help disperse the cells into a single cell suspension. Then 1 ml of FCS is added to inhibit further action of the trypsin, followed by a further 5 ml of medium to wash the cells by centrifugation at 1200 rpm for 5 min at room temperature. Cells are then resuspended in fresh culture medium and replated in fresh flasks. It is important that the stroma is not grown beyond 10 passages, because the cells may then start to transform.

Limiting Dilution Assay. Stromal cells are plated in 100 μl in 96-well plates and grown to confluence, usually within 1 week. Seeding numbers will vary with the individual cell line chosen, and an indication of the passage split ratio will usually be given when the cells are provided. One problem is that during the course of the experiment the stromal cells may strip off from the plastic, and preincubation with 0.1% gelatin to coat the plates can be used to minimize this possibility.[41] Once confluent the cultures should be used within 2 weeks, at which point they are overlaid with 100 μl of cells over a wide range of dilutions. As a rough guide 50,000 down to 100 human whole bone marrow cells per well, using 24 wells per dilution, is advisable to ascertain the correct range. Usually a 3-fold dilution factor is used for each consecutive dilution. Cultures are then maintained at 37° in a humidified atmosphere of 5% CO_2 and fed by weekly half-medium changes.

Scoring. The 96-well plates are scored using an inverted phase-contrast microscope. Each well is scanned for the presence of phase-dark cobblestone areas, which, as described by Ploemacher *et al.,*[40] are composed of at least five phase-dark cells under the stroma. Plates with human bone marrow cells are read from week 4 to week 8 after initiation, and the frequency of cobblestone-forming cells determined by Poisson statistics. Murine bone marrow is usually scored from week 1 to week 4.

Examples of the Effect of MIP-1α on Hematopoietic Progenitor Cells

Specific experimental examples of the use of these methods to investigate the activity of MIP-1α in the regulation of HPCs are provided in this section.

Protection of CFU-S by MIP-1α. The protection of HPCs from the effects of cytotoxic chemotherapeutic agents has been demonstrated by the experiments of Lord *et al.*[4] Hydroxyurea (Sigma) was dissolved with a little warming, at 100 mg/ml in HBSS. Mice were then given an HU dose of approximately 1 g/kg intraperitoneally by injecting 0.1 ml/10 g body weight followed by between 0 and 15 μg of muMIP-1α per mouse intravenously 3 and 6 hr later. The mice were then given a second dose of HU at 7 hr. Bone marrow from mice treated in this way was then harvested at time points over the following week and assayed for CFU-S activity as described above. Figure 2 shows that the CFU-S d11 in the group of mice that received MIP-1α were protected from the cytotoxic effects of HU so that by 3 days postdosing they were at or above normal levels relative to the control animals, which required a further 3–4 days to recover.

In setting up such experiments the reader should note that it may be necessary to try several concentrations of cells, for example, 0.01 and 0.02 of a femur, to get the resulting CFU-S within the range of 1–10 (described above) on the different days of harvesting. Also, the bone marrow toxicity of drug treatment can cause problems with clumping of marrow during harvesting; therefore, it is important to keep the cells on ice and to include serum in the HBSS.

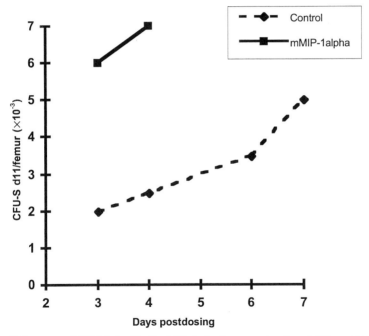

FIG. 2. Recovery of CFU-S day 11 after MIP-1α/hydroxyurea treatment. Mice were treated with hydroxyurea (900 mg/kg) at 0 and 7 hr. In the MIP-1α-treated group, 15 μg of chemokine per mouse was injected intravenously at 3 and 6 hr. The control group was injected with saline at 3 and 6 hr. Femurs were harvested at various times from 3 to 7 days after the HU injection and bone marrow assayed for CFU-S d11 activity. [Adapted with permission from Lord *et al.*, *Blood* **79,** 2605 (1992).[4]]

In vivo Protection of HPCs from Cytotoxic Effects of Hydroxyurea by MIP-1α. The murine model of repeated chemotherapy by HU has been used to show the *in vivo* myeloprotective activity of the engineered huMIP-1α variant BB-10010[9] using the methylcellulose assay to estimate survival and rate of recovery of the HPC population. Bone marrow cell suspensions were prepared from mice in each treatment group ($n = 5$ per group) to allow an assessment of the CFU per femur in individual mice (Fig. 3). On day +1, HU caused a significant reduction in mean CFU per femur compared to control. Protection by BB-10010 on day +1 posttreatment was not significant. However, by day +2, significant differences between treatment groups was apparent. In contrast to the group receiving HU alone, the groups receiving 100 and 500 μg/kg doses of BB-10010 and CFU per femur counts that were statistically indistinguishable from the control group. This shows that treatment with BB-10010 (i.e., MIP-1α) leads to an increased rate of recovery in more committed progenitors.

Mobilization of CFU-S d8, CFU-S d12, and pre-CFU-S by MIP-1α. The CFU-S and pre-CFU-S assays described in this chapter have been used to estimate the levels of HSCs and HPCs in the blood of mice after mobilization by the engineered huMIP-1α variant BB-10010.[9] Blood from control (0.4 ml) and BB-10010-treated (100 μg/kg s.c. 30 min prior to harvesting) animals (0.2 ml) was injected into irradiated recipients to estimate the levels of CFU-S and pre-CFU-S as described above. As shown in Fig. 4, elevated levels of each progenitor type were observed in the groups of animals that received BB-10010 (i.e., MIP-1α).

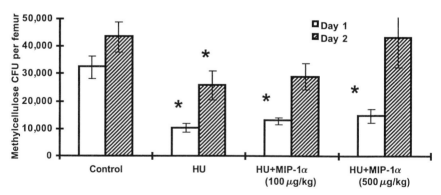

FIG. 3. Protection of HPCs by MIP-1α *in vivo*. Mice received 1 g/kg HU or saline control at 0 and 7 hr. MIP-1α at 100 or 500 μg/kg was given at 3 and 6 hr. Marrow was harvested at day 1 and day 2 postdosing and assayed for CFU activity in methylcellulose. Points showing significant suppression ($p < 0.05$) compared with controls are marked with an asterisk. [Adapted with permission from Hunter *et al., Blood* **86,** 4400 (1995).[9]]

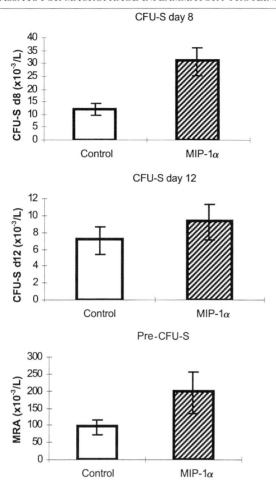

Fig. 4. Colony-forming cells in the peripheral blood of mice after treatment with MIP-1α. Mice were injected subcutaneously with 2.5 μg of muMIP-1α. After 30 min their blood was analyzed for CFU-S d8, CFU-S d12, and marrow reconstituting ability (MRA) as determined by pre-CFU-S assay. [Adapted with permission from Lord *et al.*, *Blood* **85,** 3412 (1995).[26]]

Overview of Methods

The possible involvement of MIP proteins in bone marrow dysfunction leading to disease states such as anemia and leukemia has been reviewed.[13] Assays of the type described above will continue to be crucial in expanding our knowledge of the complex mechanisms of action of MIP proteins and other chemokines and their role in bone marrow homeostasis.

Chemokines and Human Immunodeficiency Virus

The first description of the HIV suppressive activity of chemokines was a poster by Canque and Gluckman[11] in which they described the ability of huMIP-1α to suppress the infectivity of macrophage-tropic HIV-1 strains, HIV-1$_{BaL}$ and HIV-1$_{ADA}$, on primary macrophages. They also suggested that huMIP-1α may be suitable for the suppression of HIV infection within monocytes and macrophages *in vivo*. Subsequent studies by others[12,44] extended the range of C-C chemokines with HIV-1 suppressive activities to include huMIP-1β and huRANTES. The life strategy of HIV is reviewed by Haseltine,[45] and the reader can obtain mechanistic details of HIV infection and the effect of chemokines thereon in other key references.[46-54]

The chemokine-sensitive step during the infection of permissive cells by macrophage HIV-1 appears to be the interaction between the V3 loop of the viral gp120 protein and the cellular coreceptors CD4 and the chemokine receptor, of which CCR5 is the prototype. A signal transduction event is not necessary for suppression of HIV infectivity,[55] and coreceptor blockade is probably achieved by steric hindrance via the bound chemokine molecules. Studies to date have now identified the key components to be considered in this area of chemokine/HIV biology.

The interaction of HIV with chemokine receptors and the HIV sup-

[44] W. A. Paxton, S. R. Martin, D. Tse, T. R. O'Brien, J. Skurnick, N. L. VanDeventer, N. Padian, J. F. Braun, D. P. Kotler, S. M. Wolinsky, and R. A. Koup, *Nature Med.* **2,** 412 (1995).

[45] W. A. Haseltine, *Journal Acquir. Immune Defic. Syndr.* **1,** 217 (1988).

[46] Y. Feng, C. C. Broder, P. E. Kennedy, and E. A. Berger, *Science* **272,** 872 (1996).

[47] H. Deng, R. Liu, W. Ellmeier, S. Choe, D. Unutmz, M. Burhkart, P. Di Marzio, S. Marmon, R. E. Sutton, C. M. Hill, C. B. Davis, S. C. Peiper, T. J. Schall, D. R. Littman, and N. R. Landau, *Nature (London)* **381,** 661 (1996).

[48] T. Dragic, V. Litwin, G. P. Allaway, S. R. Martin, Y. Huang, K. A. Nagashima, C. Cayanan, P. J. Maddon, R. A. Koup, J. P. Moore, and W. A. Paxton, *Nature (London)* **381,** 667 (1996).

[49] G. Alkhatib, C. Combadiere, C. C. Broder, Y. Feng, P. E. Kennedy, P. M. Murphy, and E. A. Berger, *Science* **272,** 1955 (1996).

[50] H. Choe, M. Farzin, Y. Sun, N. Sullivan, B. Rollins, P. D. Ponath, L. Wu, C. R. Mackay, G. R. LaRosa, W. Newman, N. Gerard, C. Gerard, and J. Sodroski, *Cell (Cambridge, Mass.)* **85,** 1135 (1996).

[51] B. J. Doranz, J. Rucker, Y. Yi, R. J. Smyth, M. Samson, S. C. Peiper, M. Parmentier, R. G. Collman, and R. W. Doms, *Cell (Cambridge, Mass.)* **85,** 1149 (1996).

[52] E. Oberlin, A. Amara, F. Bacheterie, C. Bessia, J. Virelizier, F. Arenza-Seisdedos, O. Schwartz, J. Herd, I. Clark-Lewis, D. F. Legler, M. Loetscher, M. Baggiolini, and B. Moser, *Nature (London)* **382,** 833 (1996).

[53] C. C. Bleul, M. Farzan, H. Choe, C. Parolin, I. Clark-Lewis, J. Sodroski, and T. A. Springer, *Nature (London)* **382,** 829 (1996).

[54] P. Westervelt, D. B. Trowbridge, L. G. Epstein, B. J. Blumberg, Y. Li, B. H. Hahn, G. M. Shaw, R. W. Price, and L. Ratner, *J. Virol.* **66,** 2577 (1992).

[55] F. Arenzana-Seisdedos, J.-L. Virelizier, D. Rousset, I. Clark-Lewis, P. Loetscher, B. Moser, and M. Baggiolini, *Nature (London)* **383,** 400 (1996).

pressive activities of huMIP-1α and huMIP-1β are relatively recent discoveries, and therefore the development of assay technology in this area is still in its infancy and is consequently restricted to relatively few centers. The purpose of this section is to describe some of the techniques used to perform research in this area. However, given the practical considerations of working with a pathogen like HIV-1, it is not expected that many laboratories would initiate research with live virus solely to investigate the mechanism of chemokine suppression of HIV-1 infection. Knowledge of the viral and cellular molecules that are involved in the recognition and membrane fusion events which lead to infection has, however, allowed the development of virus-free assays such as that described by Alkhatib et al.[49] These assays are safer and more amenable, therefore, to laboratories unfamiliar with high level containment work. An example of the use of such a viral-free assay is outlined below.

Principles of Live HIV-1 Infectivity Assays

Live virus assays have three stages: (1) generation of reagents for the assay, (2) infection of the cells by the virus, and (3) readout of the assay. Reagent preparation may include isolation and growth of primary virus. However, as relatively few chemokine biologists will have access to HIV-infected individuals as a source of primary HIV-1 isolates, the focus has been on the use of established HIV-1 strains. Macrophage-tropic HIV-1 strains that have been used to study the suppressive effects of huMIP-1α and huMIP-1β include HIV-1$_{BAL}$, HIV-1$_{ADA}$, and HIV-1$_{JRFL}$. The virus must be passaged and prepared from fresh peripheral blood mononuclear cells or a suitable cell line (e.g., PM1[12]) that is permissive for macrophage-tropic HIV-1 strains. To provide consistent multiplicity of infection during a series of experiments, viral preparations are titrated using p24 antigen levels to determine concentration. The first stage of the assay might also include the preparation or sourcing of the chemokines. The infection of cells by the virus is done either with or without chemokines. This is the most flexible experimental stage, providing the option to vary dose, timing of the treatment, duration of chemokine exposure, and whether one or more chemokines are used.

Owing to the ease of measurement, quantification of p24 protein levels within the cell free culture supernatant is generally used to express assay results. The p24 concentration correlates to viral replication and release of synthesized virion. The use of p24 levels to estimate the quantity of input virus (allowing control of the multiplicity of infection) and to estimate the viral productivity of the infection may, however, be controversial because it is a measure of antigen rather than infective HIV-1 virion and the sample could include a proportion of defective virion. The simplicity and availabil-

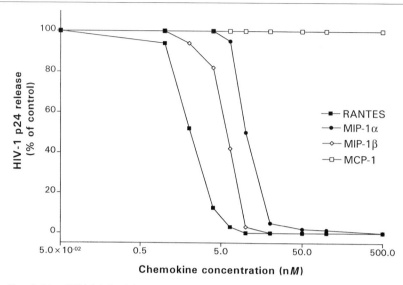

FIG. 5. Live HIV-1 infectivity assay. Dose-dependent suppression of HIV-1$_{BaL}$ infection of PM1 cells by recombinant chemokines (R&D Systems). (Reprinted with permission from Cocchi et al.[12] Copyright 1995 American Association for the Advancement of Science.)

ity of commercial enzyme-linked immunosorbent assay (ELISA) kits (e.g., Coulter Electronics, Hialeah, FL and DuPont/NEN, Boston MA), which have been used with good results, do, however, make p24 measurement the routine choice.

Infectivity Assay. Cocchi et al.[12] have used a viral infectivity assay with the HIV-1$_{BAL}$ strain and a CD4$^+$ T-cell clone, PM1, as an infectable cell line.[56] The PM1 cells (2×10^5 per assay point) were infected with HIV-1$_{BAL}$ (4 ng p24 per 10^6 cells) for 2 hr at $37°$. The cells were then washed free of excess virus three times in prewarmed phosphate-buffered saline (PBS) and finally resuspended in culture medium. The infected cells ($\sim 8 \times 10^5$ PM1 cells/ml) were then cultured in the presence of a serial dilution from 0.1 to 500 ng/ml of chemokine in complete culture. After 3 days, fresh culture medium containing the appropriate amount of chemokine was added to the cultures. Five to 7 days later the dose-dependent suppression of HIV-1$_{BaL}$ infection by recombinant chemokines, including huMIP-1α and huMIP-1β, was estimated by measuring the quantity of p24 antigen in cell free supernatants (Fig. 5). Both huMIP-1α and huMIP-1β were found to suppress HIV-1 infection in these *in vitro* cultures. Human MIP-1β was actually a more potent suppressor of HIV-1 infectivity than huMIP-1α,

[56] P. Lusso, F. Cocchi, C. Baloota, P. D. Markham, A. Louie, P. Farci, R. Pal, R. C. Gallo, and M. S. Reitz, Jr., *J. Virol.* **69**, 3712 (1995).

with an ED_{95} (dose of chemokine required to reduce HIV-1 p24 synthesis by 95%) of between 6.25 and 12.5 ng/ml versus 12.5 to 25.0 ng/ml, respectively.

Virus-Free Assays Based on Cell Fusion. Alkhatib *et al.*[49] have used a β-galactosidase-based fusion assay to investigate the effects of chemokines on HIV-1 infection. The assay used a recombinant vaccinia-based system in which the fusion between Env-expressing and CD4$^+$ CCR5-expressing cells induces the activation of β-galactosidase. The CD4- and CCR5-expressing cells (NIH 3T3) contained a vaccinia construct comprising a T7 polymerase-inducible promoter linked to the β-galactosidase gene. The Env-expressing cells (HeLa) contained a vaccinia construct with the T7 polymerase gene expressed. Thus, fusion of the two cytoplasms caused the mixing of the T7 polymerase and the T7 promoter-driven β-galactosidase gene, which induced quantifiable β-galactosidase expression. In this assay, significant, dose-dependent inhibition of cell fusion (50% reduction) was achieved with approximately 10 n*M* (about 77 ng/ml) of huMIP-1α or huMIP-1β (Fig 6). The virus-free assay is, therefore, less sensitive than the live-virus assay described above.

Overview of Methods

The live-virus assays, which first suggested the HIV-1 suppressive activity of huMIP-1α,[11,12] represent the most effective and sensitive methods.

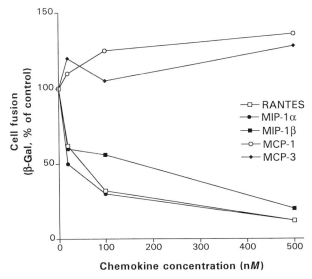

FIG. 6. Virus-free infectivity assay. Dose-dependent suppression of Env- and CD4/CCR5-dependent cell fusion by recombinant human chemokines (Peprotech). (Reprinted with permission from Alkhatib *et al.*[49] Copyright 1996 American Association for the Advancement of Science.)

They have been used to show that the suppressive effects are most pronounced if the chemokine is present prior to infection rather than postinfection and that prolonged administration is necessary to achieve complete suppression of viral replication. The virus-free assays typified by that of Alkhatib et al.[49] provide a system more suitable for general use, because high-level containment facilities are not required. The virus-free assay is also more amenable for screening chemical libraries for inhibitors of HIV-1 infection and may ultimately be more useful in dissecting the molecular interactions that are responsible for HIV-1 infection.

Conclusion

A key element to approval of a molecule for clinical trial is data supporting the theory of action and the expected therapeutic benefit. The methods described in this chapter have been critical to the acceptance of MIP-1α as a candidate biopharmaceutical for alleviation of the side effects associated with cancer chemotherapy and in controlling the progression HIV infection. As the list of chemokines continues to grow, these assays will remain central to elucidation of the basic biology and the clinical potential of this fascinating molecular family. Chapters throughout this volume detail other methods used for the characterization of chemokines, and further background information on the structure, function, and clinical utility of the chemokine family in general[57,58] and the macrophage inflammatory protein (MIP) subfamily specifically[13] can be found in extensive reviews.

[57] R. Horuk, "Chemoattractant Ligands and Their Receptors." CRC Press, Boca Raton, Florida, 1996.
[58] T. J. Schall, in "The Cytokine Handbook" (A. Thompson, ed.), 2nd Ed., Chap. 22. Academic Press, London, 1994.

[10] Gene Expression of RANTES

By Peter J. Nelson, J. M. Pattison, and Alan M. Krensky

Introduction

This chapter focuses on techniques used to study the molecular mechanisms governing RANTES gene expression in various cell types, including T cells, macrophages, endothelial cells, epithelial cells, and fibroblasts. The regulation of RANTES mRNA production in interstitial and immune cell types, along with its functional properties, afford it an important role in

0076-6879/97 $25.00

gctgcagaggatcaagacagcacgtggacctcgcacagcctctcccacaggtacc <u>ATG AAG GTC TCC GCG</u>
<u>GCA CGC CTC GCT GTC ATC CTC ATT GCT ACT GCC CTC TGC GCT CCT GCA TCT GCC</u> TCC
CCA TAT TCC TCG GAC ACC ACA CCC TGC TGC TTT GCC TAC ATT GCC CGC CCA CTG CCC
CGT GCC CAC ATC AAG GAG TAT TTC TAC ACC AGT GGC AAG TGC TCC AAC CCA GCA GTC
GTC TTT GTC ACC CGA AAG AAC CGC CAA GTG TGT GCC AAC CCA GAG AAG AAA TGG GTT
CGG GAG TAC ATC AAC TCT TTG GAG ATG AGC TAG gatggagagtccttgaacctgaacttacaca
aatttgcctgtttctgcttgctcttgtcctagcttgggaggcttcccctcactatcctaccccacccgcgcctga
agggcccagattctgaccacgacgagcagcagttacaaaaaccttccccaggctggacgtggtggctcagccttg
taatcccagcactttgggaggccaaggtgggtggatcacttgaggtcaggagttcgagacagcctggccaacatg
atgaaaccccatgtgtactaaaaatacaaaaaattagccgggcgtggtagcgggcgcctgtagtcccagctactc
gggaggctgaggcaggagaatggcgtgaacccgggagcggagcttgcagtgagccgagatcgcgccactgcactc
cagcctgggcgacagagcgagactccgtctcaaaaaaaaaaaaaaaaaaaaaaaaaatacaaaaattagccgcgt
ggtggcccacgcctgtaatcccagctactcgggaggctaaggcaggaaaattgtttcaacccaggaggtggaggc
tgcagtgagctgagattgtgccacttcactccagcctgggtgacaaagtgagactccgtcacaacaacaacaaca
aaaagcttccccaactaaagcctagaagagcttctgaggcgctgctttgtcaaaaggaagtctctaggttctgag
ctctggctttgccttggctttgcaagggctctgtgacaaggaatgaagtcagcatgcctctagaggcaaggaagg
gagaacactgcactcttaagcttccgccgtctcaacccctcacaggagcttactggcaaacatgaaaaatcgggc
ttaccaataaagttctcaatgcaaccaaaaaaaaaaaaaaaa

FIG. 1. cDNA sequence for human RANTES. The 5′ and 3′ untranslated regions are in
lowercase letters, and the signal sequence is underlined.

models of the induction and propagation of the inflammatory response.[1,2]
The techniques described here have expanded understanding of the biochemical mechanisms underlying the biology of RANTES gene expression,
the regulation of inflammatory processes, as well as aspects of peripheral
T-cell development.

RANTES cDNA

Human RANTES is encoded by a 1.2-kb mRNA containing a short 5′
untranslated region, a coding region of 275 nucleotides, and a long 3′
untranslated region that includes multiple Alu repeat sequences (Fig. 1).[2,3]
The mRNA encodes a 10-kD protein including a cleavable amino-terminal
signal sequence of 23 amino acids. The secreted protein is 68 amino acids

[1] T. J. Schall, *Cytokine* **3**, 165 (1991).
[2] P. J. Nelson, J. M. Pattison, and A. M. Krensky, *in* "Chemoattractant Ligands and Their
Receptors" (R. Horuk, ed.), CRC Press, Boca Raton, Florida, 1996.
[3] T. J. Schall, J. Jongstra, B. J. Bradley, J. Dyer, J. Jorgensen, C. Clayberger, M. Davis, and
A. M. Krensky, *J. Immunol.* **141**, 1018 (1988).

long with a predicted mass of approximately 8 kD. Mouse and rat mRNAs are approximately 560 nucleotides in length and display a much shorter 3′ untranslated region than the human mRNA.[2,4]

RANTES Locus

The RANTES gene is located on human chromosome 17 (17q11.2-q12).[5] The other C-C chemokines also localize to chromosome 17, suggesting that the C-C family members evolved by gene duplication.[1] In addition to coding sequence similarities, the genomic organization of these genes is remarkably conserved.[1] Both human and mouse RANTES genes display a three exon–two intron organization common to the C-C chemokine family.[2,6,7]

RANTES Promotor Region

The immediate upstream promotor regions of the murine and human RANTES genes have been characterized (the human DNA sequence is shown in Fig. 2).[6,7] A large number of consensus sequences for specific transcription factors have been identified within these promoter sequences.[6,7] The abundance of potential regulatory sites believed to be functional in disparate cell types, combined with the variety of tissue types known to express RANTES in response to specific stimuli, has led to the suggestion that different parts of the RANTES promoter may be operative in the different cell types that express RANTES.[2,6] This hypothesis was supported by the initial studies of RANTES promotor usage in which cells of T-lymphocyte, erythroid, and muscle lineages displayed different promotor requirements for optimal reporter gene activity.[2,6] Subsequent reports have suggested alternative mechanism for the control of RANTES transcription in T cells, monocytes, and fibroblasts.[8–10]

A comparison of the human RANTES promoter sequence to that of the mouse displays a good deal of sequence conservation, particularly in

[4] P. Heeger, G. Wolf, C. Meyers, M. J. Sun, S. C. O'Farrell, A. M. Krensky, and E. G. Neilson, *Kidney Int.* **41**, 220 (1992).

[5] T. A. Donlon, A. M. Krensky, M. R. Wallace, M. R. Collins, M. Lovett, and C. Clayberger, *Genomics* **6**, 548 (1990).

[6] P. J. Nelson, H. T. Kim, W. C. Manning, T. Goralski, and A. M. Krensky, *J. Immunol.* **151**, 2601 (1993).

[7] T. M. Danoff, P. A. Lalley, Y. S. Chang, P. S. Seeger, and E. G. Neilson, *J. Immunol.* **152**, 1182 (1994).

[8] P. J. Nelson, B. D. Ortiz, J. M. Pattison, and A. M. Krensky, *J. Immunol.* **157**, 1139 (1996).

[9] H. S. Shin, B. E. Drysdale, M. L. Shin, P. N. Noble, S. N. Fisher, and W. A. Paznekas, *Mol. Cell. Biol.* **14**, 2914 (1994).

[10] B. D. Ortiz, A. M. Krensky, and P. J. Nelson, *Mol. Cell. Biol.* **16**, 202 (1996).

5'

CTC -975
GAGGATCCCTAAAGTCCTTTGAAGCTTTCCATATTCTGTAACTTTTGTGCCCAAGAAGGCCTTACAGTGAGATGG -900
GATCCCCAGTATTTATTGAGTTTCCTCATTCATAAAATGGGATAATAATAGTAAATGAGTGATACTCGCGCTAAG -825
ACAGTGGAATAGTGGCTGGCACAGATAAGCCCTCGGTAAATGGTAGCCAATAATGATAGAGTATGCTGTAAGATA -750
GATCTTTCTCCCCTCGCTTCTCAACAAGTCTCTAATCAATTATTCCACTTTATAACAAGGAAATAGAACTCAA -675
AGACATTAAGCACTTTTCCCAAAGGTCGCTTAGCAAGTAAATGGGAGAGACCCTATGACCAGGATGAAAGCAAGA -600
AATTCCCACAAGAGGACTCATTCCAACTCATATTCTTGTGAAAAGGTTCCCAATGCCCAGCTCAGATCAACTGCCT -525
CAATTACAGTGTGAGTGTCTCACCTCCTTTGGGACTGTATATCCAGAGGACCCTCCTCAATAAAACACTTTA -450
TAAATAACATCCTTCCATGGATGAGGAAGAGGTAAGATCTGTAATGAATAAGCAGGAACTTTGAAGACTCAG -375
TGACTCAGTGAGTAATAAAGACTCAGTGACTTCTGATCCTGTCCTAACTGCCACTCCTTGTTGTCCCCAAGAAAG -300
CGGTTCCTGCTCTCTGAGGAGGACCCCTTCCCTGGAAGGTAAAACTAAGGATGTCAGCAGAGAAATTTTTCCACC -225
ATTGGTGCTTGGTCAAAGAGAGAAACTGATGAGCTCACTCTAGATGAGAGCAGTGAGGGAGAGACAGAGACTCG -150

R(C)

AATTTCCGGAGGCTATTTCAGTTTCTTTTCCGTTTTGTGCAATTTCACTTATGATACCGGCCAATGCTTGGTTG -75

LRE (MOUSE)   R(E)-NFIL6RE
CTATTTTGGAAACTCCCCTTAGGGGATGCCCCTCAACTGGCCCTATAAAGGGCCCAGCCTGAGCTGCAGAGGATCC 0
R(A)-κB   R(B)-κB   Transcriptional
Start

Fig. 2. DNA sequence of approximately the first kilobase of the immediate upstream region of the human RANTES promoter. Regions important for transcriptional regulation are underlined.

the immediate -200 nucleotides upstream of the site of transcriptional initiation.[11] This sequence similarity extends through the transcriptional start site and includes the 5' untranslated region of the mRNA, suggesting that these regions may be important in control of RANTES expression.[11]

Expression of RANTES mRNA

The temporal regulation and mode of induction of the RANTES gene varies significantly among different cell types. Megakaryocytes appear to produce RANTES constitutively, as do some tumors.[2,12] Most cell types, including fibroblasts, epithelial cells, endothelial cells, and monocytes, up-regulate RANTES mRNA within hours after simulation.[4,9,13–15] In monocytes, the peak expression is seen at approximately 6 to 8 hr following stimulation.[9] In epithelial cells, the expression peaks later, approximately 24 to 36 hr after activation.[13] In T lymphocytes, RANTES mRNA transcripts are seen several hours after activation; the mRNA is then strongly up-regulated about 3 days later.[3,16]

Regulation in Fibroblasts, Epithelial, and Endothelial Cells

An early component of the acute phase of the inflammatory response is the production of agents, such as tumor necrosis factor-α (TNF-α), interleukin-1β (IL-1β), and interferon-γ (IFN-γ), by the stressed tissue.[17] On stimulation by these agents, fibroblasts and epithelial and endothelial cells up-regulate RANTES mRNA and protein.[4,8,13–15,18] Stimulation by IL-4 appears to suppress, and IFN-γ to enhance, the TNF-α- or IL-1β-induced expression of RANTES mRNA in synovial fibroblasts.[18] Endothelial cells produce RANTES after stimulation with TNF-α, and IFN-γ, IL-4, and IL-13 inhibit this induction.[14] A human bronchial epithelial cell line up-regulates

[11] B. D. Ortiz, P. J. Nelson, and A. M. Krensky, in "Biology of the Chemokine RANTES" (A. M. Krensky, ed.), Chap. 6. Molecular Biology Intelligence Unit, R. G. Landes Company and Springer-Verlag, Austin, Texas, 1995.

[12] I. von Luettichau, P. J. Nelson, M. Vandereijin, P. Huie, J. M. Patterson, R. A. K. Stahl, C. J. Wiedermann, R. Warnke, R. K. Sibley, and A. M. Krensky, Cytokine 8, 89 (1996).

[13] C. Stellato, L. A. Beck, G. A. Gorgone, D. Proud, T. J. Schall, S. J. Ono, L. M. Lichtenstein, and R. P. Schleimer, J. Immunol. 155, 410 (1995).

[14] A. Marfaing-Koka, O. Devergne, G. Gorgone, A. Portier, T. J. Schall, P. Galanaud, and D. Emilie, J. Immunol. 154, 1870 (1995).

[15] O. Devergne, A. Marfaing-Koka, T. J. Schall, M.-B. Leger-Ravet, M. Sadick, M. Peuchmaur, M.-C. Crevon, T. Kim, P. Galanaud, and D. Emilie, J. Exp. Med. 179, 1689 (1994).

[16] L. Turner, S. G. Ward, and J. Westwick, J. Immunol. 155, 2437 (1995).

[17] P. Emergy and M. Salmon, Br. J. Hosp. Med. 45, 164 (1991).

[18] P. Rathanaswami, M. Hachicha, M. Sadick, T. J. Schall, and S. R. McColl, J. Biol. Chem. 268, 5834 (1993).

RANTES mRNA and protein by 16 hr after stimulation with TNF-α and IFN-γ. This induction was inhibited by treatment with glucocorticoids.[13]

Nelson and colleagues (and unpublished work by J. M. Pattison) suggest that, in dermal fibroblasts and epithelial cells, the induction of RANTES mRNA by TNF-α is mediated in part by Rel transcription family members, p65 and p50.[8] There are tandem κB-like binding sequences within the first 70 bases of the RANTES promotoer.[6,8] These sites bind Rel p50–p65 heterodimers rapidly after TNF-α stimulation of dermal fibroblasts[8] and epithelial cell lines.[19] In addition, the suppression of RANTES expression by dexamethasone is mediated, at least in part, through these two κB elements.[19]

RANTES Expression by Kidney Tissues

The expression of RANTES by kidney cells has been the focus of several studies.[4,20,21] Some renal diseases, including transplant rejection and interstitial nephritis, are characterized by a strong mononuclear cell infiltration of the kidney.[21] RANTES is highly expressed in renal allografts undergoing cellular rejection.[21] It is induced as an immediate early gene (2–20 hr) in murine renal tubular epithelial cells following activation with TNF-α and IL-1β, whereas no effect was reported with transforming growth factor-β (TGF-β), IFN-γ, or IL-6 stimulation.[4] Mouse mesangial cell lines up-regulate RANTES within 2 hr in response to stimulation with TNF-α or lipopolysaccharide.[20]

Monocyte and Macrophage Expression

Monocytes and macrophages up-regulate RANTES mRNA within hours after stimulation with lipopolysaccharide (LPS).[9,15] Devergne et al. found RANTES expression by monocytes in vivo and in vitro.[15] Shin and co-workers used the murine monocyte cell line RAW264.7 (ATCC, Rockville, MD; TIB 71) and reporter gene assays linking the immediate upstream region of the RANTES gene to the chloramphenicol acetyltransferase (CAT) reporter gene to study LPS induction of RANTES expression at the transcriptional level.[9] The generation of a series of promoter–deletion– reporter gene constructs allowed characterization of a LPS-responsive element in the immediate upstream region of the murine RANTES gene.[9] This region, termed LRE (lipopolysaccharide-responsive element), lies between

---

[19] J. M. Pattison, submitted for publication.

[20] G. Wolf, S. Aberle, F. Thaiss, P. J. Nelson, A. M. Krensky, E. G. Neilson, and R. A. K. Stahl, Kidney Int. **44,** 795 (1993).

[21] J. Pattison, P. J. Nelson, P. Huie, I. von Luettichau, G. Farshid, R. K. Sibley, and A. M. Krensky, Lancet **343,** 209 (1994).

nucleotides −180 and −138 relative to the start of transcription.[9] Mutational analysis indicated that the LRE consists of a TCAYR motif, located at position −172, and a second sequence at −159. Antibodies to the cAMP-responsive element-binding protein (CREB) and c-*jun* inhibit the nuclear factor complex formation on the DNA, suggesting that these proteins (or close relatives) are involved in this LPS-inducible DNA binding protein complex. The motif at −159 resembles the interferon-stimulated responsive element (ISRE). All of these elements are located within a large A/T-rich region of the promoter that is highly conserved between the murine and human genes.[9,11]

RANTES Regulation in T Lymphocytes

The expression of RANTES in T cells appears to be, in part, a developmentally controlled event. Early expression seen after mitogen stimulation of resting peripheral blood T cells is followed by a strong up-regulation occurring 3 to 5 days later.[3,16] High expression of RANTES is maintained indefinitely *in vitro* in effector T cells, such as cytotoxic and helper T lymphocytes.[3,6,8] To date, the only T-cell tumor line demonstrated to express RANTES mRNA is the Hut78 T-cell line (ATCC, TIB 161).[6,10]

Reporter gene experiments using a series of 5′ to 3′ deletions of the immediate early region of the human RANTES gene linked to the luciferase reporter gene indicate that the immediate −195 bases 5′ from the site of transcriptional initiation form the minimal region of the RANTES promoter needed for optimal reporter gene expression in the Hut78 T-cell line as well as in phytohemagglutinin (PHA)-stimulated peripheral blood lymphocytes.[6,8,10] There are at least four important transcriptional control elements contributing to RANTES expression in T cells within these 195 nucleotides (Fig. 3).[8,10]

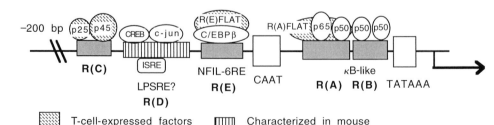

FIG. 3. Schematic model of transcription factor and control elements identified for the RANTES promoter (primarily those important for expression in T cells).

The region spanning nucleotides −78 to −42 is termed site R(A/B). This region, identified by DNase I footprinting and methylation interference, contains two Rel/κB-like consensus-binding regions, sites R(A) and R(B), separated by four nucleotides. Site R(B) efficiently binds p50–p50 homodimers and binds p50–p65 heterodimers less effectively.[8] Site R(A) has a high affinity for Rel p50–p65 heterodimers and binds p50–p50 homodimers to a lesser degree.[8] In T cells, site R(A/B) also binds a group of factors termed R(A)FLAT (RANTES site A factors of late activated T cells) specifically up-regulated 3–5 days after the activation of resting T cells.[8] These factors are also expressed in functionally mature CD8+ cytotoxic T lymphocytes (CTL).[8] The binding region for the R(A)FLAT factors includes both κB-like elements but also extends 5′ to the R(A) site to include nucleotides that lie outside the classic κB binding domain.[8]

The region R(E) was identified by sequence similarity to the consensus binding site for the C/EBP family of transcription factors.[6,10] NFIL6α/C/EBPβ positively regulates the RANTES promoter in Hut78 T cells by interacting with region R(E). Because RANTES is up-regulated early and transiently after peripheral blood lymphocyte (PBL) activation, the physiological relevance of NFIL6α/C/EBPβ to RANTES gene expression in effector T cells is still unclear.[10] This binding may help to explain why there is a low early up-regulation of RANTES mRNA in response to PHA.[16] By 2 days after activation of resting T cells with PHA, a second, as yet undefined C/EBP-like complex is up-regulated and gradually replaces NFIL6α/C/EBPβ binding to the R(E) site.[10] This C/EBP-like factor has been termed R(E)FLAT (RANTES site E factor of late activated T cells).[10]

An additional regulatory complex, termed R(C)FLAT (RANTES site C factor of late activated T cells), is a purine-rich region and is important for RANTES expression by the T-cell line Hut78.[10] This factor is induced only transiently in mitogen-stimulated PBL, with maximum expression seen 5 days after T-cell activation. This factor is not present in normal CTL lines.[10] The complex contains at least two DNA binding subunits of approximately 65 and 45 kDa as determined by UV cross-linking and polyacrylamide gel electrophoresis (PAGE), and it does not appear to be related to several transcription factor families known to bind to purine-rich sequences.[10] Preliminary data suggest that transcriptional regulation through the R(C) site may be context dependent as the R(C) site does not appear capable of transactivating a heterologous basal promoter.[10]

The study of RANTES promoter activity and the nuclear factors regulating RANTES transcription in the Hut78 T-cell line and in normal T cells has provided some explanation for the immediate early expression of RANTES in many cell types. It has also allowed the identification of several factors contributing to the late up-regulation of RANTES transcription during T-cell functional maturation.

Methods

Determination of RANTES mRNA Levels

Northern blotting is used for routine determination of steady-state levels of RANTES messenger RNA. Care should be taken to use only the coding region and 5' untranslated region as a hybridization probe because of the presence of the Alu-like repeats found in the 3' untranslated region.[3]

RNase I protection assays allow quantitative determination of specific mRNA levels. A rapid RNase I protocol used to study RANTES levels involves the hybridization of gene-specific probes to mRNA in crude cell lysates.[6,22] Sense and antisense RANTES-specific riboprobes can be prepared by cloning the coding region of the RANTES cDNA into a convenient vector (such as the pGEM vectors from Promega, Madison, WI), which permits *in vitro* transcription of RNA strands from either side of the vector, allowing production of RANTES-specific and control RNA riboprobes.[6] Following transcription, template DNA is removed by treatment with RNase-free DNase, and the riboprobe is extracted with TE [10 mM Tris-HCl (pH 7.5), 1 mM EDTA]-saturated phenol/chloroform, followed by two ammonium acetate/ethanol precipitations to remove unincorporated nucleotides. In the hybridization procedure, 1×10^7 cells are solubilized in 1 ml of lysate solution containing 5 M guanidine thiocyanate and 100 mM EDTA, pH 7. The RNA probe (2.5 μl, usually about 1–2 ng) is mixed with 10 μl of the lysate solution, heated to 60° for 5 min, and then hybridized for 18 hr at room temperature. The hybridization mixture is treated with 300 μl of a solution containing 50 μg/ml RNase A, 1 mM EDTA, 300 mM NaCl, and 30 mM Tris-HCl, pH 7.5, and incubated at room temperature for 45 min. Proteinase K at 0.2 mg/ml is then added, and samples are incubated at 37° for 15 min. The resultant digest is extracted with phenol/chloroform (1:1, v/v), coprecipitated with glycogen in ethanol, washed once with ethanol, lyophilized, and resuspended in 4 μl of gel loading buffer (90% formamide and 1 mM EDTA). After denaturation of the sample by incubating at 90° for 3 min, it can be loaded on a sequencing gel. Data are expressed as steady-state RANTES mRNA levels either per number of viable cells or relative to an internal standard such as actin mRNA.

Regulation of RANTES Promoter

Transfection of cells using promoter–reporter gene constructs allows functional characterization of *cis*-acting transcriptional regulatory components. Both stable and transient methods of transfection of RANTES promoter–reporter gene plasmids have been described.[6–10,14] One advantage in using stable transfection procedures is that the resultant transfected cell

---

[22] G. Firestein, S. M. Gardner, and W. D. Roeder, *Anal. Biochem.* **167,** 381 (1987).

line is easy to use and is very sensitive to specific stimulation (Fig. 4). As most transient transfection protocols are stressful to cells, and expression of RANTES is a stress response in many cell types, transient transfection often results in high levels of basal expression with subsequent low response to activation stimuli. In some instances, treatment of cells with the transfection procedure is sufficient to induce RANTES transcription (N. G. Miyamoto, personal communications, 1996).

Stably transfected cell lines can be generated by transfection (via electroporation, DEAE-dextran, or lipid agents; see below) of plasmids containing the RANTES promoter–reporter gene of interest together with a Geneticin (G418) resistance plasmid at a 10:1 molar ratio. At approximately 24 hr after transfection, Geneticin is added to the medium. (The level of Geneticin used must be determined empirically for each cell line prior to transfection of the cell line.) After 3 weeks, clones are pooled to dilute positional and insertional artifacts and expanded for screening by reporter gene assay. Unfortunately, experiments comparing different reporter–deletion mutations are not informative, as each stable line will show variable gene copy

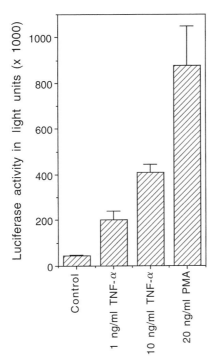

Fig. 4. One kilobase of the immediate upstream region of the RANTES promoter was stably transfected into the epithelial cell line SK-Hep-1 (ATCC, HTB 52), and the cells were then induced by addition of increasing levels of TNF-α, or PMA at 20 ng/ml (J. M. Pattison, unpublished results, 1995).

numbers and different sites of integration of the plasmid DNA. Subtle differences in reporter gene activity between constructs thus cannot be determined.

Transient Transfection Using DEAE-Dextran

Shin *et al.* used DEAE-dextran to introduce murine RANTES promoter–CAT reporter gene constructs into murine RAW264.7 monocyte cells to study LPS induction of murine RANTES at the level of gene transcription.[9] In this procedure, the RAW264.7 cells are plated at 2×10^7 cells per 10-cm tissue culture plate 1 day before transfection. The DEAE-dextran procedure uses 5 to 50 μg of plasmid DNA per dish. In this protocol, the DNA is diluted to 40 μl in Tris-buffered saline (TBS), and 80 μl of 10 ng/ml DEAE-dextran (500 kDa; Pharmacia, Piscataway, NJ) is added. After incubation at 37° for 30 min, the cell monolayer is washed two times with 1× phosphate-buffered saline (PBS) and 4 ml of TBS with 10% dextrose.[9] The DNA/DEAE-dextran suspension is added dropwise to the cell monolayer with constant mixing. The cells are then incubated at 37° for 60 min followed by addition of chloroquine (100 mM) and a further incubation for 60 min. The cells are washed again with 1× PBS, shocked for 1 min with 10% dimethyl sulfoxide in 1× PBS at room temperature, and then washed again with 1× PBS. The RANTES promoter could be induced by addition of 100 ng/ml of LPS. Resultant reporter gene activity was determined 18 hr later using established methods.[9]

Transient Transfection by Electroporation

Both suspension cell cultures and adherent cells can be efficiently transfected by electroporation.[6] Most commercial cell electroporation apparatuses can be used by following the manufacturer's specifications for transfection of eukaryotic cells. Adherent cells can be treated with trypsin, washed once in 1× PBS, and resuspended at 2.0×10^7 cells/ml before addition to electroporation cuvettes. Our laboratories routinely use standard growth medium containing 20% fetal calf serum as the electroportion "buffer," but many different buffers have been described.[6,8,10,23]

Hut78 T cells and mitogen-stimulated PBL are efficiently transfected by electroporation.[6,8,10,23] Peripheral blood lymphocytes are isolated from whole blood or from buffy coat material (obtained from a blood bank) using Ficoll–Hypaque density gradient centrifugation.[8,24] The PBL are sus-

[23] A. J. Cann, Y. Koyanagi, and I. S. Y. Chen, *Oncogene* **3**, 123 (1988).
[24] M. E. Kanof and P. D. Smith, *in* "Current Protocols in Immunology" (J. E. Coligan, A. M. Kruisbeek, D. H. Margulies, E. M. Shevach, and W. Strober, eds.) p. 7.1.2. Wiley, New York, 1994.

pended at $2–4 \times 10^6$ cells/ml in tissue culture medium (RPMI 1640 supplemented with L-glutamine, penicillin, streptomycin, and 20% heat-inactivated fetal calf serum). Adherent cells (monocytes) are reduced by incubation for 1 hr at 37° in 175-cm³ horizontal flasks (LUX, Naperville, IL, for example). Nonadherent cells are then transferred to a new 175-cm³ horizontal flask for subsequent experiments. Viability of isolated PBL should be greater than 99% as determined by trypan blue exclusion.

The PBL can then be activated with various T-cell mitogens such as PHA-M (Difco, Detroit, MI) (1–10 µg/ml), a combination of PMA (Sigma, St. Louis, MO) (0.5–10 ng/ml) and ionomycin (Sigma) (0.5 µM), or OKT3 (amount must be determined by titration).[24] For transfection of T cells only the nonadherent cells are harvested for subsequent experiments. This activated population should be greater than 95% T cells as determined by flow cytometry using monoclonal antibody to CD3 (OKT3). After incubation at 37° for 2 to 3 days, the resultant T-cell blasts can be transiently transfected.[8,23]

When using a Bio-Rad Electroporator (Bio-Rad Laboratories, Hercules, CA), the optimal electroporation voltage for Hut78 cells is approximately 240 V, and 270 V for PHA-activated PBL, at 960 µF, when using a 0.4-cm cuvette and supplemented RPMI 1640 growth medium with 20% fetal calf serum as the electroporation buffer. A control plasmid such as one-containing a cytomegalovirus (CMV) major immediate early gene promoter/enhancer–luciferase fusion construct should be used to determine empirically the optimal electroporation voltage for each cell type and each electroporation apparatus used. An aliquot of 5×10^6 cells in 250 µl of electroporation buffer is added per cuvette, and 5 to 20 mg of specific reporter gene construct is cotransfected per replicate with 3 to 10 mg of a control reporter construct (such as Rous sarcoma virus promoter driving the β-galactosidase reporter gene). Luciferase assays can be performed using commercial luciferase assay systems (such as those produced by Promega). β-Galactosidase assays are performed on an aliquot of transfected cell extracts according to the instructions accompanying the Reporter Lysis Buffer Reagent (Promega) using o-nitrophenylgalactoside (ONPG) or the Galacto-Light assay kit (TROPIX, Bedford, MA). Each experiment should be conducted at least in triplicate and the values averaged after normalization to β-galactosidase expression and protein extract levels. Cell viability should be followed through the course of the experiment using trypan blue exclusion and absolute cell number. Results routinely demonstrate 50 to 60% cell viability after the initial electroporation event, and the cells can show approximately one doubling before assaying each constructs. Only plasmids prepared at the same time should be directly compared in reporter gene assays.

Preparation of Nuclear Extracts for DNase I Footprinting,
  Electrophoretic Mobility Shift Assays, and Methylation
  Interference Assays

Nuclear proteins can be prepared by many techniques. We have followed procedures essentially according to the protocol of Durand *et al.*, except that 0.2% Nonidet P-40 (NP-40) in buffer A is used to lyse the cells instead of a homogenizer.[10,25,26] Extracts are desalted to remove ammonium sulfate using a P6DG resin (Bio-Rad) and quantitated by Bradford assay using the Bio-Rad protein assay reagent.

Electrophoretic mobility shift assay (EMSA) is performed using binding reactions (15 ml final volume) containing 10 mM Tris-HCl (pH 7.5), 80 mM NaCl, 1 mM EDTA, 1 mM dithiothreitol, 5% (v/v) glycerol, 1.5 mg of poly(dI)·poly(dC), 5–10 mg nuclear extract, and 20,000–50,000 cpm (0.1–0.5 ng) ^{32}P-end-labeled double-stranded oligonucleotide probe. After incubation for either 45 min on ice or 30 min at room temperature, the protein–DNA complexes are resolved on nondenaturing 5–7% polyacrylamide gels run in 0.5× Tris–borate/EDTA (TBE) buffer.[27] Oligonucleotides are synthesized with overhangs that can be end-labeled by fill-in using Klenow fragment as described.[27]

In EMSA competition assays, cold competitor oligomers at 1 to 1000 times molar excess of probe are added to the gel shift mixtures prior to addition of the ^{32}P-labeled oligonucleotide probe. For supershift EMSA, the antisera/monoclonal antibody reagents are added to the gel shift mixture 30 min after initiating the assay. In the case of blocking antibody reagents, the reaction mix is preincubated with the specific antibody reagent for 30 min at room temperature before the probe is added. The complete incubation time can run between 45 and 60 min and can be performed at either room temperature or 4°. The mix is then loaded onto a nondenaturing 5–7% polyacrylamide (29:1 ratio of acrylamide to bisacrylamide) gel and run in 0.25× to 1× TBE buffer. Because of the possibility of proteolytic degradation of nuclear proteins, combinations of protease inhibitors such as phenylmethylsulfonyl fluoride (PMSF, 0.1 mM), pepstatin (0.5 ng/ml), and leupeptin (0.5 mg/ml) should be added to the reaction buffer. This is especially important when determining antibody-induced blocking of specific EMSA shifts.[25,26]

[25] K. A. Jones, K. R. Yamamoto, and R. Tjian, *Cell* (*Cambridge, Mass.*) **42**, 559 (1985).
[26] D. B. Durand, J. P. Shaw, M. R. Bush, R. E. Replogle, R. Belagaje, and G. R. Crabtree, *Mol. Cell. Biol.* **8**, 1715 (1988).
[27] L. A. Chodosh, *in* "Current Protocols in Molecular Biology" (F. M. Ausubel, R. Brent, R. E. Kingston, D. D. Moore, J. G. Seidman, J. A. Smith, and K. Struhl, eds.) p. 12.2.1. Wiley, New York, 1987.

DNase I Footprinting

DNase I footprinting can be performed using a modification of the procedures described by Jones et al.[25] and Durand et al.[26] Protein binding reactions are carried out under the conditions described above for EMSA but scaled up to 50 μl and performed in protein excess. After binding, 50 μl of a 10 mM MgCl$_2$/5 mM CaCl$_2$ solution and 2 μl of an optimal DNase I (Worthington, Freehold, NJ) dilution are added and incubated for 1 min on ice. The optimal amount of DNase I must be determined for each end-labeled probe. Conditions generally range from 0.05 to 0.2 μg of DNase I per reaction. DNase I digestion is stopped by the addition of 90 μl of stop buffer (20 mM EDTA, 0.2 M NaCl), and 20 μg yeast tRNA is added as carrier. The samples are then extracted two times with an equal volume of phenol/chloroform (1:1) and precipitated by adjusting the solution to 0.3 M sodium acetate and 70% (v/v) ethanol. DNA samples can then be resuspended in 4 μl of an 80% formamide loading dye containing 1× TBE, bromophenol blue, and xylene cyanol, heated to 90° for 2 min, and loaded on 6% polyacrylamide–urea sequencing gels.

Methylation Interference

Methylation interference assays determine G residue usage for nuclear factor binding to DNA fragments necessary for formation of the complexes generated on EMSA. The results of these experiments are evidenced by the loss of piperidine cleavage product bands. An excellent protocol for methylation interference is provided by Baldwin and Sharp.[28] Briefly, partially methylated single ^{32}P-end-labeled probe representing the specific oligonucleotide probe is complexed with nuclear extracts and run out on a nondenaturing 4% polyacrylamide gel as described for EMSA (but at a 10-fold scaleup). The resultant gel is not dried, but exposed over 4 hr to X-ray film. Regions of the polyacrylamide gel containing bands of various protein–DNA complexes and free oligonucleotide probe are excised, and the modified DNA is removed from the nondenaturing acrylamide gel by electroelution (e.g., the Electro-Eluter, Bio-Rad). Following piperidine organic cleavage,[28] the DNA preparations are analyzed on 10% polyacrylamide–urea sequencing gels.[8,10,28]

Transient and stable transfection experiments using RANTES promoter–reporter gene constructs in conjunction with EMSA, EMSA competitions, and supershift assays have allowed the initial characterization of promoter elements involved in transcriptional control of RANTES gene expression in different cell types. Understanding the transcriptional regula-

---

[28] A. S. Baldwin, Jr., and P. A. Sharp, Proc. Natl. Acad. Sci. U.S.A. **85,** 723 (1988).

tion of the RANTES chemokine may prove important in a variety of diseases. The inhibition of RANTES gene expression may be therapeutic in diseases characterized by cellular infiltration. Up-regulation of RANTES expression may be useful in the therapy of cancer or acquired immunodeficiency syndrome (AIDS).

[11] Expression of Chemokine RANTES and Production of Monoclonal Antibodies

By ALAN M. KRENSKY and PETER J. NELSON

Introduction

The chemokine RANTES was first identified and characterized as a cDNA isolated by subtractive hybridization.[1] Classic biological study generally begins with a function and works toward a structure. The power of molecular biology has increasingly given rise to a new type of investigation, a type of "reverse biology" where the investigator identifies a molecule and works backward to identify its function. A required step in this process is expression of the cDNA to make protein for functional characterization. The expressed protein can then be used to make antibodies to further characterize the protein, its cell distribution, and function. This chapter describes the techniques used to characterize the RANTES protein and its function.

RANTES Protein

The chemoattractant cytokine RANTES has been implicated in the generation of an inflammatory infiltrate.[2,3] It is induced by macrophage factors, interleukin-1 (IL-1) and tumor necrosis factor (TNF), in a variety of cell types including epithelial cells and fibroblasts.[4,5] RANTES binds to

[1] T. J. Schall, J. Jongstra, B. D. Dyer, J. Jorgensen, C. Clayberger, M. M. Davis, and A. M. Krensky, *J. Immunol.* **141,** 1018 (1988).

[2] T. J. Schall, K. Bacon, K. J. Toy, and D. V. Goddell, *Nature* (*London*) **347,** 669 (1990).

[3] A. M. Krensky, ed., "Biology of the Chemokine RANTES." R. G. Landes, Austin, Texas, 1995.

[4] P. Rathanaswami, M. Hachicha, M. Sadick, T. J. Schall, and S. R. McColl, *J. Biol. Chem.* **268,** 5834 (1993).

[5] P. Heeger, G. Wolf, C. Meyers, M. J. Sun, S. C. O'Farrell, A. M. Krensky, and E. G. Neilson, *Kidney Int.* **41,** 220 (1992).

vascular endothelium where it serves as a signpost for haptotaxis (attraction along a solid matrix) and chemotaxis (attraction along a soluble gradient).[6-8] RANTES activates T cells and induces expression of important adhesion/accessory molecules of the integrin family.[9,10] T lymphocytes and monocytes attracted by RANTES enter into the site of inflammation.[7] T cells with specific receptors for the offending antigen(s) become activated and make more RANTES over the subsequent 3–5 days. This prolongs the inflammatory response in time and space. According to this model, interruption of RANTES-mediated chemoattraction may block inflammation in a variety of disease states.[11] RANTES has been implicated in inflammatory states including transplant rejection, atherosclerosis, rheumatoid arthritis, delayed type hypersensitivity, asthma, endometriosis, and cancer.[11]

The RANTES protein has also been implicated in suppression of human immunodeficiency virus (HIV) infection.[12] Cocchi and colleagues found that patients with HIV who do not progress to acquired immunodeficiency syndrome (AIDS) have high levels of RANTES, macrophage inflammatory protein-1α (MIP-1α), and MIP-1β.[12] These chemokines suppress HIV growth *in vitro*. Understanding the mechanism of RANTES suppression of HIV may lead to new therapies for this important infectious disease.

Lastly, a family of RANTES receptors have been defined.[3] RANTES binds to serpentine-7 membrane-spanning receptors, including the C-C chemokine receptors CKR-1, CKR-3, CKR-4, and CKR-5.[13-17] It also binds

[6] A. Rot, M. Krieger, T. Brunner, S. C. Bischoff, T. J. Schall, and C. A. Dahinden, *J. Exp. Med.* **176,** 1489 (1992).

[7] C. J. Wiedermann, E. Kowald, N. Reinish, C. M. Keehler, I. von Luettichau, J. M. Pattison, P. Huie, R. K. Sibley, P. J. Nelson, and A. M. Krensky, *Curr. Biol.* **3,** 735 (1993).

[8] J. M. Pattison, P. J. Nelson, P. Huie, I. von Luettichau, G. Farshid, R. K. Sibley, and A. M. Krensky, *Lancet* **343,** 209 (1994).

[9] K. B. Bacon, B. A. Premack, P. Gardner, and T. J. Schall, *Science* **269,** 1727 (1995).

[10] A. R. Lloyd, J. J. Oppenheim, D. J. Kelvin, and D. D. Taub, *J. Immunol.* **156,** 932 (1996).

[11] J. M. Pattison, P. J. Nelson, and A. M. Krensky, *Clin. Immunother.* **4,** 1 (1995).

[12] F. Cocchi, A. L. DeVico, A. Garzino-Demo, S. K. Arya, R. C. Gallo, and P. Lusso, *Science* **270,** 1811 (1995).

[13] K. Neote, D. DiGregorio, J. Y. Mak, R. Horuk, and T. J. Schall, *Cell (Cambridge, Mass.)* **72,** 415 (1993).

[14] J.-L. Gao, D. B. Kuhns, H. L. Tiffany, D. McDermott, X. Li, U. Francke, and P. M. Murphy, *J. Exp. Med.* **177,** 1421 (1993).

[15] B. L. Daugherty, S. J. Siciliano, J. A. DeMartino, L. Malkowitz, A. Sirotina, and M. S. Springer, *J. Exp. Med.* **183,** 2349 (1996).

[16] C. A. Power, A. Meyer, K. Nemeth, K. B. Bacon, A. J. Hoogewerf, A. E. I. Proudfoot, and T. N. C. Wells, *J. Biol. Chem.* **270,** 19495 (1995).

[17] M. Samson, O. Labbe, C. Mollereau, G. Vassart, and M. Parmentier, *Biochemistry* **35,** 3362 (1996).

to the Duffy blood group antigen expressed on red blood cells (DARC),[18] to a cytomegalovirus-derived protein US28,[19] and to glycosaminoglycans expressed by endothelial cells.[20] Although the functional significance of these various receptors remains poorly defined, RANTES signals T cells through C-C CKR receptors.[9] RANTES shares these receptors with other chemokines, and undoubtedly additional receptors will be defined.

The elucidation of all of this information about RANTES was dependent on expression of functional protein. The methods used to express the RANTES protein are the subject of this article.

Methods

Expression of RANTES Protein

A required step toward defining RANTES function is production of recombinant protein from the previously isolated cDNA. Three different methods have been used to express RANTES. One involves a eukaryotic expression system, and two use prokaryotes to produce the protein.

Eukaryotic Expression. Schall *et al.* have described the generation of biologically active RANTES by cloning the cDNA into the mammalian expression vector pRK5.[2] The cDNA construct is placed under control of the cytomegalovirus (CMV) immediate early promoter–enhancer and simian virus 40 (SV40) termination and polyadenylation signals. The embryonic kidney cell line 293[21] is transiently transfected with the pRK5-RANTES construct by calcium phosphate precipitation.[22] Culture supernatants containing RANTES are collected and used to test biological activity.

Prokaryotic Expression. Two methods have been used to produce functional recombinant RANTES using prokaryotic expression systems. RANTES may be prepared as a soluble peptide in *Escherichia coli* as described by Rot *et al.*[6] These investigators have linked the RANTES cDNA (minus the leader signal sequence) to the bacterial STII promoter in an expression plasmid. *Escherichia coli* expressing recombinant protein are harvested by centrifugation, the extracellular medium discarded, and the pellet stored frozen at $-70°$. Cell pastes are thawed and dispersed in 50 mM glycine, 250 mM sodium chloride, pH 3.0, using an Ultra Turrax

[18] N. Neote, W. Darbonne, J. Ogez, R. Horuk, and T. J. Schall, *J. Biol. Chem.* **268**, 12245 (1993).

[19] D. E. Kuhn, C. J. Beall, and P. E. Kolattukudy, *Biochem. Biophys. Res. Commun.* **211**, 325 (1995).

[20] D. P. Witt and A. D. Lander, *Curr. Biol.* **4**, 394 (1994).

[21] F. L. Gorman, J. Smiley, W. Russel, and R. Nairn, *J. Gen. Virol.* **36**, 59 (1977).

[22] R. E. Kingston, C. A. Chen, and H. Okayama, "Current Protocols in Molecular Biology" (F. M. Ausubel, R. Brent, R. E. Kingston, D. D. Moore, J. G. Seidman, J. A. Smith and K. Struhl, *et al.*, eds.), p. 9.1.1. Wiley, New York, 1994.

homogenizer (Tekmar, Cincinnati, OH). The cells are then mechanically disrupted in a homogenizer (1104; Microfluidics, Newton, MA) operating at 23,000 pounds per square inch (psi) cooled to room temperature, and cell debris is removed by centrifugation. The supernatant is adjusted to pH 6.0 and loaded onto a S-Sepharose fast flow column (Pharmacia, Uppsala, Sweden) equilibrated in 20 mM sodium citrate, pH 6.0. The material, including recombinant RANTES, is bound to the resin and eluted using a linear gradient of 0–1.0 M sodium chloride. The RANTES protein elutes as a peak at approximately 0.7 M sodium chloride, as determined using reversed-phase high-performance liquid chromatography (RP-HPLC) of eluted fractions. The S-Sepharose pool, containing the peak RANTES fractions, is brought to 1.8 M in ammonium sulfate and loaded onto a phenyl Toyopearl column (TosoHaas, Philadelphia, PA) equilibrated in 100 mM sodium phosphate, pH 6.0. Bound protein is eluted with a decreasing linear gradient of ammonium sulfate (1.5 to 1.0 M) in 100 mM sodium phosphate, pH 6.0. Fractions containing RANTES are selected by RP-HPLC, and ammonium sulfate is removed by diafiltration versus 10 mM sodium citrate, 450 mM sodium chloride, pH 5.0, across a 3-kD a cutoff membrane. The purity, by HPLC, is greater than 99%, and the endotoxin content is less than 1 unit/mg.

Another technique has been used by von Luettichau, Nelson, and colleagues.[23] The pET expression system yields large amounts of protein, but the protein is contained in bacterial inclusion bodies.[24] The presence of inclusion bodies greatly simplifies purification, but it requires denaturation and refolding to obtain functional protein. The RANTES cDNA (lacking the leader signal sequence) is subcloned into the pET-3c vector and transfected into the *E. coli* strain BL21(De)3 LysS.[24] Expression of RANTES protein is induced by addition of 1 mM isopropylthiogalactoside (IPTG; Sigma, St. Louis, MO). The bacteria are pelleted after an optical density of 0.9 (600 nm) is reached (2–4 hr), resuspended in 50 mM Tris-HCl, 10 mM EDTA (ethylenediaminstetraacetic acid) buffer, pH 8.0, and sonicated until an optical density of 0.2 (600 nm) is reached. The lysate is then centrifuged at 16,000 rpm at 4° for 30 min. Bacterial inclusion bodies in the pellets are washed repeatedly, treated with 2 g/100 ml sodium deoxycholate for several hours, and then digested with DNase I to remove as much debris as possible. The inclusion body pellet is then dissolved in 6 M guanidine hydrochloride in 5 mM phosphate buffer, pH 6.5, and centrifuged at 70,000g to remove insoluble debris. The denatured protein is refolded by slowly injecting it into 5 volumes of refolding buffer containing 50 mM Tris, pH 7.0, 50 mM potassium chloride, 0.1 mM EDTA, 0.75 M arginine,

[23] I. von Luettichau, P. J. Nelson, J. M. Pattison, M. van de Rijn, P. Huie, R. Warnke, C. J. Widermann, R. A. K. Stahl, R. K. Sibley, and A. M. Krensky, *Cytokine* **8,** 89 (1996).
[24] F. W. Studier and B. A. Moffatt, *J. Mol. Biol.* **189,** 113 (1986).

12 mM glutathione (reduced), and 1.2 mM glutathione (oxidized) to yield a final concentration of guanidine of 1 M. This mixture is gently stirred at 4° for 12–16 hr.

The refolded protein solution is dialyzed against 20 mM acetic acid, pH 3.0, followed by dialysis at 4°, against 10 mM MOPS, pH 7.0, containing 100 mM sodium chloride. After 30 min of centrifugation at 4°, at 70,000 g, to remove misfolded, aggregated protein, the recombinant protein is run over a Bio-Rex ion-exchange column (Bio-Rad Laboratories, Richmond, CA), equilibrated in 10 mM MOPS, pH 7.0, containing 100 mM sodium chloride, and eluted with a 10 mM MOPS, pH 7.0, 0.5 to 2 M sodium chloride linear gradient. RANTES protein elutes at approximately 1.5 M sodium chloride. Peak fractions are pooled and concentrated using polyethylene glycol (Aquacide II, Calbiochem, La Jolla, CA) and Centriprep and Centricon concentrators (Amicon, Beverly, MA), and they are further purified using fast protein liquid chromatography (FPLC) and a Superdex 75 2560 column developed with 1.5 M sodium chloride, 10 mM MOPS, pH 7.0 (Pharmacia). The recombinant protein is tested for endotoxin content by *Limulus* assay and passed over a Detoxi gel column (Sigma) if required. The purity of the protein is greater than 99%, and yields typically are between 25 and 35 mg of refolded protein per liter of starting material.[23]

Proudfoot *et al.* showed that the presence of the amino-terminal methionine in recombinant RANTES protein results in a potent and selective RANTES antagonist.[25] In many cases, however, endogenous aminopeptidases remove the amino-terminal methionine. The methods described above for the production of *E. coli*-derived RANTES protein reproducibly produce recombinant RANTES protein, which is active in the nanomolar range.[6,7,23] Horuk[25a] describes a proteolytic procedure to eliminate the terminal amino acid residues from recombinant RANTES protein.

Protein Storage

RANTES protein aggregates at concentrations over 0.5 mg/ml. Protein can be stored at 2–5 mg/ml in high salt concentrations (1.5 M sodium chloride) or at low pH (2.5).[23] Alternatively, RANTES protein can be stored in neutral HEPES buffer containing 1 mg/ml bovine serum albumin (BSA) in small aliquots at −70°.[6]

Chemotactic Activity of Recombinant RANTES Protein

The function of the recombinant protein is tested by measuring chemotaxis of human monocytes and lymphocytes to gradients of RANTES as

[25] A. E. Proudfoot, C. A. Power, A. J. Hoogewerf, M. O. Montjovent, F. Borlat, R. E. Offord, and T. N. Wells, *J. Biol. Chem.* **271,** 2599 (1996).
[25a] R. Horuk, *Methods Enzymol.* **287,** 1997 (this volume).

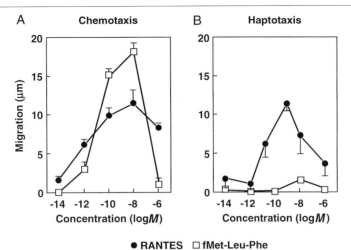

FIG. 1. Monocyte migration to (A) chemotactic and (B) haptotactic gradients of human RANTES or fMet-Leu-Phe. The chemotactic gradient is a sum of soluble and filter-bound gradients. Results shown are the means ± SEM of induced migration to test attractant minus migration to vehicle control alone after 90 min of incubation at 37°.

described by Wiedermann *et al.*[7] and Falk *et al.*[26] A 48-well microchemotaxis chamber (Neuroprobe, Bethesda, MD) in which a nitrocellulose filter separates the upper and lower chambers is used. Only actively migrating cells are able to traverse the pores in the filter. The cells (5×10^5 cells per well in RPMI 1640) are placed in the upper chamber, and RANTES protein is placed in the lower chamber. For monocytes, 5-μm pore filters are used. This size pore allows only monocytes, but not lymphocytes, to enter the filter.[27] After incubating the assay for 90 min at 37°, the nitrocellulose filters are dehydrated, fixed, and stained with hematoxylin and eosin.[28] Migration along the filter is quantified by microscopy, measuring the distance from the surface of the filter to the leading front of three cells. Data are expressed as chemotaxis, the difference between the distance of migration toward the test attractant and the distance of migration toward buffer alone.[7] Figure 1A shows representative results for monocyte migration to chemotactic gradient of RANTES compared with the positive control fMet-Leu-Phe.

Haptotactic Activity of Recombinant RANTES Protein

Chemotaxis is a classic assay but cannot account for the *in vivo* effects of RANTES. A chemokine released into the bloodstream would be washed

---

[26] W. Falk, R. H. Goodwin, Jr., and E. J. Leonard, *J. Immunol. Methods* **33,** 239 (1980).

[27] E. C. Wilkinson, *Methods Enzymol.* **162,** 38 (1988).

[28] C. J. Wiedermann, N. Reinisch, and H. Braunsteiner, *Blood* **82,** 954 (1993).

downstream immediately and would fail to attract target cells to the site of inflammation.[8] Rather, RANTES appears to mediate inflammation by haptotaxis, a process whereby cells move along the gradient of an attractant bound to a surface along which the cells move.[29] Wiederman *et al.* showed that RANTES induces a haptotactic gradient nearly equal to its chemoattractant gradient (see Fig. 1).[7] Haptotaxis is measured in 48-well chemotactic chambers across nitrocellulose filters with 5-μm pores (Sartorius, Darmstadt, Germany) by filling the bottom well with serial, threefold dilutions of RANTES and the top well with Hanks' balanced salt solution with Ca^{2+} and Mg^{2+}. The chambers are incubated at 37° for 20 min, then disassembled, and the filters are removed and washed in Hanks' balanced salt solution.[7] The filters are blotted using a paper towel, air dried, and placed in a new chemotactic chamber. The top wells are filled with 5×10^4 mononuclear cells isolated by Ficoll–Hypaque density gradient centrifugation and washed in Hanks' balanced salt solution three times. The migration of monocytes is measured after 30–90 min of incubation at 37° in 5% (v/v) CO_2, and compared to migration with buffer control (35–70 min).[7]

Generation of Monoclonal Antibodies

Protein made using the pET-3c expression system described above is used to generate anti-RANTES monoclonal antibodies.[23,30] Five BALB/c retired breeders are injected with purified, denatured RANTES protein. Mice are immunized two times per week for 3 weeks with 5 μg of protein. Popliteal, inguinal, and axillary lymph nodes are removed from the mice. Thymus is removed to be used for feeder cells. The nodes and thymus are placed in phosphate-buffered saline (PBS) or RPMI, teased apart on glass slides or dishes, and filtered through Nytex. Clumps of tissue are allowed to settle and are not used for the fusion. The resultant cell suspension is transferred into 15-ml tubes, washed, and the cells counted. The Sp2 fusion partner hybridoma (ATCC) is washed in a 50-ml tube and cells counted. The Sp2 cells and lymph node cells are combined at a ratio of 1:1 and spun together in a 50-ml tube. The supernatant is decanted, the pellet resuspended, and 1 ml of fusion media [40% polyethylene glycol, 10% dimethyl sulfoxide (DMSO), RPMI (no fetal calf serum)] is added over 30 sec with gentle tapping to suspend the cells. The mixture is allowed to stand for 2 min, and 0.25 ml of 5% DMSO, RPMI (no fetal calf serum) is added every 15 sec for 2 min with gentle tapping. An additional 0.5 ml of 5% DMSO in RPMI is added every 15 sec to a total of 10 ml. Then PBS or DMSO is added to a total of 50 ml, and the tubes are spun at room temperature. The supernatant is decanted and excess medium removed with a

[29] A. Rot, *Eur. J. Immunol.* **23**, 303 (1993).
[30] K. Ozato and D. H. Sachs, *J. Immunol.* **126**, 317 (1981).

sterile cotton tip applicator. Ten mililiters of RPMI followed by 30 ml of fusion medium are combined and used to fill two 96-well Linbro flat bottom plates. Approximately one thymus per plate of thymic cells are added as feeders. Next, 200 μl/well are plated. Each well contains approximately 2 $\times$ 10^5 fusion cells/well. Cells are fed at 7 days with hypoxanthine-containing medium. Positive fusions are generally testable within 2 weeks.

Fusion hybrids are screened by enzyme-linked immunosorbent assay (ELISA) on Nunc Immuno Plates (MaxiSorp F96) using refolded recombinant RANTES as capture antigen.[20,27] Positive hybridomas are transferred to larger wells, additional feeder cells are provided and the samples are retested and subcloned by limiting dilution. Hybridomas positive in ELISA are evaluated further for Western blotting and cell staining as described below. The isotypes of four selected monoclonal antibodies are determined using a commercial mouse monoclonal antibody typing kit (Amersham, Amersham, UK). Data regarding the four anti-human recombinant RANTES monoclonal antibodies produced in this manner are summarized in Table I and detailed in von Luettichau et al.[23]

Inhibition of Recombinant RANTES-Induced Chemoattraction by Anti-RANTES Antibodies

Inhibition of recombinant RANTES-induced chemoattraction by anti-RANTES antibodies is tested by adding various dilutions (10 to 100-fold) of the hybridoma supernatants or control antibodies to the lower compartment of Boyden chambers as described above. Inhibition of monocyte and lymphocyte migration is expressed as a percentage of migration in the presence of control antibody.[7]

Enzyme-Linked Immunosorbent Assay

Cross-reactivity of the anti-RANTES monoclonal antibodies with other chemokine family members is tested using a plate ELISA. Titrations from

TABLE I
ANTI-RANTES MONOCLONAL ANTIBODIES GENERATED AGAINST RECOMBINANT HUMAN PROTEIN

Monoclonal antibody	Isotype	Western blot	Immunoprecipitation	ELISA	Cytospin	Frozen tissue
VL1	G2bκ	++++	++++	++++	+++	+++
VL2	G2bκ	−	−	++++	+++	++++
VL3	G2bκ	++++	++++	++++	−	+
VL4	G2bκ	++++	++++	++++	−	+

[a] From von Luettichan et al.[23]

1 μg to 1 ng of RANTES, human (Hu)MIP1α, HuMIP-1β, human macrophage chemotactic protein-1 (HuMCP-1), and HuIL-8 protein are bound to Nunc Immuno Plates (MaxiSorb) and developed using horseradish peroxidase-conjugated goat anti-mouse IgG$_{2b}$ antibody (Southern Biotechnology, Birmingham, AL) as the secondary antibody. The assays are developed using a commercial peroxidase detection reagent (Kirkegaard & Perry Laboratories, Gaithersburg, MD).[23]

Western Blot Analysis

Western blotting is carried out using a Bio-Dot (Bio-Rad) apparatus.[31] Purified recombinant RANTES protein and control chemokines and cytokines (100 ng) are transferred to Immobilon P membrane (Millipore, Bedford, MA) according to the manufacturer's directions. Transfer membranes are blocked overnight at 4° using TBST buffer (50 mM Tris-HCl, pH 7.5, 0.9% sodium chloride, 0.05% Tween 20) and 3% powdered milk. The membranes are then washed three times using TBST buffer and incubated with a 1:50 dilution of the spent hybridoma supernatants in TBST for 1 hr at room temperature. The membranes are washed three times with TBST buffer. Horseradish peroxidase-conjugated goat anti-mouse IgG$_{2b}$ antibody (Southern Biotechnology) is used as the secondary antibody at a 1:2000 dilution. The secondary antibody is applied for 1 hr at 25°, and the membranes are developed using ECL (enhanced chemiluminescence, Amersham).

Immunoprecipitation

Cells are incubated at 5 × 10^6 cells/ml in cysteine- and methionine-free RPMI 1640 (GIBCO-BRL, Grand Island, NY) supplemented with 5% dialyzed and heat-inactivated fetal calf serum, 2 mM glutamine, and 50 mM 2-mercaptoethanol (2-ME) for 3 hr at 37°. ^{35}S-Labeled cysteine and methionine (Translabel, 1225 Ci/mmol, Amersham, Arlington Heights, IL) and [^{35}S]cysteine (Amersham) are added at 1 mCi per 5 × 10^7 cells and incubated for 4 hr at 37°. After washing with 1× PBS, cells are lysed with modified RIPA lysis buffer containing 1% Nonidet P-40 (NP-40), 0.1% sodium deoxycholate, 1 mg/ml leupeptin (Sigma), 1 mg/ml pepstatin (Sigma), and 2 mM phenylmethylsulfonyl fluoride (PMSF) (Sigma), in 100 mM Tris-HCl, pH 7.2, 140 mM sodium chloride, 0.025 mM sodium azide, at 0.5 ml buffer per 5 × 10^7 cells. After two freeze–thaw cycles, the lysate is centrifuged for 30 min at 20,000g, and the supernatant is prepared and immunoprecipitated using the immunoisolation techniques described by Tamura *et al.*[32]

[31] W. N. Burnette, *Anal. Biochem.* **112,** 195 (1981).
[32] G. S. Tamura, M. O. Dailey, W. M. Gallitan, E. C. Butcher, M. S. McGrath, I. L. Weissman, and E. A. Pillemer, *Anal. Biochem.* **136,** 458 (1984).

For indirect solid-phase immunoisolation, 100 μl of affinity-purified anti-immunoglobulin at a concentration of 100 μg/ml in PBS with 1 mM azide is added to each well of a polyvinyl chloride microtiter plate (Dynatech, Chantilly, VA) and incubated overnight at 4°. These wells can be saved for later use and the antiimmunoglobulin reagent reused. The wells are washed three times with PBS, aspirating each time with a clean, dry Pasteur pipette. A 100-μl aliquot of specific reagent (antisera diluted 10-fold or monoclonal antibody supernatants or purified antibody) is added to the wells, and plates are incubated at 4° for 4 hr and then again washed three times with PBS. Nonspecific protein binding is blocked by adding 200 μl of PBS with 1 mM azide and 1% normal serum to each well for 2 hr at room temperature. The serum is from the same species (human, mouse) as the specific reagent.

One hundred microliters of cell lysate is added to each well and incubated at 4° overnight. The lysate is removed and the well washed three times with phosphate lysis buffer (0.1% SDS, 0.1% sodium deoxycholate, 1% NP-40, 100 mM NaCl, 10 mM sodium phosphate, pH 7.1, and 1 mM sodium azide). For sodium dodecyl sulfate–polyacrylamide gel electrophoresis (SDS–PAGE), 100 μl of sample buffer containing 1% SDS, 1% 2-mercaptoethanol, 50 mM Tris, pH 6.8, 10% (v/v) glycerol, and 0.01% bromphenol blue is added and incubated for 5 min. Sample buffer is removed, and the samples are heated in a boiling water bath for 3–5 min. Samples are then run on polyacrylamide gels as described by Laemmli.[33]

Cell Staining Experiments

Human fetal tissue is obtained from discarded therapeutic abortion material. Adult tonsil, kidney, and other tissues are obtained from surgical procedures. All biopsy specimens are embedded in OCT medium, snap-frozen, and stored at −70°. Immunostaining is done on 6-μm cryostat sections with the antirecombinant human RANTES monoclonal antibodies described above.[7,8,34] A four-layer peroxidase–antiperoxidase–immuno-peroxidase approach is used as described by McCaughan et al.[35] Tissue is incubated with monoclonal antibody for 1 hr at room temperature followed by sequential application of rabbit antimouse immunoglobulin (1:500, Dakopatts A/S, Glostrup, Denmark), swine antirabbit immunoglobulin (1:75, Dakopatts A/S Z196), and rabbit peroxidase–antiperoxidase (1:150, Dakopatts). Peroxidase activity is revealed using 3,3'-diaminobenzidine tetrahy-

[33] U. K. Laemmli, *Nature (London)* **227**, 680 (1970).
[34] J. M. Bindl and R. A. Warnke, *Am. J. Clin. Pathol.* **85**, 490 (1986).
[35] G. W. McCaughan, J. S. Davies, J. A. Waugh, G. A. Bishop, B. M. Hall, N. D. Gallagher, J. F. Thompson, A. G. Sheil, and D. M. Painter, *Hepatology* **12**, 1305 (1990).

drochloride and developed over 10 min, at which time the solution is rinsed off the sections. The sections are counterstained in Mayer's hematoxylin, rinsed, dehydrated, and mounted in ProtexX (American Scientific Products, McGaw Park, IL). Immunofluorescence is performed using a fluorescein-conjugated $F(ab')_2$ fragment of rabbit immunoglobulin as a secondary antibody. Controls included the omission of the primary antibody, the use of irrelevant isotype-matched antibodies as primary antibodies, and the staining patterns of tissues with known positive or negative reactivity to the reagents used.

In Situ Hybridization

For *in situ* hybridization, 6-μm sections of tissue are placed on baked glass microscope slides and fixed in 4% paraformaldehyde in PBS for 10 min at room temperature.[7,8,36] Slides are washed three times with 5-min changes of $1\times$ PBS/10% (v/v) ethanol, and then 100% ethanol for 5 min and air dried. The sections are prehybridized in hybridization solution [5 ml $2\times$ hybridization solution (Sigma), 4 ml deionized formamide, 1 ml diethyl pyrocarbonate (DEPC) water, and 125 mg tRNA (Sigma), adjusted to pH 8–9 with sodium hydroxide] for 1 hr at 42°. Prehybridization solution is replaced with hybridization solution containing the appropriate probes at a concentration of 1.5 ng/μl and incubated overnight at 42°. The RANTES probe is a cocktail of two digoxigenin-labeled 21-residue antisense RANTES-specific oligodeoxynucleotides (RGB, 5' GGCACGGGG-CAGTGGGCGGGC 3'; RGD, 5' CAAAGAGTTGATGTACTCCCG 3').[36,37] Controls included omission of an oligomer and use of a nonsense probe and a poly(T) labeled probe. Slides are washed twice for 45 min in $5\times$ SSC and then for 30 min each in $2\times$ SSC, $1\times$ SSC, $0.5\times$ SSC, and $0.25\times$ SSC. Tissues are then incubated in 0.1 M Tris-HCl, 0.15 M sodium chloride, pH 7.5, containing 0.2% normal sheep serum and 0.3% Triton X-100 for 1 hr at room temperature and then anti-digoxigenin-conjugated alkaline phosphatase is added for 2 hr. Unbound antibody is removed by washing the tissue in 0.1 M Tris-HCl, 0.1 M sodium chloride containing Triton X-100, pH 7.5, and then in Tris-HCl, 0.15 M sodium chloride, pH 8.5, for 5 min each. Nitro blue tetrazolium/5-bromo-4-chloro-3-indoyl phosphate (GIBCO-BRL) are used as substrates for the alkaline phosphatase color reaction. The slides are washed with water and mounted using mounting medium (Glycergel, Dako).

We have found that fluorescein isothiocyanate (FITC) conjunction

---

[36] C. M. Chleq-Deschamps, D. P. LeBrun, D. P. Besnier, P. Huie, R. K. Sibley, and M. L. Cleary, *Blood* **81,** 293 (1993).

[37] J. Pattison, P. J. Nelson, P. Huie, R. K. Sibley, and A. M. Krensky, *J. Heart Lung Transplant.* **15,** 1194 (1996).

(Pierce, Rockford, IL) of the VL1 and VL3 monoclonal antibodies works well with commercial permeabilizing reagents (Fix and Perm, Caltag Laboratories, South San Francisco, CA) for visualization of intracellular RANTES. This approach is suitable for quantifying protein using fluorescence activated cell sorting (FACS) and permits study of RANTES expression in lymphocyte subpopulations.

TABLE II

RANTES EXPRESSION IN NORMAL, INFLAMED, TRANSFORMED, AND FETAL TISSUES

Cell type	Protein expression	Message expression
Normal tissues		
Adult kidney	Rare scattered cells	Rare cells
Adult spleen	Scattered cells	Not tested
Adult tonsil	Scattered mantle cells; subepithelial nononuclear cells; rare scattered germinal center cells	Same pattern as with antibody
Inflamed tissues		
Spleen (extramedullary hematopoiesis)	Megakaryocytes	Megakaryocytes
Kidney transplants	Epithelial cells, fibroblasts, endothelial cells, inflammatory cells	Epithelial cells, fibroblasts, endothelial cells, inflammatory cells
Lung epithelium (asthma)	Epithelial cells, fibroblasts, endothelial cells, inflammatory cells	Epithelial cells, fibroblasts, endothelial cells, inflammatory cells
Tumors		
Lymphomas	Scattered positive cells including infiltrating and dendritic cells	Not tested
$\gamma\delta$ T-cell lymphoma	Positive	Positive
Renal cell carcinoma	Positive	
Wilm's tumor	Positive	
Rhabdomyosarcoma	Some positive	
Fetal tissues		
Kidney	Glomerular mesangial cells; proximal tubules; subcapsular blastema (weak)	S bodies, proximal tubules, glomerular mesangial cells
Spleen	Extramedullary hematopoiesis	Extramedullary hematopoiesis
Thymus	Rare scattered positive cells in cortex	
Lung	Rare scattered positive cells	
Heart	Negative	
Brain	Negative	
Muscle	Negative	

Cellular Distribution of RANTES Protein

The cell distribution of RANTES was first characterized by Northern blots using the original cDNA.[1] The development of monoclonal antibodies made it possible to confirm the expression of RANTES and to differentiate the presence of protein from message.[21] A summary of the cellular expression of RANTES protein in shown in Table II. These studies confirmed the expression of RANTES in a variety of tissues, including T lymphocytes in a variety of normal tissues and sites of inflammation, selected tumors, and fetal tissues. In addition, they suggested a role for RANTES in development. In kidney, for example, RANTES message and protein are widely expressed in a variety of cell types.

Summary and Conclusions

RANTES was first identified as a cDNA in a search for genes expressed late (3–5 days) after T-cell activation. Definition of RANTES function depended on the generation of protein. This chapter describes the various techniques used to make recombinant RANTES protein, to test its activity, and to generate monoclonal antibodies to assess RANTES protein cell distribution.

[12] Biological Responses to C-C Chemokines

By ROBERT C. NEWTON and KRISHNA VADDI

Introduction

Chemokines from the β family have been shown to target monocytes, basophils, eosinophils, mast cells, and T lymphocytes. The cellular specificity of a given chemokine is due mainly to restriction of the expression of the responsive chemokine receptor (CCR) to particular cell types. This review of methods to assess the biological activity of chemokines focuses on assays using human cells, because that is the major area of work by the authors, but they should also generally apply to cells from other species or tissue culture transfectants.

Cell Isolation

Various methods are available for the purification of human blood cells for *in vitro* assays, and it is not the objective of this chapter to detail these

0076-6879/97 $25.00

methods. Instead, a brief outline of the generally preferred methods for each cell type is presented, and the investigator is referred to the literature for more detailed methods. It should be noted that a major goal of whatever procedure is used should be to obtain the desired cell population in as high a degree of purity as possible without risk of inadvertent cell activation. Many of the methods used to characterize biological activity represent whole population responses, and the researcher needs to ensure that the effect seen is not due to a high response by a contaminating population, which masks a negative or absence of response from the intended study population. For example, myeloid cells contain significantly higher levels of intracellular esterases than lymphocytes. Therefore, techniques which use dyes that accumulate intracellularly following esterase action will cause dye to accumulate preferentially in myeloid cells, and the signal per cell will be higher than with lymphocytes. Care should be exercised when interpreting such assays, and alternative methods such as single cell analysis may be used if contamination with unwanted cells is unavoidable. Whatever method is used, care should also be taken to avoid inadvertent cell activation, as activation may lead to receptor desensitization or the engagement of other feedback mechanisms that block subsequent biological responses.

Monocytes. Monocytes can be readily obtained from human peripheral blood by a combination of Ficoll–Hypaque density centrifugation with centrifugal elutriation.[1] Lymphocytes are the primary contaminant. Alternately, human cell lines such as U937, THP-1, and MonoMac6 can be utilized, although it should be noted that there are subclones of these cell lines currently being grown in various laboratories that are not designated as such but that differ widely in their responses.

Lymphocytes. T lymphocytes can be isolated from human peripheral blood by Ficoll–Hypaque density centrifugation followed by adherence to plastic culture dishes and collection of the nonadherent cells. Further purification can be obtained by negative selection of monocytes and B lymphocytes, respectively, with anti-CD14 and anti-CD19 antibodies.[2]

Basophils. Basophils can be enriched to 80–95% by a combination of discontinuous Percoll gradient centrifugation with negative selection using anti-CD3, CD4, CD8, CD14, CD16, and CD19 monoclonal antibody-coated magnetic beads.[3] Lymphocytes are the primary contaminant.

[1] K. Vaddi and R. C. Newton, *J. Leukocyte Biol.* **55**, 756 (1994).
[2] S. Mattoli, V. Ackerman, E. Vittori, and M. Marini, *Biochem. Biophys. Res. Commun.* **209**, 316 (1995).
[3] S. C. Bischoff, M. Krieger, T. Brunner, A. Rot, V. Tscharner, M. Baggiolini, and C. A. Dahinden, *Eur. J. Immunol.* **23**, 761 (1993).

Eosinophils. Eosinophils can be purified from the venous blood of healthy donors to >99% purity by Percoll density gradient centrifugation of dextran-sedimented leukocytes followed by negative selection with anti-CD16 monoclonal antibody-coated magnetic beads.[4] Neutrophils are the primary contaminant.

Mast Cells. Mast cells can be obtained to 65–75% purity from enzymatically dispersed lung tissue by isolation on Percoll gradients and negative selection with anti-CD2, CD14, and CD20 antibodies.[5]

Chemotaxis Assay

It is superfluous to say that chemotaxis is the definitive method for characterizing chemotactic factors. Since the days of Metchnikoff, who originally described chemotaxis, a number of assay variations have been developed to examine chemotactic factors from a variety of sources. However, with the advent of the Boyden chamber in the early 1960s, the assays have become fairly well standardized around the use of two chambers separated by a porous polycarbonate filter. Current variations may involve different geometries developed for ease of use or higher sample numbers, and they may include cell monolayers or extracellular matrix as relevant surrogates for *in vivo* cell migration.

The chemotaxis assay we utilize is performed using a 48- or a 96-well chemotaxis chamber (Neuroprobe, Cabin John, MD) with protocol modifications according to the method of Junger *et al.*[6] The cells are labeled with a fluorescent marker, calcein AM (Molecular Probes, Eugene, OR), a polyanionic fluorescein derivative widely used as an indicator of cell viability. The ester form of the dye facilitates cell penetration, and intracellular accumulation of dye occurs following cleavage by cell esterases. The dye is well retained in cells and is not appreciably pH sensitive. Another dye that is also used to label cells, BCECF (Molecular Probes, Eugene, OR), is primarily used for monitoring intracellular pH changes and has the disadvantages of not being as well retained by cells and exhibiting fluorescence changes on cell activation that result from changes in intracellular pH.

Ten milliliters of cells (1×10^6 cells/ml in 10 mM phosphate-buffered saline (PBS), pH 7.4) are incubated with 50 μg of calcein AM for 45 min at room temperature in the dark and then washed twice to remove the extracellular dye. Cells are kept in the dark for most of the procedures by

[4] T. T. Hansel, I. J. M. de Vries, T. Iff, S. Rihs, M. Wandzilak, S. Betz, K. Blaser, and C. Walker, *J. Immunol. Methods* **145**, 237 (1991).

[5] T. Jinquan, B. Deleuran, B. Gesser, H. Maare, M. Deleuran, C. G. Larsen, and K. Thestrup-Pedersen, *J. Immunol.* **154**, 3742 (1995).

[6] W. G. Junger, T. A. Cardoza, F. C. Liu, D. B. Hoyt, and R. Goodwin, *J. Immunol. Methods* **160**, 73 (1993).

wrapping tubes and plates with aluminum foil. Cells are resuspended to 1×10^6 myeloid cells/ml or 1×10^7 lymphocytes/ml in Hanks' balanced salt solution containing calcium and magnesium and 0.05% bovine serum albumen (BSA; Sigma, St. Louis, MO). We have found 0.05% BSA to be optimal for chemotaxis with β-chemokines, with BSA concentrations in the range of 0.5–1.0% showing a significant dampening of the chemotactic response. Chemokines are prepared in the same solution and warmed to 37° before use. Each of the quadruplicate wells of a 96-well microtiter plate are filled with the chemokine solution to form a slight positive meniscus, and the plate is placed in the recess of the bottom part of the chamber. A framed uncoated polycarbonate filter with a pore size of 5 μm (Nucleopore, Pleasanton, CA) is carefully placed over the plate and the chamber assembled. For lymphocytes, some investigators report precoating the upper surface of the filter for 1 hr at room temperature with mouse type IV collagen (Life Technologies, Gaithersburg, MD). One hundred microliters of cell suspension is placed into the upper wells, and the chamber is placed into an incubator at 37°. Assays normally incubate for 60 min with monocytes, eosinophils, and basophils and for 4 hr with T lymphocytes (owing to the slower rate of migration).

After incubation, the cell suspension is aspirated, and the upper side of the filter is wiped with PBS to remove unmigrated cells. After the filter is inverted, the fluorescence is quantitated using a Cytofluor fluorescence reader (Millipore, Bedford, MA). Results are expressed as relative fluorescence units, and a typical dose–response curve obtained with macrophage chemotactic protein-1 (MCP-1) on human monocytes is shown in Fig. 1. Alternately, the membrane may be fixed and stained with Diff-Quick (Harleco, Gibbstown, NJ) or Giemsa (Sigma) and the lower surface of the filter examined microscopically. Typically in this analysis, a representative area is counted, and the total cells migrated is obtained by calculation. For example, counting a 1-mm^2 area in a 48-well plate (3.2 mm well diameter) would mean multiplying the value obtained by 8. A caveat to these experiments is that cells may not migrate homogeneously over the entire membrane (usually greater at the center, less at the edges), and not all cells remain attached to the membrane following migration but may drop into the chamber. With the dye-loaded cells, the latter value can be easily measured by reading the wells in the fluorescence reader. Techniques using microscopic evaluation should ensure that the cells that drop off the membrane are proportional in all instances.

The background for the assay should not exceed 5% of the input cells. Migration of up to 50% of the total cell input can be seen with some myeloid cells, although 10–30% is typical. For a positive control, a sample with a final concentration of 100 nM formylmethionylleucylphenylalanine (fMLP) added to the lower chamber should serve to give a strong positive signal.

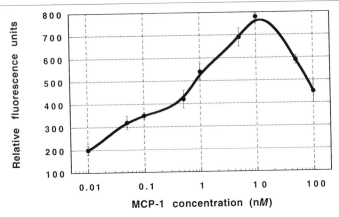

Fig. 1. Dose–response curve for chemotaxis of human monocytes to MCP-1. Cells were placed into the upper chamber, MCP-1 into the lower chamber, and chemotaxis allowed to proceed for 30 min at 37°. The upper surface of the membrane was wiped clear of cells, and the lower surface was measured using a fluorometer. Each point represents the average ± standard deviation of three separate determinations within the same assay.

In addition, a control for chemokinesis (accelerated but nondirectional migration) needs to be included. The ideal control is a well with the same concentration of chemokine in the upper and the lower chamber. The investigator also needs to be aware of the typical bell-shaped curve seen in chemotaxis. With β-chemokines the peak usually occurs around 10 nM protein. Responses below this concentration tend to be linear in magnitude with increasing concentration of chemokine, probably because of a direct relationship of increasing receptor occupancy to rate of chemotaxis. Above the peak concentration, rapid breakdown of the gradient in the effective range of sensitivity, receptor desensitization, and feedback inhibition of the various biological mechanisms at play in the process of locomotion are some of the factors that probably serve to inhibit the response.

Calcium Mobilization Assay

The other assay widely used to characterize the activity of chemokines is the mobilization of intracellular calcium. All chemotactic factors described on all cell types have been shown to elevate the level of intracellular free calcium ion. Normally cells are loaded with a vital dye that acquires new or altered fluorescent properties after interaction with free calcium. Intracellular free calcium is held at a low level in most cells ($\sim$50–100 nM) by maintaining a calcium gradient through the action of ion pumps in the plasma membrane, by sequestration into intracellular compartments known as calciosomes, and by the association of intracellular calcium with cell membranes. Receptor activation normally leads to release of some of the membrane-bound calcium through actions on phospholipases and the release of inositol phosphate (IP_3). By binding to specific receptors on intracellular vesicular membranes, IP_3 induces the release of bound calcium into the cytosol. The elevated intracellular calcium can, in turn, lead to the opening of calcium channels on the plasma membrane and the additional influx of extracellular calcium.

There is often a dissociation seen of about 1 log between the dose–response curve for chemotaxis and that for calcium mobilization. This is most likely due to the effects on chemotaxis being a local membrane event, which occurs continually over time, and the fact that the calcium studies measure whole cell (i.e., nondirectional) calcium mobilization over a relatively short time. In addition, the calcium response appears to be directly related to the receptor occupancy, and cells with higher receptor numbers will generate a signal of greater magnitude. There is no high dose inhibition as observed with chemotaxis, but cross-desensitization of receptors has been noted, probably due to the use of common signaling pathways. In addition, the calcium response is tightly regulated in cells and is generated transiently over a period of about 60 sec before the free calcium is either pumped out of the cell or rebound by membranes and the signal returns to baseline.

Calcium mobilization is measured using the acetoxymethyl (AM) esters of the fluorescent dyes Indo-1 and Fura-2. These dyes are calcium chelators that form a cagelike trap for the calcium ion and undergo a change in fluorescence properties on binding. For the dye molecule, both tightness and specificity of binding are essential. Chelation complexes are reversible and follow the rules of chemical equilibrium. In reversible equilibrium systems, the amount of bound chelate–ion complex will depend on the total concentration of target ion and the affinity of the chelator for this ion. A very important property of the molecule chosen is the dissociation constant, which is a measure of the ability of the molecule to associate itself to the target ion (i.e., its binding strength). By definition the dissociation constant (K_d) is the half-saturation point, and the K_d is expressed in units of concentration. The smaller its value the more powerful and selective a chelation agent. Most of the change in the magnitude of the chelator–ion binding (about 75%) occurs over a 1 log concentration range, which therefore determines the dynamic range for the dye. Almost 99% of the change in binding occurs over a 2 log concentration range. For example, on the basis of a K_d of 224 nM for Fura-2 association with calcium, the effective measurable range of calcium concentration extends from 22.4 to 2240 nM. From this discussion it should be evident that the homogeneous distribution of the chelator dye, its K_d for calcium, and the magnitude of the signal generated by the complex are all important parameters for selection. For more information on these topics, the investigator should visit two very informative Internet sites set up by Molecular Probes (http://www.probes.com) and Perkin-Elmer (http://www.perkin-elmer.com:80/uv/fural/html).

To monitor the changes in intracellular free calcium levels, cells are loaded with 1.25 μM Indo-1AM (Molecular Probes) in the dark in 10 mM phosphate-buffered saline, pH 7.2, without calcium and magnesium and containing 50 μl of Pluronic F-127 (Molecular Probes, low UV absorbance grade) for 1 hr at room temperature. Pluronic F-127 is used as a dispersing agent to facilitate the dye loading. Indo-1 is used in these studies because of its large change in fluorescent emission wavelength on calcium binding and its high affinity (250 nM K_d) at 37°, which allows determination of relatively small changes in intracellular concentrations of free calcium. Fura-2, with an affinity of 224 nM, can also be used. The ester form allows cell penetration and intracellular accumulation of dye following cleavage by cell esterases. About 20% of the dye can be loaded into cells during a typical loading, facilitated by the fact that myeloid cells contain high levels of intracellular esterases (e.g., relative to lymphocytes). Compartmentalization of Indo-1 can occur within myeloid cells,[7] which may place the dye in a location that is inaccessible to cytoplasmic calcium. This will lead to a reduced signal and will become more apparent on continued incubation at 37°. Indo-1 is nontoxic to cells, but overloading of cells will lead to a buffering of the release of free calcium by the dye. Buffering is manifest as a delayed rise in signal and a slowed rate of return to baseline. Higher levels of loading will also reduce the peak height obtained. The level of cell loading or the presence of compartmentalization can be quickly checked by fluorescence microscopy.

After loading, the cells are washed twice with PBS with calcium to remove extracellular dye, resuspended in the same buffer, and transferred to a thermostatically controlled cuvette set at 37° (the cellular response is temperature dependent) and equipped with a magnetic stirring bar. If not used immediately, cells are maintained at room temperature in the dark. Intracellular calcium measurements are carried out with a Perkin-Elmer (Foster City, CA) or similar spectrofluorometer, set to collect emission data at 390 nm after excitation with 360 nm. Greater sensitivity can be obtained by using a ratio method, where emissions are collected at both 402 (bound dye) and 486 nm (free dye). If Fura-2 is used, excitation is at 334 nm with emissions collected at >490 nm. Calibration for both dyes is done with 5 μM ionomycin for F_{max} measurement with subsequent addition of 0.5 mM MnCl$_2$ for F_{min} measurement. Ionomycin is used preferentially over the ionophore A23187 because of its selective affinity for calcium and

---

[7] C. H. June and P. S. Rabinovitch, *in* "Flow Cytometry" (Z. Darzynkiewicz, J. P. Robinson, and H. A. Crissman, eds.), 2nd Ed., pp. 149–174. Academic Press, New York, 1994.

low fluorescence. The use of manganese to quench the fluorescence of the Indo-1 will also indicate the presence of residual nondeesterified dye, as the ester form is insensitive to changes in anions. A typical set of traces obtained for the calibration and stimulation of human blood monocytes by MCP-1 are shown in Fig. 2.

For all samples, the data are converted to average intracellular calcium (nM) using the following formula:

$$[Ca^{2+}]_i = K_d \left(\frac{F - F_{min}}{F_{max} - F} \right)$$

in which 250 nM is used as the K_d for the Ca^{2+}/Indo-1 complex and F_{max} and F_{min} are the fluorescence intensity ratios obtained from the calibration with ionophore and $MnCl_2$, respectively. The calibration is independent of

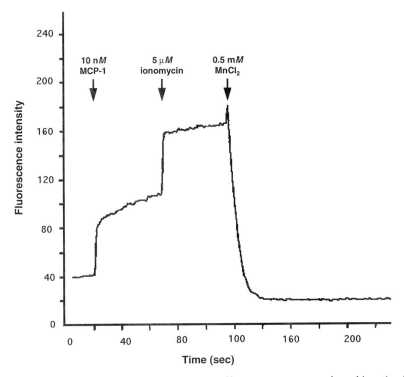

FIG. 2. Standardization of the calcium response of human monocytes to chemokine stimulation. Monocytes, preloaded with Indo-1, were activated at 37° by the addition of 10 nM final concentration of MCP-1. To calibrate the maximum level of response, the cell suspension was treated with ionomycin followed by quenching to the baseline by addition of manganese chloride. The fluorescence intensity values are used in the equation described in the text to determine the intracellular free calcium concentration.

the cell size and the amount of dye loaded. Depending on the cell type and the stimulus used, quiescent cells have an intracellular level of 50–100 nM free calcium (actively proliferating cell lines are usually higher than primary cells), and fully activated cells can reach levels in the 1–2 μM concentration range. For a positive control, a sample with a final concentration of 100nM fMLP added should serve to give a strong positive signal with the expected kinetics. With experiments that extend over long periods (>1 hr), cells can be maintained at room temperature in the dark, but calibration before and after the experiment is necessary to ensure that the system has remained consistent throughout the evaluation period. As noted above, changes can occur due to cell death, dye leakage, or dye redistribution. Methods also exist for measurement of cell fluorescence by flow cytometry,[7] which may be preferred when mixed populations are used because the population of interest may then be tagged with a specific surface marker.

Cell Polarization and Actin Polymerization Assays

One consequence of the generation of inositol phosphate, the mobilization of intracellular free calcium, and the activation of the small GTPases subsequent to chemokine interaction with receptor is the polymerization of actin. Activation of these second messengers results in the local formation of actin filaments, which is regulated through a large set of positive and negative regulating actin binding proteins. The actin filament network generally determines the shape of the cell and the local mechanical properties. Despite the apparent simplicity of this model, the chemotaxis of cells is very difficult to explain at the molecular level in terms of actin polymerization because the polymerization is dynamic, involves focal adhesions to cell surfaces, involves the formation and dissolution of lamellapodia, etc. However, it seems likely that actin polymerization at the leading edge results in a cell protrusion (lamellapod), which extends the cell in the direction of the chemotactic gradient. Therefore, measures of cell polarization (indicative of direction) and actin polymerization can be determined as an event downstream of the calcium response and directly related to chemotaxis.

Cells are suspended in Hanks' balanced salt solution (HBSS) with 0.1% BSA at 1×10^6 cells in a 100-μl volume and treated with chemokines for 10 min at 37°. To measure cell polarization, the activation is terminated by addition of 100 μl of 3.7% formaldehyde, and incubation is continued for 10 min. Cell polarization is determined by measuring forward scatter on a flow cytometer such as a FACScan (Becton Dickinson), and the data are presented as the percent polarized cells.

To measure the extent of actin polymerization, the activation is terminated by addition of 100 μl of 3.7% formaldehyde containing 100 μg of lysophosphatidylcholine (Sigma) and incubation continued for 10 min. The cells are centrifuged and resuspended in 100 μl of 10 mM PBS containing 1.65 μM 7-nitrobenz-2-oxa-1,3-diazole (NBD)-phallacidin, a fluorescent analog of phallacidin, and incubated for 20 min at room temperature. The cells are washed twice and resuspended in 1 ml of 10 mM PBS without calcium and magnesium. Fluorescence detection is acquired by a FACScan flow cytometer, and the total F-actin is represented by the relative fluorescence of stimulated cells divided by the fluorescence of prestimulated or unstimulated control cells. For a positive control, a sample with a final concentration of 100 nM fMLP added should be included.

For time course analysis, the stimulation can be stopped at 1-min intervals and samples fixed, processed, and data collected as outlined above.

Respiratory Burst Assay

As is evident by the assays described above, closely linked to chemotactic events is the further activation of cells. Myeloid cells can also be characterized to undergo a respiratory burst, accompanied by the generation of oxygen radicals. The NADPH oxidase system in myeloid cells generates superoxide anion (O_2^-), which then is metabolized to hydrogen peroxide and hydroxyl radical, both of which possess microbicidal properties. Although the respiratory burst is not directly related to chemotaxis, it is hypothesized that when the cell reaches a sufficient level of chemotactic stimulation (or chemotactic stimulation in concert with other activation stimuli), the cell is triggered to undergo a respiratory burst.

Culture determination of the respiratory burst utilizes the ability of culture supernatants to reduce electron accepting compounds such as ferric (Fe^{3+}) cytochrome c to the ferrous form (Fe^{2+}). In this assay, cells at 1 $\times$ 10^6 to 1 $\times$ 10^7 cells/ml (depending on the cell type used) in phenol red-free HBSS are placed into wells of a 96-well microtiter plate at 100 μl per well. Each well receives 50 μl of a 1.35 mg/ml solution of ferric cytochrome c in HBSS (horse heart, type VI, Sigma), and the plate is equilibrated to 37°. An additional 50 μl of HBSS at 37° containing the chemokine at four times the final desired concentration is added. For a positive control a sample with a final concentration of 100 nM fMLP added should serve to give a strong positive signal, and a sample with 25 μl of a 5 mg/ml solution of superoxide dismutase (bovine erythrocyte, Sigma) should be used as the negative control. A final concentration of 50 ng/ml phorbol ester can also be used as a positive control but the kinetics and the maximum response will not be reflective of a typical chemokine response. Superoxide dismutase works specifically with O_2^- as a substrate and is used to determine specificity because a number of reactive molecules can reduce ferric cytochrome c. After 30 min, the plate is read in a 96-place spectrophotometer at 550 nm.

Alternately, for more accurate kinetic measurements, the plate may be read at regular intervals (2-min intervals are suggested for initial studies). Following the final reading, 50 μl of a 1 mg/ml solution of sodium dithionate (Sigma) in HBSS is added per well to reduce all the ferric cytochrome c and obtain the maximum value for each well. Data can then be expressed as percent of the total signal possible for each well.

The respiratory burst can also be measured on a cellular basis using dihydrorhodamine 123. Dihydrorhodamine 123 is a colorless, cell-permeant dye that is oxidized by superoxide anions to the cell-permeant, relatively nontoxic and fluorescent rhodamine 123. This dye has been used to study the neutrophil respiratory burst as well as mitochondrial respiration.

Cells are washed twice, suspended at 1×10^7 cells/ml in 10 mM PBS, pH 7.2, containing dihydrorhodamine 123 (Molecular Probes) at 1 μg/ml, and incubated at 37° for 10 min. The baseline fluorescence level is obtained on a FACScan flow cytometer with excitation at 488 nm using an argon laser, and the emission is monitored at 530 nm with a long-pass filter. The cells are then subjected to chemokine stimulation for 15 min at 37°. At the end of incubation, the mean fluorescence of the cell suspension is acquired and used in analysis. Stimulation with phorbol esters can be used as a positive control to ensure that the assay is functioning properly. However, the investigator needs to be aware that the cell response to phorbol esters is much more gradual and that the peak response is greater than typically seen with chemokine stimulation. For a positive control, a sample with a final concentration of 100 nM fMLP added should serve to give a strong positive signal. In addition, chemokines may induce only a minimal respiratory burst unless the cells have been primed as with degranulation (see next section). Finally, the impact of contamination of reagents (especially proteins like cytochrome c and superoxide dismutase) with endotoxin, which can act as a priming and activation agent with cells such as monocytes, needs to be considered by the investigator.

Degranulation

Associated with elevated intracellular calcium and often accompanying a respiratory burst, degranulation is also seen in many myeloid cells. Degranulation can be enhanced in most myeloid cells by preincubation with 5 μg/ml cytochalasin B (Sigma) for 5 min. In some instances, cells must be primed or presensitized before undergoing degranulation in response to chemokines. For example, basophils need to be treated with interleukin-3, interleukin-5, or granulocyte–macrophage colony-stimulating factor (GM-CSF) to become responsive to RANTES-induced degranulation.[3]

In general, degranulation is a rapid cellular event closely following the calcium transient and can be measured by incubating $1–3 \times 10^5$ myeloid cells for 10 min in the presence of the chemotactic stimulus. Supernatants should be separated from the cells and quickly frozen if the analysis is to be done at a later time. Cells and serum contain factors that may either break down or inactivate the mediator or enzyme of interest. Two freeze–thaw cycles of the cells after the time course can be used to obtain the maximum release of granule contents. Again, for a positive control, a sample with a final concentration of 100 nM fMLP added should serve to give a positive signal. Many enzymes and small molecular weight molecules are released by degranulation, and the granule contents tend to be cell selective. Some common assays used include the following.

Histamine Assay. Histamine is released by mast cells and basophils. Levels can be measured by radioimmunoassay using commercially available kits.

Leukotriene C$_4$ Assay. Leukotriene C$_4$ is released by eosinophils. Levels can be measured by enzyme-linked immunosorbent assay (ELISA) assay or radioimmunoassay using commercially available kits (Amersham, Arlington Heights, IL; DuPont-NEN, Boston, MA).

Peroxidase Assay. Peroxidase is released by eosinophils. Levels can be assayed using kits available commercially (Sigma).

β-Glucuronidase Assay. β-Glucuronidase is released by mast cells and basophils. A premixed kit with protocol to assay this enzyme using a chromogenic substrate is available from Sigma.

Cell Surface Receptor Expression and Cell Adhesion

Often chemokine activation can also be manifest by a change in the expression of a surface marker or by enhanced adhesion. These events are not associated with new protein synthesis but with the activation of integrins or the mobilization of intracellular stores of receptors to the cell surface. As such, the events involved occur over a short time (10–30 min), are transient, and correlate well with other intracellular events such as calcium mobilization.

A cell surface receptor that has been useful in the determination of cellular activation is Mac-1 ($\alpha_M\beta_2$) on monocytes.[1] Upregulation can be quantitated using monoclonal antibody staining coupled with fluorescence activated cell sorting (FACS) analysis and quantitation by measuring the percent positive cells or examining the mean fluorescent channel. Alternately, an activation antigen recognizing antibody, such as that described by Elemer and Edgington,[8] can be utilized.

Functional adhesion is usually measured using endothelial cells or the endothelial-like cell line ECV109. Normally, confluent cultures of cells are

[8] G. S. Elemer and T. S. Edgington, *J. Immunol.* **152,** 5836 (1994).

grown in 96-well plates and then treated overnight with 10–100 ng/ml interleukin-1, tumor necrosis factor-α, or interleukin-4 before assay to enhance adhesion molecule expression. The following day the supernatant is removed, the cells gently washed twice, and the medium replaced with HBSS. The adhering cell is activated for 10 min with chemokine at 37° in HBSS and then added to plates. After 30 min, the wells are filled with HBSS to provide a small meniscus and the plate sealed with a plate sealer. It is important to avoid trapping air bubbles within the wells in this step. The plates are then inverted and placed into a tabletop centrifuge using a 96-well plate carrier rotor and spun for 1 min at 100 rpm. This forces detachment of the nonadherent and loosely adherent cells. Following centrifugation, the sealer is removed and the remaining fluid gently tapped from the well. Cell number can then be assayed by (1) direct staining and visualization, (2) assay utilizing fluorescent dye-containing cells (see section on chemotaxis), (3) staining using a vital marker such as tetrazolium blue, or (4) monoclonal antibody staining using a whole cell ELISA protocol.

The investigator needs to consider that adhesion is a temperature-dependent, avidity-based event, and therefore the number of interactions (both background and induced) will increase with time of contact. The researcher needs to work out the timing for the assay on the bases of the two cell types interacting, the stimulus used with both cell types, and the method used to separate bound from unbound cells.

[13] Gene Targeting Strategies to Study Chemokine Function *in Vivo*

By Donald N. Cook

Introduction

The chemokines (<u>chemo</u>tactic cyto<u>kines</u>) comprise a large superfamily of structurally similar, low molecular weight, proinflammatory proteins that induce directed migration (chemotaxis) of cells. There are two major subfamilies of chemokines, termed C-X-C and C-C, based on the presence or absence of an intervening amino acid between the first two of four conserved cysteine residues. The majority of known chemokine genes are clustered within two relatively small chromosomal regions located on mouse chromosome 11 for the C-C chemokines, and chromosome 5 for the C-X-C chemokines. Most of what is currently known of chemokine

METHODS IN ENZYMOLOGY, VOL. 287   0076-6879/97 $25.00

function has been learned from studies *in vitro*. Several of these studies have suggested that chemokines within a subfamily have biological activities that overlap with one another; however, the conditions used in these assays cannot reproduce exactly the microenvironment of inflamed tissues, and chemokines with similar activities *in vitro* may function differently *in vivo*. One of the major challenges of future chemokine research is to identify the *in vivo* functions of individual chemokines in health and in disease.

Gene targeting in embryonic stem (ES) cells provides a powerful approach to study chemokine function *in vivo*. This technology combines two separate technological advances: the derivation of murine ES cell lines capable of repopulating the mouse blastocyst stage embryo,[1,2] and the ability to generate planned genetic alterations in mammalian cells by homologous recombination.[3] Gene targeting in ES cells,[4,5] which is the combination of these two technologies, allows targeted mutations to be generated *in vitro* and then transferred into the mouse germ line,[6] where their effects can be studied *in vivo*. In principle, the functions of any cloned gene can be determined by studying the phenotype of mice homozygous or heterozygous for the gene altered appropriately by targeting.

The methods involved in generating mutations in ES cells and transferring those mutations to the mouse germ line have been previously described in detail in several books and comprehensive review articles. The present review does not provide a condensed account of these methods, but rather gives an overview of gene targeting and discusses some problems likely to be encountered when using this technology to study chemokine function *in vivo*.

We have previously used gene targeting to generate MIP-1α (macrophage inflammatory protein-1α)-deficient mice.[7] The targeting of the MIP-1α gene is used here as an example to illustrate general problems associated with chemokine gene targeting.

[1] M. J. Evans and M. H. Kaufman, *Nature (London)* **292,** 154 (1981).
[2] G. R. Martin, *Proc. Natl. Acad. Sci. U.S.A.* **78,** 7634 (1981).
[3] O. Smithies, R. G. Gregg, S. S. Boggs, M. A. Koralewski, and R. S. Kucherlapati, *Nature (London)* **317,** 230 (1985).
[4] K. R. Thomas and M. R. Capecchi, *Cell (Cambridge, Mass.)* **51,** 503 (1987).
[5] T. Doetschman, R. G. Gregg, N. Maeda, M. L. Hooper, D. W. Melton, S. Thompson, and O. Smithies, *Nature (London)* **330,** 576 (1987).
[6] S. Thompson, A. R. Clarke, A. M. Pow, M. L. Hooper, and D. W. Melton, *Cell (Cambridge, Mass.)* **56,** 313 (1989).
[7] D. Cook, M. Beck, T. Coffman, S. Kirby, J. Sheridan, I. Pragnell, and O. Smithies, *Science* **269,** 1583 (1995).

Reagents and Materials

All the reagents, materials, and methods needed to target ES cells and to transfer the mutations into the mouse germ line have been well described in previous publications. For example, *Manipulating the Mouse Embryo: A Laboratory Manual* (1986)[8] focuses primarily on manipulating one-cell embryos to generate transgenic mice, although protocols for genetic manipulation of blastocyst stage embryos are also described; the second edition of that book (1994) contains an expanded section on ES cell culture. *Teratocarcinomas and Embryonic Stem Cells: A Practical Approach* (1987)[9] provides a comprehensive description of ES cell culture and the use of these cells to generate chimeric mice. More recent descriptions of the procedures used in gene targeting[10] and generating chimeras can be found in Volume 225 of this series, *Guide to Techniques in Mouse Development* (1994).[11] Most of these procedures are relatively straightforward in principle, although considerable practice may be required before an acceptable degree of success in their use is achieved. Anyone wanting to perform the entire procedure is encouraged to learn firsthand from an experienced investigator, or to participate in a practical course on gene targeting in ES cells.

Embryonic Stem Cell Culture

Embryonic stem cells are used in gene targeting experiments because of their ability to repopulate the mouse blastocyst, thereby allowing transfer of the targeted gene to future generations. Many ES cell lines now exist, and some are available through the American Type Culture Collection (ATCC, Rockville, MD). Obtaining a good ES cell line is critical to a successful experimental outcome. The number of passages the ES cells have undergone since their isolation may vary considerably, and the investigator should ensure that the ES cell line chosen has been used previously at or close to the available passage number and has generated germ line chimeras. It is equally important that the ES cells be maintained in an undifferentiated state throughout the course of the experiments. This is accomplished by culturing the cells with leukemia inhibitory factor (LIF), which can be supplied either as a recombinant protein, by mitotically inactivated LIF-

[8] B. Hogan, F. Costantini, and E. Lacy, "Manipulating the Mouse Embryo: A Laboratory Manual," 1st and 2nd Eds. Cold Spring Harbor Laboratory Press, Cold Spring Harbor, New York, 1986, 1994.

[9] E. J. Robertson, "Teratocarcinomas and Embryonic Stem Cells: A Practical Approach." IRL Press, Oxford, UK, 1987.

[10] R. Ramírez-Solis, A. C. Davis, and A. Bradley, *Methods Enzymol.* **225,** 855 (1993).

[11] C. L. Stewart, *Methods Enzymol.* **225,** 823 (1993).

expressing cell lines, or by primary fibroblast feeder cells. Our laboratory most frequently uses the BK4 ES cell line [a subclone of the hypoxanthine phosphoribosyltransferase (HPRT)-negative cell line E14TG2A], which we grow on primary fibroblasts. However, different ES cell lines may have been adapted to grow under specific conditions, and it is recommended that the investigator use reagents and conditions that have proved successful for the particular cell line they have obtained.

Even under ideal culture conditions, ES cells having abnormal karyotypes will arise. Such cells often have a selective growth advantage and eventually become the predominant cell type in the culture. Unfortunately, these aneuploid cells are rarely able to transmit the mutation to the mouse germ line, and the following precautions should be taken to minimize their frequency. (1) Freeze as many vials of early passage cells as is convenient for later use in targeting experiments. (2) Do not passage ES cells extensively prior to their use in electroporations. (3) Do not to allow the ES cell colonies to become too crowded or too sparse; they should be passaged at a dilution of 1 : 10 approximately every 3 days. (4) Medium should be replaced if its color indicates acidification.

Gene Targeting

Overview

The first step in gene targeting experiments is to assemble a recombinant DNA targeting vector that contains several kilobases (kb) of DNA having the same sequence as the gene of interest. The gene targeting vector can recombine with its homologous gene in ES cells, resulting in structural changes to the locus that depend on the type of the targeting vector used. The targeting efficiency of linear DNA is severalfold higher than that of supercoiled DNA. For this reason, a restriction endonuclease is used to linearize the vector DNA prior to its introduction into ES cells by electroporation. There are two major types of targeting vectors, termed Ω or replacement type, and O or insertion type. In Ω type vectors, the two horizontal arms of the Ω represent regions of DNA sequence identity with the gene of interest (Fig. 1A). Each of these two regions can undergo homologous recombination with the target gene. The loop of the Ω represents nonhomologous DNA, typically a positive selectable marker gene whose expression enables ES cells that incorporate the targeting vector to survive and form colonies in selective media; the most commonly used positive selectable marker gene is *neo*, the gene encoding resistance to the aminoglycoside G418. After recombination with an Ω-type vector, cellular DNA is replaced by DNA from the vector.

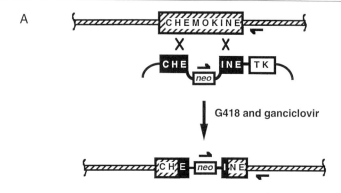

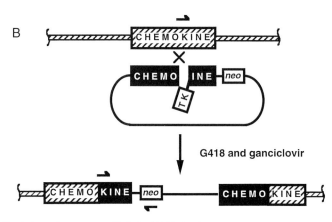

FIG. 1. Basic targeting vectors. Cellular DNA is shown by hatching; regions of homology in the targeting vector are in black. Small arrows represent primers for use in PCR analysis of candidate cell lines. Plasmid sequences are not drawn to scale. (A) Ω-type vector. The *neo* gene is inserted into the chemokine gene, and the sequence "MOK" is simultaneously deleted. (B) O-type targeting. The entire chemokine gene is duplicated, accompanied by an insertion of *neo* between the two copies. Note that gap repair will fill in sequences missing in O-type vectors (illustrated here by the absent letter K, which is filled in during the recombination).

Although *neo* expression allows targeted cells to survive in selective media containing G418, the majority of drug-resistant colonies arise from cells that have incorporated the DNA in a nonhomologous (random) manner. To reduce the frequency of these nonhomologous recombinants, most investigators use a positive–negative procedure in which a herpes simplex virus thymidine kinase (HSV TK) gene is included in the targeting construct, and the drug ganciclovir is added to the culture medium.[12] Most cells that randomly incorporate the targeting vector will express the TK gene, and these cells will be killed by ganciclovir. In contrast, the TK gene is not incorporated into targeted cells, and they will be resistant to ganciclovir. However, even the double selection is not completely restrictive, so that the DNAs of colonies resistant to both G418 and ganciclovir must be screened for a correctly targeted gene by southern blotting and hybridization, or by using the polymerase chain reaction (PCR) with one primer corresponding to DNA in the targeting vector and the other primer corresponding to a region in the locus not contained in the targeting construct (Fig. 1).

With O-type vectors, the entire targeting construct, including *neo* and the bacterial plasmid sequences, is inserted into the host chromosome (Fig. 1B). This type of targeting event is mediated by a single crossover that results in the duplication of the regions of homology. Gap repair will fill in sequences missing in O-type vectors (K in Fig. 1B).

After a targeted ES cell line has been identified, the cells are microinjected into the blastocoel cavity of 3.5-day-old mouse blastocysts, which are then transferred to the uterus of a pseudopregnant mouse for their continued development. Properly maintained ES cells can contribute to most tissues of the embryo that develops into a chimeric mouse. If the ES cells contribute to the germ tissue in the chimera, the targeted gene can be transmitted to future generations.

Vector Design

Gene Disruptions. The most frequent application of gene targeting is gene disruption, often referred to as gene knockout. In this type of experiment, an Ω-type targeting vector is typically used to insert *neo* into a region of the gene required for its function. The site of *neo* insertion depends on the borders of the regions of homology adjacent to *neo* in the targeting construct. It is desirable to generate a deletion in the target gene at the site of *neo* insertion by omitting DNA from the targeting construct (MOK in Fig. 1A), as this will help ensure functional inactivation of the target gene. If specific regions required for the function of the gene of interest have not been identified, it is prudent to delete as much of the gene as possible to avoid producing a partially functional product.

The targeting strategy that we used to disrupt the MIP-1α gene is shown in Fig. 2. The regions of homology in this targeting construct lack several features of the MIP-1α gene that were predicted to be required for its

[12] S. L. Mansour, K. R. Thomas, and M. R. Capecchi, *Nature (London)* **336**, 348 (1988).

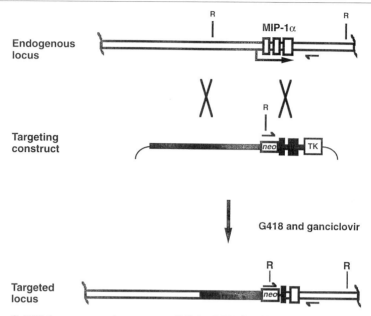

Fig. 2. MIP-1α gene targeting strategy. Cellular DNA is white, and regions of homology in the targeting construct are black. Boxes represent exons. The direction of transcription of the MIP-1α gene is indicated by the large horizontal arrow. Small arrows denote primers used in PCR analysis of candidate clones. Suitable restriction enzyme sites for Southern blot detection of targeted colonies are indicated by an R. The large Xs represent arbitrary points of DNA crossover.

function: the first exon, half of the second exon, the mRNA start site, and 250 base pairs (bp) of DNA upstream of this site. Northern blot analyses of mRNA from cells derived from mice homozygous for the disruption revealed very little if any MIP-1α-specific mRNA.[7] Similar approaches should be suitable for inactivating other chemokine genes.

Gene Modifications and Eliminating Marker Gene DNA. As chemokine function *in vivo* becomes better understood, it may become useful to modify regions of chemokine genes in more subtle ways. These modifications could be directed either to the coding region of the chemokine gene or to its transcriptional regulatory regions. One example of an interesting coding region modification that might be attempted is a 2-amino acid substitution in MIP-1α that abolishes its ability to bind proteoglycans *in vitro*.[13] A

[13] G. J. Graham, P. C. Wilkinson, R. J. Nibbs, A. Lowe, S. O. Kolser, A. Parker, M. G. Freshney, M. L.-S. Tsang, and I. B. Pragnell, *EMBO J.* **15**(23), 6506 (1996).

mouse carrying this mutation would be useful to test the hypothesis that proteoglycan binding is important to MIP-1α function *in vivo*.

Modifications might be introduced into the regulatory regions of chemokine genes with the aim of altering their transcription either in all cells or in only specific cell types or tissues. It should also be possible to insert exogenous drug- or hormone-responsive transcription elements into the regulatory regions of chemokine genes so that their transcription can be induced at will.

Several strategies have been developed that allow the introduction of specific gene modifications into ES cell DNA. The simplest strategy employs an Ω-type vector in which the desired modification is contained within one of the arms of homology (N* in Fig. 3A) and *neo* is inserted upstream or downstream of the chemokine gene of interest. Depending on the point of crossover between the incoming and host DNAs, the desired modification will or will not become incorporated into the targeted gene, and an appropriate PCR assay is therefore needed to screen targeted cells for the presence of the modification.

One disadvantage of using a simple Ω-type vector to introduce gene modifications is that *neo* can affect the transcription of the modified gene or neighboring genes[14] and complicate analysis of the desired modification. Retention of the marker in the targeted locus is therefore undesirable even for gene disruption experiments. This may be a particularly serious problem in mice carrying disruptions of chemokine genes, because most chemokine genes are situated near other chemokine genes having related activities. For example, we have found that mice carrying the disrupted MIP-1α gene have an approximately 50-fold reduction in mRNA specific for the closely linked C-C chemokine MIP-1β.[14a] Therefore, part of the difference in the virus-induced inflammatory response between the MIP-1α-deficient mice and wild-type mice could be due to reduction in the expression of MIP-1β. Simple gene disruptions of other chemokine genes could likewise affect transcription of their respective neighboring genes.

The disadvantages of retaining marker genes in targeted loci have led investigators to develop strategies to eliminate the selectable marker DNA from the targeted locus prior to the generation of animals. One of these strategies employs the Cre/*lox* system. Cre is a P1 bacteriophage-encoded recombinase that catalyzes sequence-specific DNA recombination between

---

[14] E. N. Olson, H.-H. Arnold, P. W. J. Rigby, and B. J. Wold, *Cell* (*Cambridge, Mass.*) **85**, 1 (1996).

[14a] Cook *et al.*, in preparation (1997).

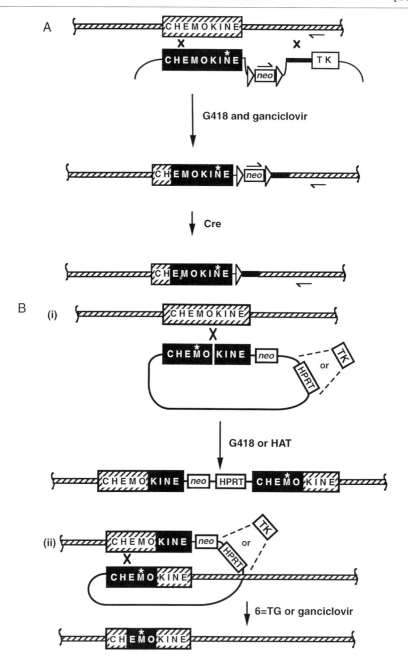

34-bp repeats denoted *loxP*.[15,16] If *loxP* sites are engineered into the targeting vector at both ends of *neo,* Cre can be transiently expressed in targeted cells to catalyze recombination-mediated deletion of the marker[17] (lines 2 and 3 in Fig. 3A). Cells having undergone this second recombination event can be identified by screening for sensitivity to G418.

An alternative strategy to target genes and subsequently remove the selectable marker from the targeted locus is shown in Fig. 3B. This strategy, termed hit and run[18] or in–out,[19] employs an O-type vector carrying a gene modification in one of the regions of homology together with both a positive and a negative selectable marker gene. A single HPRT gene can be used for both of these purposes in conjunction with HPRT$^-$ ES cells because ES cells expressing the HPRT gene will be positively selectable using media containing hypoxanthine, aminopterin, and thymidine (HAT) and negatively selectable in media containing 6-thioguanine (6-TG). (The use of the HPRT gene for this dual selection requires a male ES cell line, such as E14TG2A, having a nonrevertable mutation in its endogenous X-chromosome encoded HPRT gene.) Alternatively, *neo* can be used as a positive marker and HSV TK as a negative selectable marker. As with all O-type vectors, part or all of the target gene is duplicated following the recombination, but in this case one of the duplicated regions is designed to contain the desired modification. The first recombination event (hit or in) is positively selected in media appropriate for the marker gene used: *neo* or HAT. The second event (run or out) requires a second, intrachromosomal recombination event that excises one of the two duplicated regions together with the negatively selectable marker gene. Cells that have undergone this second event can be selected by culture in media containing a drug appropriate for the negative selectable gene: 6-TG for HPRT and ganciclovir for the HSV TK gene. It is essential to confirm that the targeted

[15] N. Sternberg and D. Hamilton, *J. Mol. Biol.* **150,** 467 (1981).
[16] B. Sauer and N. Henderson, *Proc. Natl. Acad. Sci. U.S.A.* **85,** 5166 (1988).
[17] B. Sauer, *Methods Enzymol.* **225,** 890 (1993).
[18] P. Hasty, R. Ramirez-Solis, R. Krumlauf, and A. Bradley, *Nature* (*London*) **350,** 243 (1991).
[19] V. Valancius and O. Smithies, *Mol. Cell. Biol.* **11,** 1402 (1991).

FIG. 3. Gene modification strategies. (A) An Ω-type vector is used to change the N in the chemokine gene to N*. The *neo* gene is inserted downstream of the gene to minimize transcriptional disregulation of the locus. *LoxP* sites (shown as triangles flanking *neo*) and Cre recombinase can be used to remove *neo* from the targeted locus. (B) An in–out or hit-and-run strategy is shown. A gene modification (M*) is introduced in a duplicated chemokine gene using an O-type targeting vector in the first step. A second, intrachromosomal recombination event follows, resulting in loss of the wild-type gene together with the negative selectable marker (HPRT or TK) gene.

locus has the correct structure because cells resistant to the selective drugs can also arise by somatic mutations to the selectable genes. It is also necessary to verify by PCR or Southern blotting that the specific mutation (M*) has been retained after the second, excision event.

Repeated Targeting of a Single Locus

For some purposes, it may be desirable to analyze the effect of several different modifications of the same chemokine gene. For example, various modifications to the noncoding region of a single chemokine gene could be used to affect its transcription in different cell types. Normally, considerable effort is required to screen the large number of drug-resistant cells generated in these kinds of experiments. However, Detloff *et al.* have designed a "plug-and-socket" strategy for use with HPRT− ES cells that allows homologous recombination events to be directly selected.[20] In this strategy, a socket is first generated using a gene replacement vector to insert *neo* and an incomplete copy of the HPRT gene downstream (as exemplified by ΔPRT in Fig. 4) or upstream (HPRΔ) of the chemokine gene of interest. This socket provides a region of homology for recombination with a second plasmid, termed the plug, that carries a different incomplete (HPRΔ in Fig. 4) but complementary copy of the HPRT gene. Homologous recombination between the overlapping parts of the mutant HPRT genes provides the only means that a functional HPRT gene can be restored, and it confers on cells the ability to grow in media containing HAT. A single socket can be used together with many different plugs to generate a variety of gene-targeted ES cells, all of which can be directly selected. If desired, *loxP* sites can be engineered onto the two ends of the HPRT gene in the socket-and-plug vectors so that Cre recombinase can be used later to eliminate the HPRT gene from the targeted locus.

Analysis of Embryonic Stem Cell DNA

Southern Blotting. Southern blotting is the most reliable method for analyzing the DNA of candidate clones. The DNA should be cut with a restriction enzyme(s) that recognizes a site(s) that differs in the untargeted and targeted loci. If the resulting targeted locus is of a different size than the endogenous locus, it may be convenient to use an enzyme whose sites span the selectable marker. Ideally, the Southern blot should be probed with DNA from a part of the locus that is not contained in the targeting vector, so that nonhomologous recombinants will not be confused with

[20] P. J. Detloff, J. Lewis, S. W. M. John, W. R. Shehee, R. Langenbach, N. Maeda, and O. Smithies, *Mol. Cell. Biol.* **14,** 6936 (1994).

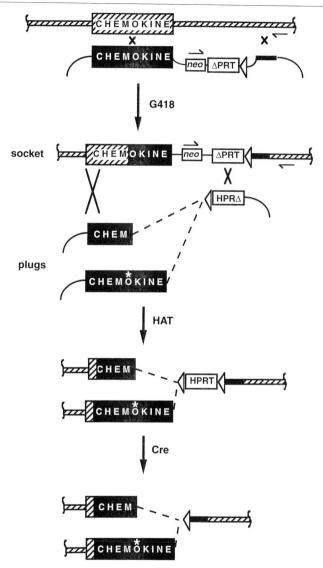

Fig. 4. Plug-and-socket repeat targeting strategy. In the first step, a socket is generated, in which *neo* and a 5′ deletion mutant of the HPRT gene (ΔPRT) are inserted downstream of the chemokine gene. In the second step, a plug completes the HPRT gene and allows direct selection with HAT-containing media. Various 5′ regions of homology can be used in different plugs to generate different mutations. The two plugs shown respectively generate a 3′ deletion and a gene modification in the chemokine gene.

correctly targeted cells. If a probe from outside the regions of DNA identity is unavailable or unsuitable because it yields a smeared pattern on Southern blots of genomic DNA, a probe that hybridizes to a region within the targeting construct can be used, but this type of probe will also hybridize to nonhomologously (randomly) integrated vector DNA. Consequently, DNA from cells that have randomly incorporated the targeting vector will occasionally give a hybridizing fragment that by chance is the same size as that expected for a targeted cell line. This will lead to a false-positive unless further analysis of the candidate clone is carried out using one or more additional restriction enzymes. Analysis by PCR can also serve as an independent means of confirming the targeting event.

Polymerase Chain Reaction Analysis. Many investigators grow sufficient numbers of cells so that each candidate cell line can be frozen prior to its analysis by Southern blotting. For experiments having a low targeting frequency, this can result in the expenditure of considerable time. In such cases, PCR may be preferable to Southern blotting as a means of identifying targeted clones, because targeted cells can sometimes be identified[21] prior to their need to be passaged from the well into which the original colony was placed. If the targeting frequency of a vector is very low, larger numbers of G418-resistant colonies can be screened by pooling several colonies (typically five) in the same PCR reaction. Once a positive pool is identified, reactions can be carried out on the individual colonies that make up the positive pool to identify the targeted cell line. Two PCR primers are necessary: one complementary to a region of the target gene that is not contained in the targeting construct, and the other complementary to a region in the targeting construct, typically *neo* or the adjacent polylinker DNA, that is not in the locus of interest. Examples of appropriate locations of PCR primer binding sites are indicated by small arrows in Figs. 1–4.

A disadvantage of using PCR to detect targeted genes is that correctly targeted cells may be missed because their DNA did not yield a visible band in ethidium bromide-stained gels of PCR products. To reduce the frequency of these false-negatives, each PCR experiment should ideally include DNA from a positive control cell whose amplification ensures that the procedure at any particular time was sufficiently sensitive to detect targeted cells. To generate this positive control cell, ES cells should be electroporated with a DNA that is related to the final targeted locus and therefore has binding sites for both primers used in the PCR reactions (see Kim *et al.*[22] for details). G418-resistant colonies resulting from this electroporation should be screened for a cell line that contains only a single

[21] H. S. Kim and O. Smithies, *Nucleic Acids Res.* **16**, 8887 (1988).
[22] H. S. Kim, B. Popowich, W. R. Shehee, E. Shesely, and O. Smithies, *Gene* **103**, 227 (1991).

insert of the positive control DNA. Colonies of this cell line should then be treated in parallel with the colonies that are candidates for targeted cell lines. If necessary, the sensitivity of the PCR screening procedure can be increased by southern blotting of gels on which the PCR products were electrophoresed, followed by hybridization of the membranes with a probe corresponding to a region between the two primers.

A rapid method for isolating DNA from candidate colonies is as follows: (1) Place one-half of the G418-resistant colony in a 24-well plate containing feeder cells and place the other half in an Eppendorf tube containing 0.5 ml phosphate-buffered saline (PBS; up to five colonies can be added to that tube). Centrifuge the tube at 12,000 rpm for 30 sec, pour off the supernatant, and freeze the cell pellet. (2) Resuspend the frozen cell pellet in 20 μl deionized water, incubate at 95° for 10 min, and then allow the tube to cool to room temperature. (3) Add 1 ml of 10 mg/ml proteinase K and incubate for 1 hr at 55°. (4) Heat the tube again to 95° for 5 min. One-half of this crude DNA preparation can be used directly in a PCR reaction.

Breeding of Mice

Chimeras. The strain of mouse chosen as a source of the recipient blastocysts used to generate chimeric mice is irrelevant to the genetic background of the mice that eventually carry the targeted gene, because the chromosomes of the targeted mice inherited from the chimera are derived from the injected ES cells, not from the recipient blastocysts. However, because chimeric mice are most easily identified by their coat color, the recipient blastocyst should be from a strain that has a coat color different from that of the strain from which the ES cells were derived.

Once obtained, chimeric mice are bred to an inbred strain of mouse to derive mice heterozygous for the targeted gene. The choice of inbred strain with which to breed the chimera is important because this strain will contribute genetic material to all subsequent generations. If the ES cells were derived from strain 129 and the chimeras are bred to strain 129 mice, the genetic background of all subsequent generations will be inbred and therefore uniform with the exception of the targeted gene. Therefore, any phenotypic differences between wild-type 129 mice and the gene-targeted 129 mice can be attributed solely to the targeted gene. Unfortunately, strain 129 mice are difficult to breed, and they are prone to infections. For these reasons, the chimeras are usually mated to a hardier strain of mice such as C57BL/6.

The phenotype to be studied is an important factor in deciding which mouse strain(s) should be bred with the chimeras, because there are some inflammatory diseases that occur only in certain strains of mice. For example, strain SJL mice are susceptible to experimental allergic encephalomyelitis (EAE), and strain DBA mice are susceptible to collagen-induced rheumatoid arthritis (RA). To study these diseases in gene-targeted mice, the targeted chemokine gene must be backcrossed for 6 to 10 generations onto the genetic background of a susceptible mouse strain.

F_1 *Generation*. If the chimera was bred to a C57BL/6 mouse, germ line transmission of the 129 genome is indicated by the agouti coat color of the pups. The offspring from this mating, the F_1 generation, have half of their chromosomes derived from the 129 strain and half from the C57BL/6 strain. The hybrid vigor conferred by such a genetic composition is obvious from the typically large and healthy appearance of the mice. Each mouse in the F_1 generation is genetically identical to the others, with the exception of the targeted gene. Therefore, any differences in the F_1 generation phenotype between wild-type mice and mice heterozygous for the targeted gene can be attributed solely to the mutation.

F_2 *Generation*. The F_2 generation, derived by interbreeding F_1 mice heterozygous for the targeted mutation, is the first generation in which homozygous mutant mice can be obtained. Heterozygous and wild-type mice, which can serve as littermate experimental controls, are also obtained in this generation. Unlike the F_1 generation, phenotypic differences between the three target locus genotypes ($+/+$, $+/-$, and $-/-$) in F_2 mice cannot necessarily be attributed to the targeted mutation. This is because individual F_2 mice differ from one another not only at the targeted locus but also at genetic loci that are linked to the target gene and at loci that are unlinked. Differences at the unlinked loci arise from the random segregation of the C57BL/6- and 129-derived chromosomes that occurs during meiosis in the germ cells of the F_1 mice. Depending on the phenotype under study, these strain-specific differences can give rise to considerable variation between individual littermates. For this reason, it may be necessary to use a larger number of experimental F_2 mice than of F_1 mice to obtain statistically significant data relating to phenotypic differences between the various genotypes. The variability among individual F_2 mice resulting from unlinked genes can be progressively reduced in future generations by backcrossing the mutation onto a mouse strain appropriate to the phenotype of interest.

When F_1 heterozygous mice are interbred to generate wild-type, heterozygous, and homozygous mutant mice, these mice will differ from one another at genes that are linked to the targeted gene. This is because the F_1 heterozygous mice inherit the targeted gene (together with DNA sequences linked to it) from the 129 strain-derived ES cells. In contrast, the F_1 mice inherit the wild-type copy of the chemokine gene (together with DNA sequences linked to it) from the mouse strain that was bred to the chimera, C57BL/6 in our example. Therefore, F_2 mice that are homozygous for the targeted gene are also homozygous for the 129-derived chemokine genes that are linked to the targeted locus, whereas F_2 mice homozygous for the wild-type copy of the chemokine gene are also homozygous for the C57BL/6-derived chemokine genes linked to the wild-type chemokine gene (Fig. 5A). This strain-specific difference surrounding the targeted chemokine locus will remain even after the targeted gene has been backcrossed onto an inbred strain, because the frequency of crossovers between the targeted gene and other closely linked neighboring chemokine genes will be extremely low—the average size of a centimorgan (the length of DNA for which the frequency of crossover during meiosis is 1%) is approximately 3×10^6 bp, a length of DNA that is much larger than the multigene chemokine loci. Although important functional differences between chemokine genes derived from the 129 and C57BL/6 strains have not yet been described, any such differences could cause phenotypic differences between the targeted and wild-type mice that are unrelated to the mutation that was introduced into the targeted gene.

One of the few ways of avoiding this complication that results from linked genes is to carry out specific matings to generate the proper control mice (Fig. 5). This requires that the investigator first identify a restriction fragment length polymorphism (RFLP) or simple repeat polymorphism (SRP) that is different between the two mouse strains at or closely linked to the locus of interest. Wild-type F_1 animals can then be interbred, and the polymorphism can be used to identify offspring homozygous for the 129-derived wild-type gene. These wild-type mice serve as good controls for homozygous mutant mice because both types of mice will have strain 129-derived chemokine genes linked to the gene of interest. Mice carrying the wild-type 129 locus and those carrying the targeted 129 locus can subsequently be backcrossed to the strain of mouse appropriate for the particular phenotype under study. This breeding strategy will eliminate most of the complications associated with both linked and unlinked genes.

Mice in which the wild-type 129 chemokine locus has been backcrossed to an inbred strain can also be used to determine whether strain-specific differences for any phentoype of interest exist in the same general region. Thus, because the only difference between the inbred wild-type mice and the congenic backcrossed mice will be at or linked to the chemokine locus, any differences in their phenotypes must be due to genes within that region. If there are no differences in phenotype between these two strains, then any strain-specific DNA sequence differences in or linked to that chemokine locus are irrelevant to that particular phenotype.

Even wild-type mice carrying 129 chemokine genes backcrossed onto a C57BL/6 background are not perfect controls for comparably backcrossed targeted mice because the two backcrossed strains will differ from one

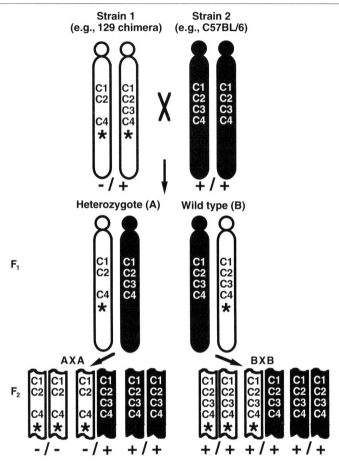

FIG. 5. Breeding strategies for F_1 mice. A hypothetical chemokine gene, C3, is deleted in one of the 129-derived ES cell chromosomes, which are shown in white; C57BL/6-derived chromosomes are shown in black. Note that the typical breeding strategy of using $+/-$ F_1 mice (A × A) generates F_2 mice of all three target gene genotypes: $-/-$, $+/-$, and $+/+$. However, the $-/-$ mice have strain 129-derived genes linked to the target locus, whereas the $+/+$ F_2 mice carry C57BL/6-derived linked genes. Control $+/+$ mice that carry 129-derived chemokine genes can be generated only by interbreeding $+/+$ F_1 heterozygotes (B × B). A polymorphism (shown here by an asterisk) in or near the chemokine locus is used to determine the strain (129 or C57BL/6) of wild-type loci. [Adapted from O. Smithies and N. Maeda, *Proc. Natl. Acad. Sci. U.S.A.* **92,** 5266 (1995).]

another in the exact points of crossover between the 129- and C57BL/6-derived chromosomes. In fact, there are no perfect genetic controls for gene-targeted mice after the first generation (F_1) unless the chimera is bred to a mouse of the strain from which the ES cells were derived. Embryonic stem cells have been derived from C57BL/6 mice,[23] and these and similar kinds of ES cell lines should eventually eliminate the complicated breeding schemes that have been required to generate good controls for gene targeted mice.

Analysis of Gene-Targeted Animals

Inflammation can be induced experimentally in mice by a wide variety of chemical and biological reagents. The type of leukocytes in the inflamed tissue depends on the nature of the proinflammatory stimulus. *In vitro*, chemokines are chemotactic for specific types of leukocytes, suggesting that expression of specific chemokines *in vivo* may recruit specific leukocyte subsets to sites of infection. This hypothesis can be tested using mice carrying disruptions of individual chemokines. For example, MIP-1α is chemotactic *in vitro* for monocytes and lymphocytes, cells that protect the animal from intracellular pathogens. These observations led us to experiments that demonstrated that MIP-1α $-/-$ mice have a reduced inflammatory response to coxsackie virus B3 (CVB3) and influenza virus.[7] Similar lines of reasoning can be used to design experiments using mice carrying disruptions of other chemokine genes. Thus, mice lacking members of the C-X-C chemokine subfamily could be tested for their response to extracellular pathogens, since these chemokines have been shown to be generally chemotactic *in vitro* for neutrophils.

Injection of blocking antibodies into mice has been useful for studying chemokine function *in vivo*. However, certain types of biological questions are not readily answered by this approach. For example, the cell types that express chemokines and are important to the inflammatory response cannot be easily identified using blocking antibodies because the function of the chemokine will be blocked regardless of its cellular source. However, it may be possible to identify these cells by cell transfer experiments between wild-type and gene-disrupted mice. For example, we have been studying this type of problem in MIP-1α $-/-$ mice, which are resistant to CVB3-mediated myocarditis. Wild-type and MIP-1α $-/-$ recipient mice have both been irradiated to deplete their bone marrow and have then been reconstituted with bone marrow cells from either wild-type or MIP-1α $-/-$ donor mice. We anticipate that analysis of the response of these mice to

[23] B. Ledermann and K. Burki, *Exp. Cell Res.* **197,** 254 (1991).

infection with CVB3 will determine the relative importance of hematopoietic and nonhematopoietic cell expression of MIP-1α to CVB3-induced myocarditis. Similarly, the importance to chemokine expression in other, more specific cell types such as T lymphocytes can be determined by isolating these cells from wild-type mice and transferring them into gene-disrupted recipients prior to inducing the inflammatory response.

Perspectives

Gene targeting provides a powerful genetic approach to analyze the function of chemokines *in vivo*. It is now possible to introduce into the mouse genome virtually any desired modification. Occasionally, no phenotypic differences are evident between mice carrying targeted genes and their wild-type littermates, but this does not not necessarily imply that the gene has no biological function. Thus, the mice that carry a disrupted chemokine gene may have a normal inflammatory response because of an increased expression of some other proinflammatory genes that compensates for the chemokine gene disruption. A search for alterations in the expression of other proinflammatory genes may then provide useful clues to the function of the targeted gene.[24]

Mice lacking more than one chemokine should be useful to study the extent of redundancy or synergy among various chemokines. The usual approach to generating mice carrying more than one disrupted gene is to interbreed mice carrying individual gene disruptions. Unfortunately, this approach is unsuitable for generating mice homozygous for more than one chemokine gene disruption because genes within a chemokine subfamily are so closely linked that combining the mutations by a crossover between the two targeted genes within the complex is virtually impossible. However, it is possible to first generate ES cells having gene disruptions in two closely linked genes and then to use these cells to derive mice. This can be accomplished by carrying out two sequential gene targeting experiments on the same ES cells using vectors carrying different selectable markers, such as *neo* and *hyg,* the gene encoding resistance to hygromycin.[25] In this type of experiment it must be established that the second (*hyg*) targeting event is on the same chromosome as was targeted in the first (*neo*) experiment, and not on the other copy of that chromosome. Thus, ES cells carrying the two mutations on the same chromosome should be identified by pulsed-field gel electrophoresis or by fluorescent *in situ* hybridization

[24] O. Smithies and N. Maeda, *Proc. Natl. Acad. Sci. U.S.A.* **92,** 5266 (1995).
[25] H. te Riele, E. R. Maandag, A. Clarke, M. Hooper, and A. Berns, *Nature* (*London*) **348,** 649 (1990).

(FISH) before proceeding. An alternative is to inject several independently derived double targeted ES cell lines into blastocysts and then to establish chromosomal locations of the two targeted genes in each ES cell line by analysis of the F_1 progeny of the chimeras generated from that cell line. Because the F_1 mice carry only one copy of each of the ES cell-derived chromosomes, ES cells that have both disruptions on the same chromosome will give rise to F_1 mice that have either both or neither targeted genes.

A different approach that allows the deletion of multiple chemokines within a subfamily is to generate a single large deletion in a locus that encompasses more than one gene. A 20-kb deletion has been achieved in ES cells using conventional targeting vectors,[26] but larger deletions have not been reported, presumably because targeting events of this sort are very infrequent. The plug-and-socket strategy may be a useful way of generating large deletions[27] because correctly targeted cells can be directly selected. A combination of the Cre/*lox* system and HPRT selection to generate large deletions has been described by Ramírez-Solis *et al.*[28] In this strategy, a *loxP* site together with an incomplete HPRT gene was inserted into one region of the *HoxB* locus using a conventional targeting vector. A second targeting vector carrying a different positive marker was then used to introduce a second *loxP* site together with a different incomplete but complementary HPRT into a different *Hox* gene 3–4 cM distant from the first insertion. Cre recombinase was then transiently expressed to catalyze recombination across the two *loxP* sites, thereby deleting the 3–4 cM of intervening DNA. The recombination event also restored HPRT function so that the targeting event was directly selectable. A similar strategy might be suitable to generate large deletions in a chemokine locus.

Inducible gene targeting[29] in mature mice provides a valuable alternative to conventional gene targeting. Such mice are generated in two steps. In the first, conventional gene targeting in ES cells is used to insert *loxP* sites at both ends of the gene(s) of interest. Mice carrying this modified gene are then bred to animals carrying a Cre transgene whose expression can be induced by agents such as α/β interferon[29] or tetracycline.[30] The offspring of such a breeding scheme are then induced to delete the target gene by recombination across the *loxP* sites.

[26] H. Zhang, P. Hastey, and A. Bradley, *Mol. Cell. Biol.* **14**, 2404 (1994).

[27] B. Yang, S. Kirby, J. Lewis, P. Detloff, N. Maeda, and O. Smithies, *Proc. Natl. Acad. Sci. U.S.A.* **92**, 11608 (1995).

[28] R. Ramírez-Solis, P. Liu, and A. Bradley, *Nature (London)* **378**, 720 (1995).

[29] R. Kuhn, F. Schwenk, M. Aguet, and K. Rajewsky, *Science* **269**, 1427 (1995).

[30] P. A. Furth, L. St. Onge, H. Böger, P. Gruss, M. Gossen, A. Kistner, H. Bujard, and L. Henninghausen, *Proc. Natl. Acad. Sci. U.S.A.* **91**, 9302 (1994).

Tissue-specific targeting may be a useful approach in cases where a gene has different functions in different cell types. Gu *et al.* have used the Cre/*lox* system to restrict targeting of a DNA polymerase gene to T lymphocytes by first generating mice in which the *loxP* sites flank the polymerase gene and then breeding these mice to animals expressing the Cre recombinase gene under control of the T-cell-specific *lck* promoter.[31] Only a fraction of the cells examined in this prototype experiment carried the expected targeting event, but the efficiency of the technique will likely be improved in the future. It may also be possible to devise approaches that allow gene targeting to be induced in a specific cell type at a chosen time. As these new targeting technologies develop, we will be able to use them to address very specific questions about chemokine function *in vivo*. Meanwhile, the more conventional targeting strategies, such as gene disruption and gene modification, can be expected to continue to reveal both interesting and surprising aspects of chemokine function *in vivo*.

Acknowledgments

This work was supported by National Institutes of Health Grant GM 20069 to Oliver Smithies. The author thanks Suzanne Kirby, John Krege, and Oliver Smithies for critical reading of the manuscript.

[31] H. Gu, J. D. Marth, P. C. Orban, H. Mossman, and K. Rajewsky, *Science* **265,** 103 (1994).

[14] Lymphotactin: A New Class of Chemokine

By Joseph A. Hedrick and Albert Zlotnik

Introduction

The chemokines have classically been divided into two families depending on the relative positions of the first two of the four invariant cysteines found in these proteins. Those chemokines whose first two cysteines are separated by a single variable amino acid are known as the C-X-C or α-chemokines. Those chemokines whose first two cysteines are directly adjacent to one another are designated the C-C or β-chemokines. Interestingly, the majority of the genes encoding the β-chemokines cluster to chromosome 17 in humans (chromosome 11 in mouse), whereas the genes encoding the α-chemokines generally cluster to chromosome 4 (chromosome 5 in mouse). These two groups can also be distinguished on the basis of the general cell types that are attracted by them, with monocytes responding to the C-C chemokines and neutrophils responding to the C-X-C chemokines.[1–4] Members of both chemokine classes have been variously observed to attract lymphocytes.[1–4]

The cloning of a novel chemokine designated lymphotactin (Lptn) has prompted the addition of a third chemokine subfamily, the C or γ-chemokines.[5,6] This unique chemokine possesses only two of the four cysteines present in other chemokines (cysteines 2 and 4). Furthermore, the gene encoding Lptn localizes to chromosome 1 in both human and mouse.[5–8] Finally, Lptn attracts neither monocytes nor neutrophils, but is instead specific for lymphocytes.[5,8] These features clearly distinguish lymphotactin from all other chemokines. In this chapter, we review what is already known about lymphotactin; its genetics and characterization of the protein, cell, and tissue expression; and its function in lymphocyte biology. Finally, we briefly discuss the therapeutic potential of lymphotactin and provide some discussion of what the future might hold for this interesting molecule.

Genetics and Protein Structure

The gene encoding murine lymphotactin (mLptn) has been localized to the distal region of mouse chromosome 1 by interspecific backcross analysis.[5] Genes linked to *lptn* include Fas ligand, antithrombin 3, selection endothelium, and octamer-binding transcription factor 1.[5] The gene for human lymphotactin (hLptn) has been generally localized to human chromosome 1 by somatic hybrid analysis[6,8] and specifically to chromosome 1q23 by fluorescence *in situ* hybridization.[7]

The Lptn gene also seems to be present in a variety of other mammalian species. Zoo blot analysis using mLptn cDNA as a probe reveals a signal in monkey, dog, cow, rabbit, and rat DNAs, but not in chicken or yeast DNA.[6,8] Interestingly, although no signal for Lptn was detected in chicken,

[1] T. J. Schall, *in* "The Chemokines" (A. Thomson, ed.), p. 419. Academic Press, New York, 1994.

[2] M. Baggiolini, B. Dewald, and B. Moser, *Adv. Immunol.* **55,** 97 (1994).

[3] T. J. Schall and K. B. Bacon, *Curr. Opin. Immunol.* **6,** 865 (1994).

[4] J. A. Hedrick and A. Zlotnik, *Curr. Opin. Immunol.* **8,** 343 (1996).

[5] G. S. Kelner, J. Kennedy, K. B. Bacon, S. Kleyensteuber, D. A. Largaespada, N. A. Jenkins, N. G. Copeland, J. F. Bazan, K. W. Moore, T. J. Schall, and A. Zlotnik, *Science* **266,** 1395 (1994).

[6] T. Yoshida, T. Imai, M. Kakizaki, M. Nishimura, and O. Yoshie, *FEBS Lett.* **360,** 155 (1995).

[7] S. Müller, B. Dorner, U. Korthäuer, H. W. Mages, M. D'Apuzzo, G. Senger, and R. A. Kroczek, *Eur. J. Immunol.* **25,** 1744 (1995).

[8] J. Kennedy, G. S. Kelner, S. Kleyensteuber, T. J. Schall, M. C. Weiss, H. Yssel, P. V. Schneider, B. G. Cocks, K. B. Bacon, and A. Zlotnik, *J. Immunol.* **155,** 203 (1995).

a cDNA encoding what appears to be a chicken version of Lptn (cLptn) has been isolated in our laboratory.[9] This molecule is similar to both mLptn and hLptn in that it lacks cysteines 1 and 3, but it is otherwise only 33% identical to mLptn at the amino acid level (Fig. 1A). Utilizing the cLptn cDNA to reprobe the Zoo blot revealed a detectable signal only in chicken and perhaps yeast.[9] This result indicates that although the Lptn protein structure has been maintained through evolution, the nucleic acid sequence has diverged considerably.

Comparison of the hLptn and mLptn amino acid sequences shows 58% overall sequence identity (Fig. 1A), with many of the differences clustered in the region immediately preceding the extended tail. Computer predictions of the secondary structure for both hLptn and mLptn suggest the presence of typical chemokine structural motifs, including a series of β sheets lying between the two cysteine residues, which are presumably involved in a disulfide bridge, and a carboxyl-terminal helix (Fig. 1B). As previously mentioned, both hLptn and mLptn have an unusually long carboxyl terminus. This feature is not unique among murine chemokines, as mMCP-1 (murine monocyte chemotactic protein-1) also has an extended carboxyl terminus (although its human homolog does not), but it is the first to be identified among the human chemokines. In fact, the amino acid sequences of hLptn and mLptn are quite similar in the last 25 amino acids, showing 68% identity in this region (Fig. 1A).[6,8] The fact that this long tail is conserved across species suggests that it may be important for the function of Lptn. The hLptn mRNA encodes a protein of 114 amino acids that includes a signal peptide with a predicted cleavage site between Gly-21 and Val-22 (Fig. 1B). Cleavage of the hLptn precursor at this site would result in a mature protein of 93 amino acids with a predicted molecular mass of 10.2 kD. Müller et al.[7] have proposed two alternative forms of hLptn, one with Gly-21 as its N-terminal amino acid and another with Gly-23 in this position. However, neither of these two hLptn variants was found to be active in chemotactic assays. In addition, Yoshida et al.[10] have reported that N-terminal amino acid sequencing of purified hLptn secreted from insect cells shows the predicted cleavage; thus, it seems very likely that valine is the N-terminal amino acid. The amino-terminal sequence of native hLptn was recently shown to be Val-21, Gly-22.[10a]

The genes for the C-X-C chemokines are generally organized into either

[9] D. Rossi, A. Zlotnik, and J. Kennedy, in "Lymphotactin: A Third Type of Chemokine" (B. Aggarwal and J. Gutterman, eds.), in press.
[10] T. Yoshida, T. Imai, S. Takagi, M. Nishimura, I. Ishikawa, T. Yaoi, and O. Yoshie, FEBS Lett. 395, 82 (1996).
[10a] B. Dorner, S. Müller, F. Entschladen, J. M. Schröder, P. Franke, R. Kraft, P. Friedl, I. Clark-Lewis, and R. A. Kroczek, J. Biol. Chem. 272, 8817 (1997).

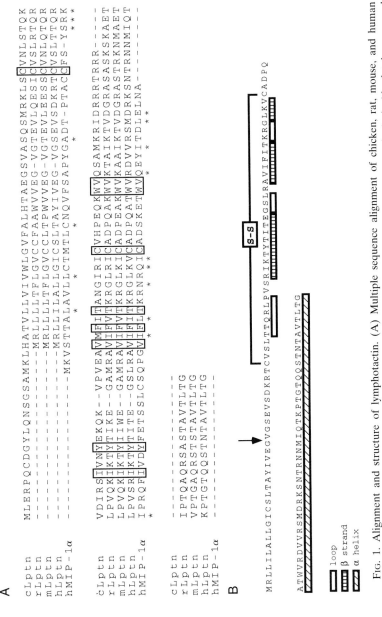

Fig. 1. Alignment and structure of lymphotactin. (A) Multiple sequence alignment of chicken, rat, mouse, and human lymphotactin compared with the human β-chemokine MIP-1α (macrophage inflammatory protein 1-α). Absolutely conserved residues are boxed, whereas residues of similar characteristic are denoted with asterisks. The multiple sequence alignment was generated using the ClustalW program [J. D. Thompson, D. G. Higgins, and T. J. Gibson. *Nucleic Acids Res.* **22**, 4673 (1994)]. (B) Amino acid sequence of human lymphotactin with various structural features as noted in the figure key. The arrow indicates the signal peptide cleavage site, and the disulfide bond between the cysteine residues is shown. Secondary structure is as predicted by PredictProtein mail server [B. Rost, *Methods Enzymol.* **266**, 525 (1996)] and is based on the known structure of other chemokines (overall three-state accuracy of 72.1%).

A

```
tttcctgtgg agtcagcagg gccacggggc tgatttgtgg ttgtcgacgc cttcagtccc
aggcagcagg tgggatggac caccgtgacc acaccacctc atttcctttg cccagttatc
acactgagcc atgacacgga ctggcttcag gatggaagat ccgaccttag cttcacaact
gtataaaaaa aaaaaaagac aactacactt catctcatgc tcacatttag tttcaatgac
acattaacca cacagggaga agaaaaggcc acaaggtctc atgttggtgt gaacaaggca
aaggaacttg tgtttaaatt taaaaaaaaa aaagagagag agagacagaa aatatcatca
tcattgcaaa gactttccgt gattccacac ccacccccgaa agccctcgc aacccaggaa
gtgtgctgac cattgagggt aataaaaggg gctcctgggg agtctgctcc acattcttct
tgcacagccc agcaagacct cagccatgag acttctcctc ctgactttcc tgggagtctg
                          M   R   L   L    L   T   F   L    G   V   C
ctgcctcacc ccatgggttg tggaag ..(intron 1).. gtgtggggac tgaagtccta
 C   L   T    P   W   V   V    E   G                 V   G   T    E   V   L
gaagagagta gctgtgtgaa cttacaaacc cagcggctgc cagttcaaaa aatcaagacc
E   E   S   S    C   V   N    L   Q   T    Q   R   L   P    V   Q   K    I   K   T
tatatcatct gggaggggggc catgagagct gtaat  ..(intron 2).. ttttgtcacc
Y   I   I   W    E   G   A    M   R   A    V   I                   F   V   T

aaacgaggac taaaaatttg tgctgatcca gaagccaaat gggtgaaagc agcgatcaag
K   R   G   L    K   I   C    A   D   P    E   A   K   W    V   K   A    A   I   K
actgtggatg gcagggccag taccagaaag aacatggctg aaactgttcc cacaggagcc
T   V   D    G   R   A   S    T   R   K    N   M   A   E    T   V   P    T   G   A
cagaggtcca ccagcacagc gataaccctg actgggtaac agcctccagg acaatgtttc
Q   R   S    T   S   T   A    I   T   L    T   G   Stop
ctcactcgtt aagcagctca tctcagttcc caaacccatt gcacaaatac ttatttttat
ttttaacgac attcacattc atttcaaatg ttataagtaa taaatattta ttattgatga
                                     poly A signal
tggccctata tttgatttat tatataaagg gaaaggtttt ctttatcatt tgttttattt
ctctctcatt gatctctctg tctcctctct ctctccttgt ctctctctgt atgtctgtct
ctgtctctgt ctctctctct ctctgtctct ctctctcgga cctcagatc
```

B

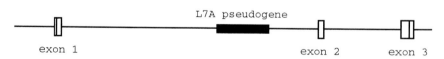

SCM-1α

exon 1   L7A pseudogene   exon 2   exon 3

SCM-1β

exon 1   deletion   L7A pseudogene   exon 2   exon 3

a three exon/three intron pattern [e.g., interleukin-8 (IL-8) and growth related oncogene (GRO)] or a three exon/two intron pattern [e.g., platelet factor 4 (PF4) and connective tissue activating peptide-III (CTAP-III)/ neutrophil-activating peptide-2 (NAP-2)].[2] The genes for the C-C chemokines are similarly organized into a three exon/two intron pattern.[2] The mouse and human Lptn genes are organized along a similar pattern of three exons and two introns (Fig. 2).[10] Interestingly, a second gene for mLptn has been reported that also localizes to human chromosome 1q23.[10] This gene, designated SCM-1β, is identical to the known gene SCM-1α, but it has a deletion in the first intron (Fig. 2B) and encodes a slightly different form of Lptn with two amino acid differences in the amino terminus, Asp-Lys in SCM-1α and His-Arg in SCM-1β. Transcripts for both genes were observed by reverse transcription-polymerase chain reaction (RT-PCR), suggesting that both genes express protein; however, all of the original reports for human lymphotactin have identical sequences[6-8] that correspond to the transcript produced from SCM-1α. Furthermore, sequence analysis of a number of hLptn clones from a variety of cell sources shows only transcripts from the SCM-1α gene to be present. Thus, the significance of the SCM-1β gene and whether, in fact, any protein is produced which originates from it remain to be demonstrated.

Cell and Tissue Expression

The expression of mLptn message appears to be restricted to activated CD8[+] thymocytes, that is, CD8[+] peripheral blood T cells.[5,7,8] It is also of particular interest to note that two specialized T-cell populations, the CD4[+]NK1.1[+] T cells and the TCR$\alpha\beta$[+]CD4[-]8[-] T cells, express mLptn mRNA.[5,8] These two populations have similar characteristics in that they are both biased toward V$_\beta$8.2 in their T-cell receptor (TCR) repertoire, both produce large amounts of interleukin-4 (IL-4) (but are not of the T helper type 2 subclass), and are both class I major histocompatibility complex (MHC)-restricted. Northern blot analysis of murine tissue reveals no expression of mLptn in heart, brain, lung, liver, kidney, testis, or skeletal muscle.[5] Low-level expression can be detected in spleen and, more rarely, in thymus.[5] This is consistent with the hypothesis that expression of Lptn

FIG. 2. Gene structure of lymphotactin. (A) Nucleic acid sequence of the mLptn exons (based on GenBank accession numbers U28491, U28492, and U28493) along with the appropriate amino acid translation. The start and stop codons are underlined as is the single polyadenylation signal sequence. Message destabilization sequences are indicated in boldface type. (B) Overall structure of the two human Lptn genes, SCM-1α and SCM-1β. Exons are shown as open boxes, introns are represented by lines, and the L7A pseudogene is shown as a black box. The deletion present in SCM-1β is indicated. Figure is not drawn to scale.

is dependent on activation.[5–8] The expression of Lptn in human tissue seems more widespread, although still primarily restricted to lymphoid organs. Northern analysis of multiple tissue blots shows expression of Lptn in resting thymus, spleen, peripheral blood lymphocytes (PBL), and small intestine, with a low level of expression detected in lung, colon, ovary, and testis.[6–8] The expression of hLptn by resting PBL, however, was not observed in Northern blots prepared either by our laboratory or by other groups.[6–8] In fact, the expression of hLptn in these blots was observed only on activation and was primarily found in the CD8$^+$ subset of cells.[7,8] Although low levels of Lptn mRNA could be detected in activated CD4$^+$ cells, it is very likely that this signal arises from the human equivalent of murine CD4$^+$NK1.1$^+$ T cells.[7,8] Given these results it seems likely that the detection of Lptn in the PBL mRNA present on the multiple tissue blots may be due to the particular source of mRNA. Likewise the expression of hLptn in spleen, small intestine, lung, and colon may be attributable to the presence of activated lymphocytes in these tissues. Müller *et al.* similarly found expression of Lptn only in activated human CD8$^+$ cells, whereas activated human B cells, monocytes, and fibroblasts failed to express hLptn mRNA.[7]

The expression of Lptn mRNA by murine intraepithelial $\gamma\delta$ T cells ($\gamma\delta$ IEL) has been reported.[11] These cells are found in epithelial tissues and normally reside in close association with epithelial cells. The $\gamma\delta$ IEL are distinguishable from the $\gamma\delta$ T cells found in lymphoid organs by their expression of restricted tissue-specific TCR. It has been proposed that the $\gamma\delta$ IEL may play a role in the early inflammatory response to damage or disease and that production of chemotactic factors might in turn be involved in the response of these cells. Indeed, a subset of the $\gamma\delta$ IEL, the dendritic epidermal T cells (DETC) have been previously shown to produce MIP-1α.[12] A report by Boismenu *et al.* shows that these cells can indeed make a variety of chemokines and that, in fact, Lptn mRNA was the most abundant chemokine mRNA detected.[11] They further demonstrated the production of Lptn protein by the DETC line 7–17 and that the supernatant of these cells possessed T cell chemotactic activity that was neutralized by anti-mLptn antibody. As is true of $\alpha\beta$ T cells, the expression of Lptn mRNA by murine DETC and other $\gamma\delta$ IEL was also found to be dependent on activation.[11]

Findings in our own laboratory have extended the range of lymphocytes known to express Lptn to include natural killer (NK) cells.[13] Like the $\gamma\delta$

[11] R. Boismenu, L. Feng, Y. Xia, J. Chang, and W. Havran, *J. Immunol.* **157**, 985 (1996).
[12] H. Matsue, P. Cruz, P. Bergstresser, and A. Takashima, *J. Invest. Dermatol.* **101**, 537 (1993).
[13] J. Hedrick, V. Saylor, D. Figueroa, L. Mizoue, Y. Xu, S. Menon, J. Abrams, T. Handel, and A. Zlotnik, *J. Immunol.* **158**, 1533 (1997).

IEL, NK cells are believed to play an important role in early inflammatory responses, particularly to viral and bacterial pathogens. Consistent with this idea and with the hypothesis that Lptn is important in the early phases of an immune response, we have found that human NK cell clones express Lptn mRNA and that activated, but not resting, murine NK cells produce Lptn protein. Furthermore, protein could be observed within 2 hr of NK cell activation by IL-2.

Analysis of the full-length hLptn mRNA shows multiple polyadenylation signal sequences, raising the possibility of alternative splicing in the mRNA.[6–8] Indeed, examination of hLptn Northern blots shows a broad band of transcripts ranging in size from 0.5 kb to approximately 0.8 kb, and rare messages of more than 1.0 kb have been isolated.[6–8] The significance of these various transcripts has yet to be elucidated. In contrast to hLptn, mLptn seems to be encoded by only a single transcript of about 0.5 kb.[14] Consistent with this finding is the fact that the mLptn mRNA contains only a single polyadenylation signal sequence (Fig. 2A).

In summary, the expression of Lptn mRNA thus appears to be limited to activated lymphocytes and more specifically to pro-T cells, class I MHC-restricted T cells (CD8$^+$ T cells, CD4$^+$NK1.1$^+$ T cells, and TCR$\alpha\beta^+$CD4$^-$8$^-$ T cells), NK cells, and intraepithelial (but not lymphoid) $\gamma\delta$ T cells. Furthermore, expression of Lptn mRNA is rapidly up-regulated on activation, suggesting that Lptn plays a central role in the early phase of an inflammatory or immune response.

Biological Function

The hallmark characteristic of chemokines is, of course, their ability to direct the migration of leukocytes to the sites of inflammation. Most other chemokines attract either monocytes or neutrophils, depending generally on whether they are C-C or C-X-C chemokines.[1–4] Lymphotactin is once again unusual among chemokines in that it attracts neither neutrophils nor monocytes, but is instead a potent chemoattractant for lymphocytes.[5,8] Human CD4$^+$ and CD8$^+$ PBL are both strongly attracted by hLptn.[8] The same is also true for murine CD4$^+$ and CD8$^+$ splenocytes and thymocytes.[5] In addition, we now also know that both human[13,15] and murine NK cells respond to Lptn.[13] These activities have been characterized *in vitro* using the standard Boyden chamber microchemotaxis assay and *in vivo*, in the case of mLptn, through intraperitoneal injection of recombinant mLptn. Both T cells and NK cells were found to be responsive to nanomolar

[14] G. Kelner and A. Zlotnik, *J. Leukocyte Biol.* **57**, 778 (1995).
[15] G. Bianchi, S. Sozzani, A. Zlotnik, A. Mantovani, and P. Allavena, *Eur. J. Immunol.* **26**, 3238 (1996).

concentrations of Lptn,[5,8,13,15] although there is some indication that CD8[+] T cells may be responsive to doses as low as 10^{-10} M.[5,8] Although, as discussed earlier, $\gamma\delta$ IEL have been shown to produce Lptn, it is not presently known whether this or any $\gamma\delta$ T cell population responds chemotactically to Lptn.

The activity of chemokines is well known to be mediated through their interaction with a variety of G-protein-coupled, seven-transmembrane domain receptors. One consequence of this interaction is the induction of an intracellular calcium flux. Although a receptor for Lptn has not yet been identified, hLptn and mLptn have been shown to induce an intracellular calcium flux in PBL and CD4[+]-depleted thymocytes, respectively.[5,8] Furthermore, it appears that the Lptn receptor is distinct from the receptors for IL-8 and C-C chemokines, as no cross-desensitization was observed when Lptn was added prior to the introduction of these chemokines.[16] In addition, Lptn is unable to compete with a variety of C-C and C-X-C chemokines for binding to the Duffy antigen promiscuous chemokine receptor.[16]

Therapeutic Application

The potential application of chemokines as therapeutics is currently an area of extreme interest and activity, much of it brought about by findings regarding the role of chemokines and chemokine receptors in the binding of human immunodeficiency virus (HIV) to its target cells. Although lymphotactin has not been shown to have any role in HIV pathogenesis, human T-lymphotropic virus type 1 (HTLV-1)-positive T-cell lines have been shown to express a number of chemokines including Lptn.[17] It was further demonstrated that the retroviral Tax protein was responsible for induction of Lptn as well as several other chemokines.[17] What role Lptn and other chemokines play in HTLV-1 pathogenesis is unclear, but this is clearly an area that deserves further scrutiny.

One of the more promising therapeutic applications for chemokines is their use as antitumor agents. Several chemokines have been used with some success in various animal antitumor models,[18,19] but lymphotactin is a particularly attractive chemokine to use in such applications because of its ability to specifically attract both effector cells (NK cells and CD8[+] T cells) and helper cells (CD4[+] T cells). Indeed, lymphotactin was shown

[16] M. C. Szabo, K. S. Soo, A. Zlotnik, and T. J. Schall, J. Biol. Chem. **270**, 25348 (1995).
[17] M. Baba, T. Imai, T. Yoshida, and O. Yoshie, Int. J. Cancer **66**, 124 (1996).
[18] A. D. Luster and P. Leder, J. Exp. Med. **178**, 1057 (1993).
[19] J. Laning, H. Kawasaki, E. Tanaka, Y. Luo, and M. E. Dorf, J. Immunol. **153**, 4625 (1994).

to synergize with IL-2 in retarding the growth of a preexisting tumor.[20] In this model, Lptn alone was observed to cause the infiltration of CD4+ cells into the tumors, and the use of both Lptn and IL-2 caused the infiltration of tumors by both CD4+ and CD8+ cells. Furthermore, depletion of either CD4+ cells or CD8+ cells abrogated the protection provided by the combination of IL-2 and Lptn.[20] Lymphotactin alone had very little antitumor activity, despite its ability to cause tumor infiltration by CD4+ cells. This is consistent with the fact that, other than promoting chemotaxis, Lptn has not been observed to have any additional effects on lymphocytes.[13,21] In fact, this focused function may be one of the more attractive aspects of lymphotactin used as a therapeutic agent, suggesting that lymphotactin will have few if any side effects.

Conclusion

Murine lymphotactin remains the sole example of the γ-chemokine family. Homologs have been cloned from human, rat, and chicken, but no additional family members have been identified. The original activity ascribed to lymphotactin, chemotaxis, remains its only known function, although the range of cells that produce and respond to it has been expanded to include NK cells and $\gamma\delta$ T cells. In addition, the *in vivo* efficacy of lymphotactin as a lymphocyte attractant has been demonstrated. What we know about Lptn at this point suggests that it is made early in an immune response and that it attracts both effector and helper cells. This suggests that Lptn has an important role in the early phases of an immune response, perhaps representing the immunological equivalent of an emergency "911" call, after which other chemokines are produced to recruit other cell populations.

Despite what has been learned about Lptn, many questions remain to be answered. For example, what factors regulate the expression of Lptn? Is Lptn indeed the only γ-chemokine? What molecule serves as the Lptn receptor? Does Lptn have additional, nonchemotactic biological functions? Among the lymphocyte populations that have been shown to respond to Lptn, are some subpopulations more or less responsive? How does the presence of other chemokines or cytokines affect the activity of Lptn? How does Lptn affect the activity of other chemokines or cytokines? These are questions that will not doubt be addressed in the near future, to be the subject of a future review.

[20] D. Dilloo, K. Bacon, W. Holden, W. Zhong, S. Burdach, A. Zlotnik, and M. Brenner, *Nat. Med.* **2,** 1090 (1996).
[21] J. Hedrick, unpublished observations (1996).

[15] Isolation and Purification of Neutrophil-Activating Peptide-4: A Chemokine Missing Two Cysteines

By JENS-MICHAEL SCHRÖDER

Introduction

Platelets play a critical role in hemostasis and are important in initiating tissue repair after injury. Wound healing is associated with the presence of mainly neutrophils in injured areas. Thus, it has been suggested that platelets play a role as inflammatory cells, which can release, either from α granules or through perturbation of membrane phospholipids, a wide range of growth factors and inflammatory mediators including chemokines.

Chemokines identified in platelets include the C-X-C chemokines platelet factor 4 (PF4) and platelet basic protein, which is the precursor of connective tissue activating peptide-III (CTAP-III), β-thromboglobulin, and neutrophil-activating peptide-2 (NAP-2), as well as the C-C chemokine RANTES (regulated and normal T-lymphocyte expressed and presumably secreted). Apart from the release of chemokines from platelets, a number of different chemokine mRNAs could be identified to be expressed in platelets.[1]

In previous investigations in our laboratory we raised the question of whether platelets are capable of releasing the C-X-C chemokine interleukin-8 (IL-8). During purification of neutrophil chemotactic activity seen in platelet lysates we identified a novel chemokine with close similarity to PF4. Because this activity was identified as a neutrophil attractant, this chemokine was termed neutrophil-activating peptide-4 (NAP-4).[2]

Purification of Platelets

As yet the only cellular source of NAP-4 appears to be platelets. To get homogeneous starting material it is important to isolate and purify platelets obtained from human blood.

Freshly drawn venous blood (100 ml) is immediately mixed with 10 ml of a prewarmed (37°) acidic dextran and anticoagulants [65 mM citric acid, 85 mM sodium citrate, 10 mM EDTA, and 20 g/liter dextran T-70 (Sigma, Munich, Germany)]. Care should be taken to hold the temperature at 37°

---

[1] C. Power, J. M. Clemetson, K. J. Clemetson, and T. N. C. Wells, *Cytokine* **7**, 479 (1995).

[2] J.-M. Schröder, M. Sticherling, N.-L. M. Persoon, and E. Christophers, *Biochem. Biophys. Res. Commun.* **172**, 898 (1990).

using a polystyrene box. The blood is centrifuged for 20 min at 37° and 500g. Then the plasma and the white buffy coat are removed and centrifuged again for 10 min at 37° and 200g. The remaining blood cell sediments can be used for isolating neutrophils as described elsewhere in this series. Platelet-containing supernatants are centrifuged at 1000g for 20 min at 37°. Sediments are washed two times with warm (37°) phosphate-buffered saline (PBS), pH 7.4 (without $CaCl_2$ and $MgCl_2$), containing 10 mM EDTA. The final recovery is usually 4–8 × 10^{10} platelets.

It is important to keep the temperature at 37° during purification of platelets. Otherwise drastic losses of platelet-derived chemokines can be seen. This release of material is not inhibited by adding anticoagulants such as EDTA. Platelets are either stored at −70° until use of total extracts or stimulated for release of granule constituents.

Preparation of Platelet Extracts

Frozen platelet sediments are thawed and suspended in 2 M NaCl and acidified to pH 3.0 with formic acid. The suspensions are frozen/thawed three times, clarified by centrifugation (1000g at 4°), and finally clarified by filtration through a 0.1-μm filter. The extract is stored at −70° until further use. Alternatively, freshly purified platelets are suspended in PBS containing 0.6 mM $CaCl_2$ and 0.8 mM $MgCl_2$ to a density of 1 × 10^{10} platelets/ml and subsequently are stimulated with thrombin (2 U/ml, Sigma) for 15 min at 37°. Platelets are then spun down (1000g at 4°), and the supernatant are stored at −70° until further use.

Bioassays for Neutrophil-Activating Peptide-4

Neutrophil-activating peptide-4 can be detected by the use of a neutrophil chemotaxis or degranulation assay. Neutrophils are purified from freshly drawn blood as described elsewhere in this series.[3] It is important to note that neutrophils isolated from buffy coats taken from local blood centers usually do not respond well to NAP-4 or other C-X-C chemokines. Therefore, neutrophils isolated from freshly taken blood should be used within 1–2 hr after final purification of the cells. For detection of neutrophil attractants the endogenous chemotaxis compound Boyden chamber assay system is used. Details are described elsewhere.[3] The lower parts of the Boyden chambers are filled with high-performance liquid chromatography (HPLC) fractions (5–40 μl) to be tested, which previously were lyophilized in a microtiter plate after adding 10 μl of PBS containing 0.1% (w/v) bovine

[3] J.-M. Schröder, *Methods Enzymol.* **288**, [17] (1997).

serum albumin (BSA) and subsequently are dissolved in 300 μl complete (c) PBS [PBS containing 0.6 mM CaCl$_2$, 0.8 mM MgCl$_2$, and 0.1% (w/v) BSA].

Alternatively, a degranulation assay system can be used for detection of NAP-4. Prewarmed (37°) neutrophils, suspended in cPBS, are preincubated with cytochalasin B [5 μg/ml from a stock (1 mg/ml) in dimethyl sulfoxide (DMSO, Sigma)] for 5 min at 37°. The HPLC fractions (treated as described above) are added to the wells of a microtiter plate and are prewarmed to 37° (5 min). Then 100 μl of a neutrophil suspension treated with cytochalasin B (10^7 cells/ml) is added to each well, and plates are incubated for 30 min at 37°. After centrifugation 100 μl of the supernatants taken from each well is incubated in another microtiter plate with 100 μl of β-glucuronidase substrate solution [10 mM p-nitrophenyl-β-D-glucuronide (Sigma) in 0.1 M sodium acetate, pH 4.0] for 18 hr. The enzymatic reaction is terminated by adding 200 μl of 0.4 M glycine buffer, pH 10,0, and finally p-nitrophenolate is determined at 405 nm.

As a total control, one well should contain 100 μl of 0.2% (v/v) Triton X-100 (Sigma) for lysis of cells. As a positive control use both 10^{-8} M N-formylmethionylleucylphenylalanine (Sigma) and 3 × 10^{-8} M IL-8 (Peprotech, London). It is important to use IL-8 as a positive control because we sometimes have observed neutrophil preparations that show very low or nearly absent enzyme-releasing activity when IL-8 is used as stimulus. Because cellular responses toward NAP-4 are less than those elicited by IL-8, such neutrophil preparations easily can lead to the failure to detect NAP-4 in HPLC fractions using this bioassay.

Purification of Neutrophil-Activating Peptide-4: Strategy I

Platelet lysates frozen below −70° are thawed, centrifuged at 1000g, and then applied to a Sephadex G-75 gel column (2.6 × 65 cm, Superdex, Pharmacia, Uppsala, Sweden), which previously was equilibrated with 0.1 M ammonium formate, pH 5.0. Proteins are eluted at 4° with equilibration buffer at a flow rate of 10 ml/hr. Twenty microliters of each fraction are diluted with 300 μl cPBS and then tested for polymorphonucleocytic (PMN)-chemotactic activity using the Boyden chamber technique described above. Fractions containing PMN-chemotactic activity (Fig. 1), which correspond to relative molecular masses between 5 and 40 kDa, are pooled and concentrated in Amicon chambers using YM2 filters (Amicon, Danvers, MA).

For further purification, pooled neutrophil-activating peptide-containing fractions are diafiltered against PBS, pH 7.4, and then applied to an anti-IL-8 affinity column, which is prepared as follows. Two milliliters

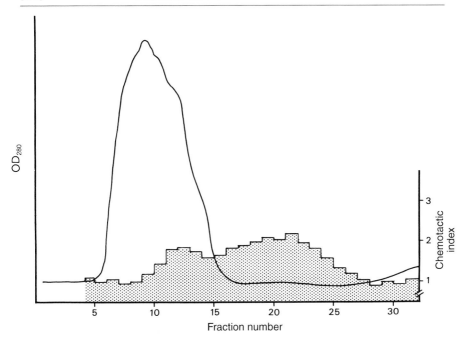

FIG. 1. Platelet lysates contain 10-kDa neutrophil chemotactic peptides. Platelet lysates were separated with a Sephadex G-75 size-exclusion column, and neutrophil chemotactic activity was measured in column fractions. Note the presence of activity in fractions 15–25 corresponding to relative molecular masses between 5 and 40 kDa.

preactivated Sepharose (Affi-Gel 10, Bio-Rad, Munich, Germany) is mixed with 2 ml purified monoclonal anti-IL-8 antibody 14E4,[4] dissolved in sodium hydrogen carbonate buffer, pH 8, and gently shaken for 30 min at room temperature. Thereafter, the gel is centrifuged and the nonbound antibodies containing supernatant is collected. The gel is then incubated with 2 ml of 1 M diethanolamine in water (pH 9.0) for a further 20 min. Antibody-containing beads are washed twice with PBS and finally purged into a syringe, which had previously been filled with glass wool. This column should be stored at 4° in PBS containing 0.1% (w/v) of sodium azide to prevent bacterial growth.

For isolation of NAP-4 the affinity column is depleted of sodium azide-containing PBS by exchange against PBS. Thereafter the pool from the Sephadex G-75 chromatography step containing neutrophil chemotactic activity (and NAP-4) is applied to the affinity column at a flow rate of 1

[4] M. Sticherling, J.-M. Schröder, and E. Christophers, *J. Immunol.* **143**, 1628 (1989).

ml/min. This procedure is repeated twice with the effluent. After washing the anti-IL-8 column with 3 ml PBS, proteins bound to the column are stripped (flow rate 1 ml/min) from the affinity column with 5 ml of 2 M NaCl in PBS, adjusted to pH 7.4. Finally remaining proteins are extracted with 5 ml of 0.2 M glycine buffer, pH 2.0 (flow rate 1 ml/min). The low affinity-bound material (which was eluted with 2 M NaCl) and high affinity-bound material (which was eluted at pH 2.0) should be further processed separately.

In our hands, the majority of NAP-4 is present in the 2 M NaCl extract. The acidic effluent of the affinity column apart from low amounts of NAP-4 often contains unrelated proteins, which are not present in the 2 M NaCl extract. Thus, it is necessary to separate these proteins by additional HPLC steps. The 2 M NaCl eluate is acidified to pH 2 and then diafiltered against 0.1% (v/v) trifluoroacetic acid (TFA) in water using an Amicon YM2 filter (cutoff of 2 kDa). Using these conditions nonspecific absorption to the ultrafilter is minimized. The material (2–5 ml) is then applied to a narrow-pore reversed-phase (RP-18) HPLC column [Nucleosil, 5 μm octadecylsilyl column with end-capping, 250 × 4.6 mm (Bischoff, Leonberg, Germany)], which was previously equilibrated with 0.1% (v/v) aqueous TFA containing 10% (v/v) acetonitrile. Proteins are eluted from the column by the use of a gradient of increasing concentrations of acetonitrile. A typical result is shown in Fig. 2. Using these conditions NAP-4 is the major protein eluting from the column, and it coelutes with neutrophil chemotactic activity.

Electrophoretic Analysis of Purified Neutrophil-Activating Peptide-4

For sodium dodecyl sulfate–polyacrylamide gel electrophoresis (SDS–PAGE) analysis of NAP-4, the Tricine system described by Schägger and von Jagow is used.[5] Proteins are analyzed in the presence of 8 M urea. Electrophoresis is performed at 30 V and room temperature overnight (~18 hr). Proteins in the gel are fixed with a mixture of water/2-propanol/acetic acid/glutaraldehyde in the ratio 170/100/31/1 (v/v/v/v) for 30 min at room temperature. The gel is then washed three times with water for 10 min followed by a 30-min incubation with aqueous silver nitrate solution (1 g AgNO$_3$/liter). Gels are rinsed for 1 min with water and then for 1 min with developer solution [10% saturated aqueous Na$_2$CO$_3$ solution containing 0.27% (v/v) of a saturated formaldehyde solution]. When metallic silver forms in the gel, it is necessary to change the developer solution. Thereafter

---

[5] H. Schägger and G. V. von Jagow, *Anal. Biochem.* **166,** 368 (1987).

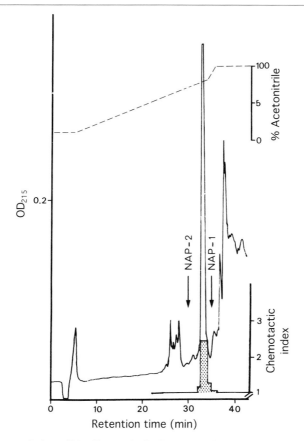

FIG. 2. Reversed-phase (RP-18) HPLC of a 2 M NaCl eluate obtained from a 14E4 anti-IL-8 affinity column. PMN chemotactic activity (shaded area) of 10-μl aliquots of each HPLC fraction was measured in the Boyden chamber system. Elution times of NAP-1 [IL-8]$_{72}$ (72 amino acids long, starting with Ser [NH$_2$-terminus]) and NAP-2 are indicated by arrows. [From J.-M. Schröder, M. Sticherling, N.-L. M. Persoon, E. Christophers, *Biochem. Biophys, Res. Commun.* **172,** 898 (1990).]

the gel is incubated with developer solution until bands are visible and background is not too high. Reaction is stopped by exchanging the developer solution with 3% acetic acid.

We perform SDS–PAGE analysis in the absence of a reducing reagent, because in the case of NAP-4 no change in mobility can be seen. A typical experiment is shown in Fig. 3. It is interesting to note that the electrophoretic mobility of NAP-4 in this system is slightly higher than that of PF4 and Gro α but much higher than that of [Ser-IL-8]$_{72}$. The electrophoretic mobility of NAP-4 indicates a relative molecular mass near 8000.[2]

M_r

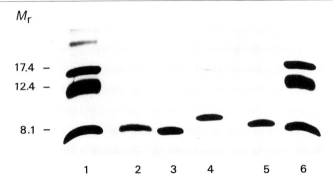

FIG. 3. Electrophoresis of NAP-4. NAP-4 as well as some other structurally related polypeptides were analyzed by SDS–PAGE using the Tricine/SDS system. Lanes 1 and 5 contained myoglobin (17.4 K), cytochrome c (12.4 K), as well as ubiquitin (8.1 K) as standards. Whereas in lane 2 70 ng platelet factor 4 was applied, lane 3 was loaded with 100 ng NAP-4. Lane 4 was loaded with 50 ng natural [Ser-IL-8]$_{72}$. In lane 5, 90 ng of authentic natural NAP-3/Gro α [J.-M. Schröder, N. Persoon, and E. Christophers, *J. Exp. Med.* **171**, 1091 (1990)] was loaded. [From J.-M. Schröder, M. Sticherling, N.-L. M. Persoon, and E. Christophers, *Biochem. Biophys. Res. Commun.* **172**, 898 (1990).]

This material (3.2 μg), which shows a single band on SDS–PAGE, has been N-terminally sequenced using an Applied Biosystems (Foster City, CA) gas phase sequencer with on-line HPLC analysis of the phenylthiohydantoin derivatives of amino acids. The following unambiguous sequence of 31 residues can be obtained: Glu-Ala-Glu-Gln-Leu-Gln-Asp-Leu-Gln-Val-Lys-Thr-Val-Lys-Gln-Val-Ser-Pro-Val-His-Ile-Thr-Ser-Leu-Glu-Val-Asp-Lys-Ala-Gly-Arg.[2] Data search using the Micro Genie Protein Data Bank (Beckman Instruments, Columbia, MD) revealed that NAP-4 is a unique protein. However, as shown in Fig. 4, NAP-4 shows strong (71%) homology with human platelet factor 4 as well as some homology to tumor necrosis factor-α starting at residue 21. The most striking phenomenon is the lack in NAP-4 of two cysteines, which are known to be present in all other chemokines. In all Edman degradation steps, sufficient amounts of the phenylthiohydantoin derivatives of amino acids can be obtained. There is no evidence for the absence of an Edman degradation product within the first 31 residues, which can be interpreted either as the absence of a cysteine (which does not form a stable thiohydantoin) or as the absence of derivatized (glycosylated) amino acids.

Measurement of Neutrophil-Activating Peptide-4 Biological Activity

Fractions from HPLC containing NAP-4 express chemotactic activity for neutrophils *in vitro*. Dose–response studies reveal half-maximum re-

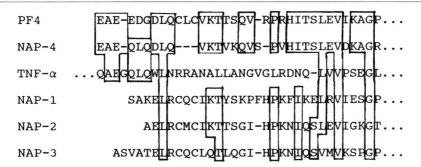

FIG. 4. Sequence alignment of NAP-4 with other closely related polypeptides. Sequences of PF4 (platelet factor 4) [T. F. Deuel, P. S. Keim, M. Farmer, and R. L. Heinrikson, *Proc. Natl. Acad. Sci. U.S.A.* **74**, 2256 (1977)], TNF-α (tumor necrosis factor-α) [D. Pennica, G. E. Nedwing, J. S. Hayflick, P. H. Seeburg, R. Derynck, M. A. Palaldino, W. J. Kehr, B. B. Aggarwal, and D. V. Goeddel, *Nature* (*London*) **312**, 724 (1984)], NAP-1 (IL-8) [T. Yoshimura, K. Matsushima, S. Tanaka, E. A. Robinson, E. Appella, J. J. Oppenheim, and E. J. Leonard, *Biochem. Biophys, Res. Commun,* **149**, 755 (1987)], NAP-2 [A. Walz and M. Baggiolini, *Biochem. Biophys, Res. Commun.* **159**, 969 (1989)] and NAP-3 (Gro α) [A. Richmond, E. Balentien, H. G. Thomas, G. Flaggs, D. E. Barton, J. Spiess, R. Bardoni, U. Francke, and R. Derynck, *EMBO J.* **7**, 2025 (1988)] are shown. The single letter code for amino acids is used. Gaps are indicated by dashes (–). Residues conserved in NAP-4 and the other cytokines are boxed with a solid line. [From J.-M. Schröder, M. Sticherlin, N.-L. M. Persoon, and E. Christophers, *Biochem. Biophys. Res. Commun.* **172**, 898 (1990).]

sponses at a concentration near 300 ng/ml, which is nearly 100-fold higher than necessary for [Ser-IL-8]$_{72}$ (Fig. 5). The chemotactic index (percentage of input migrating cells) is less than that of IL-8 and comparable to that of NAP-2 or Gro α. This raises the question of whether the biological activity of NAP-4 originates from traces of PMN chemotactic impurities. This hypothesis is rather unlikely because known PMN chemotactic factors can be easily separated by reversed-phase HPLC (see Fig. 2) and, using a different strategy to purify NAP-4 (see below), the same specific activity can be obtained. Furthermore, IL-8 is absent in purified platelet preparations, and NAP-2 is only formed by truncation of the major platelet products CTAP-III and β-thromboglobulin with proteases from PMN or monocytes, which are absent in the platelet preparations we use for isolation of NAP-4.

The purified NAP-4 also shows some myeloperoxidase-releasing activity in PMN, which is dose dependent (Fig. 5). The low specific activity, however, does not allow the use of PMN enzyme release as a detection system for NAP-4 in HPLC fractions. Even the dose–response curve of PMN chemotactic activity shows that high protein amounts are necessary to detect NAP-4 via a bioassay.

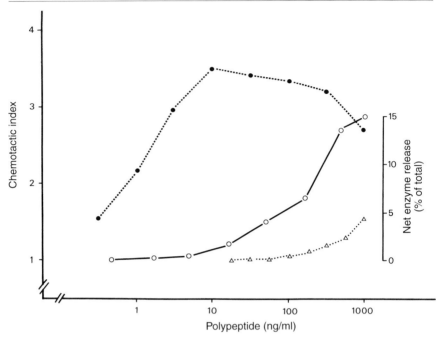

FIG. 5. PMN-activating properties of NAP-4 preparations. The PMN chemotactic activity (○) as well as the degranulating activity (myeloperoxidase release) (△) of NAP-4 preparations were determined using established assays. Chemotactic activity is expressed as the chemotactic index, which represents the quotient of the number of migrating cells under stimulatory conditions divided by the number of migrating cells migrating toward the buffer control. Degranulation is expressed as the percentage of release by a total control [prepared by lysis of PMN with 0.1% (w/v) hexadecyltrimethylammonium bromide]. The means of three duplicate experiments are shown. For comparison the dose–response curve of [Ser-IL-8]$_{72}$-released PMN chemotaxis (●) is included. [From J.-M. Schröder, M. Sticherling, N.-L. M. Persoon, and E. Christophers, *Biochem. Biophys. Res. Commun.* **172,** 898 (1990).]

In the case of affinity purification by the use of the cross-reacting anti-IL-8 antibody 14E4, concentrations of NAP-4 are usually high enough to detect NAP-4 via PMN chemotactic activity. When a different strategy is used to purify NAP-4 (see below), it is essential to use sufficient material for detection of NAP-4 bioactivity in HPLC fractions. Usually up to 10% of each HPLC fraction needs to be used for testing when NAP-4 has to be purified from extracts of platelets obtained from 200 ml blood.

Purification of Neutrophil-Activating Peptide-4: Strategy II

The original aim of our study was to investigate whether platelets contain biologically active IL-8. By the use of a simple anti-IL-8 affinity chromatography step we were able to isolate NAP-4 instead of IL-8, when platelet extracts were analyzed. Two conclusions are drawn from this finding. First, platelets do not contain IL-8, and, second, the 14E4 monoclonal anti-IL-8 antibody, in addition to IL-8, recognizes a novel chemokine. Because we obtained NAP-4 only by chance, we describe the purification of NAP-4 without any affinity chromatography in a more classic manner using HPLC techniques.

Neutrophil chemotactic activity-containing fractions of Sephadex G-75 gel chromatography of platelet lysates obtained from the platelets isolated from 200 ml blood (see Fig. 1) are acidified to pH 2–3, concentrated in Amicon chambers (YM2 filters, Amicon), and diafiltered against 0.1% TFA. Samples then are applied to a preparative wide-pore reversed-phase (RP-8) HPLC column (C_8 Nucleosil with end-capping, 250 × 12.6 mm, 7-μm particles; Macherey and Nagel, Düren, Germany), previously equilibrated with water/acetonitrile/TFA (90:10:0.1, v/v/v). Proteins are eluted with an increasing gradient of acetonitrile containing 0.1% TFA at a flow rate of 3 ml/min.

A typical example is shown in Fig. 6. For testing of PMN chemotactic activity, aliquots of fractions (50–100 μl) are mixed with 10 μl cPBS and then lyophilized in a microtiter plate. Owing to the use of volatile solvents, testing of high volumes of fractions is possible without any interference with the assays. Fractions (10–30 μl) can also be tested for the presence of 14E4 (anti-IL-8) immunoreactivity by the use of a solid-phase enzyme-linked immunosorbent assay (ELISA) as described. The 14E4 immunoreactivity usually occurs in separate fractions (Fig. 6), which is the result of a specific cross-reactivity of the 14E4 anti-IL-8 antibody. Using different monoclonal anti-IL-8 antibodies, we see either no immunoreactivity in these HPLC fractions or, more often, cross-reactivity with CTAP-III and its fragments, which elute early from the HPLC column (Fig. 6).

For detection of 14E4 immunoreactivity using the solid-phase ELISA it is important to use amounts of HPLC fractions that allow complete binding of proteins present in the aliquot. Thus, it is important to repeat the ELISA experiment with lesser amounts of fractions when at high concentrations no 14E4 immunoreactivity is detectable. Otherwise, false-negative results may occur. It is our experience that NAP-4 nearly coelutes in the system used for purification with RANTES[6] as well as PF4 (Fig. 6). Usually neutrophil chemotactic activity coelutes with 14E4 ELISA reactivity. In some cases no PMN chemotactic activity can be seen, which is the result of inadequate amounts of NAP-4 in crude extracts. In addition,

[6] Y. Kameyoshi, A. Dörschner, A. I. Mallet, E. Christophers, and J.-M. Schröder, *J. Exp. Med.* **176,** 587 (1992).

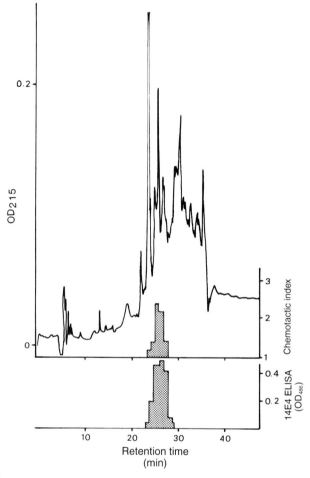

FIG. 6. Preparative RP-8 reversed-phase HPLC of a platelet lysate. Platelet lysates obtained from 2×10^{10} platelets were separated on a preparative RP-8 HPLC column. Proteins were eluted using a gradient of increasing concentrations of acetonitrile containing 0.1% TFA. Absorbance was monitored at 215 nm. Fractions were tested for neutrophil chemotactic activity (*top*) and 14E4 anti-IL-8 immunoreactivity (*bottom*), and active fractions are shaded.

care should be taken when choosing the blood donors: the method for preparation of granulocytes described above does not allow a separation of eosinophils from neutrophils, and some donors have higher blood eosinophil counts. This may result in high (>30%) contamination of PMN preparations with eosinophils and thus may lead to the detection of eosinophil attractants in HPLC fractions. In the special case of NAP-4 peak biological

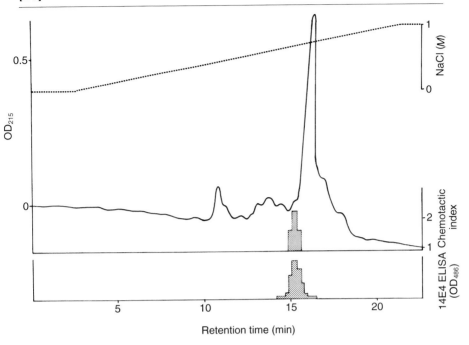

FIG. 7. Cation-exchange HPLC of partially purified NAP-4 preparations. NAP-4-containing fractions from preparative RP-8 HPLC were separated by micro-Mono S HPLC using an increasing gradient of NaCl. Absorbance was monitored at 215 nm. Neutrophil chemotactic activity present in HPLC fractions is in top shaded area. 14E4 and anti-IL-8 immunoreactivity is demonstrated in lower shaded area.

activity would be artificially extended to fractions containing RANTES, which is a highly efficient eosinophil attractant[6] that nearly coelutes with NAP-4.

Further purification of NAP-4 can be done by microcation-exchange HPLC. Fractions containing NAP-4 from preparative RP-8 HPLC are lyophilized and dissolved in 50–100 μl of 50 mM ammonium formate, pH 4.0, containing 20% (v/v) acetonitrile and are applied to a micro-Mono S HPLC column (Smart-System, Pharmacia) previously equilibrated with the same buffer. Proteins are eluted with a gradient of increasing concentration of NaCl (maximum of 1 M) in equilibration buffer using a flow rate of 100 μl/min. Fractions are taken manually and tested for PMN chemotactic activity. Figure 7 shows a representative run.

The major component present in the HPLC run is PF4, which can be identified by N-terminal sequencing. NAP-4 usually elutes slightly earlier as a minor peak. It can be identified by its immunoreactivity on 14E4 solid-phase ELISA, its neutrophil chemotactic activity, or its slightly lower migratory activity, as seen for PF4 on Tricine–SDS–PAGE analysis.

Final purification is achieved by the use of microreversed-phase HPLC. Fractions from Mono S HPLC containing NAP-4 are mixed with TFA to give a final concentration of 1% (v/v). Samples are then applied to a microreversed-phase RP-18 HPLC column (Sephasil C_{18}, 2.1 × 100 mm, particle size 5 μm; Pharmacia) attached to a Smart-System (Pharmacia), previously equilibrated with 10% (v/v) acetonitrile in 0.1% (v/v) aqueous TFA. NAP-4 is eluted with an increasing gradient of acetonitrile. Aliquots (1 μl) are tested for either PMN chemotactic activity by direct dilution in 300 μl complete PBS and/or for 14E4 immunoreactivity by solid-phase ELISA.

In nearly 50% of the experiments NAP-4 can be obtained in a pure form, giving the known N-terminal sequence. In cases where NAP-4 cannot be obtained as a pure product (the major impurity usually was found to be PF4), one should try to use an additional HPLC step prior to Smart Mono S HPLC. Fractions from the preparative RP-8 HPLC step are lyophilized, and the residue is dissolved in 200–400 μl of 0.1% (v/v) aqueous TFA. The sample is then applied to a cyanopropyl HPLC column (4.6 × 250 mm, cyanopropyl-silica with end-capping, pore size 5 μm; J. T. Baker, Gross Gerau, Germany) previously equilibrated with 0.1% aqueous TFA. We find that it is essential to use either lyophilizated samples or samples which have been diafiltered against 0.1% TFA, because with this HPLC column the majority of proteins elute in the void volume when too much acetonitrile is present in the sample. Thus, to be on the safe side, the samples are evaporated prior to cyanopropyl HPLC.

Proteins are eluted from the column with a gradient of increasing concentrations of 1-propanol containing 0.1% TFA, and fractions again are tested for neutrophil chemotactic activity. A typical example is shown in Fig. 8. The total recovery of pure NAP-4 is low when compared with that of CTAP-III or PF4, thus indicating that NAP-4 appears to be a minor protein in platelets. Usually we obtain with purification strategy I 10–15 μg NAP-4 per 10^{11} platelets. Using strategy II only 1–4 μg pure NAP-4 per 10^{10} platelets could be obtained.

Outlook

In this chapter, the isolation of a chemokine that is lacking the motif Cys-X-Cys (Fig. 4), which is a conserved structural element of the C-X-C branch of chemokines, is described. The striking similarity of NAP-4 to PF4 indicates that its biological role may be similar. The methods described here should allow purification of sufficient material of this unique chemo-

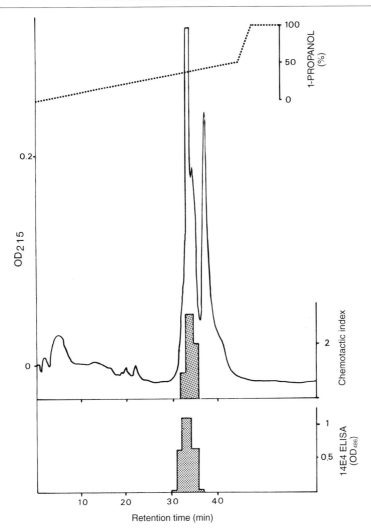

FIG. 8. Cyanopropyl reversed-phase HPLC of partially purified NAP-4 preparations. NAP-4-containing fractions from preparation RP-8 HPLC were separated with a cyanopropyl HPLC column using a gradient of increasing concentration of 1-propanol. The effluent was monitored at 215 nm. Fractions with PMN chemotactic activity and 14E4 anti-IL-8 immunore-activity are indicated by shading.

kine for further biological and biochemical characterization, including determination of the complete amino acid sequence.

Acknowledgments

This work was supported by the Deutsche Forschungsgemeinschaft. The author thanks Dr. Michael Sticherling for providing the 14E4 anti-IL-8 monoclonal antibody, Christine Gerbrecht-Gleissmann, Jutta Quitzau, Marlies Brandt, and Birgit Radtke for technical assistance, and Gabriele Tams for editorial help.

Section III

Other Methods

[16] Chemical Synthesis, Purification, and Folding of C-X-C and C-C Chemokines

By IAN CLARK-LEWIS, LUAN VO, PHILIP OWEN, and JENNIFER ANDERSON

Introduction

Chemokines can be readily produced by chemical synthesis. As of 1996, the limit for stepwise solid-phase peptide synthesis methods is about 110 residues, and because most chemokines contain between 70 and 80 amino acids, they fall well below this size limit. In the solid-phase synthesis method, as first outlined by Merrifield,[1] the COOH-terminal residue is linked to a solid support, and residues are added one at a time until the entire polypeptide[2] is formed. The polypeptide is then removed from the solid support and purified. Standard laboratory-scale methods allow the synthesis of up to 100 mg of homogeneous protein.[3] We have synthesized more than 600 chemoattractant cytokines (chemokines) and their analogs, encompassing both the C-X-C and C-C chemokine subfamilies. The synthetic chemokines have been employed for investigations of their functions,[4-7] structure–activity relationships,[8-11] and three-dimensional structure.[12-15] These stud-

---

[1] R. B. Merrifield, *J. Am. Chem. Soc.* **85,** 2149 (1963).

[2] Only the folded form of a polypeptide is termed a protein. Prior to folding the molecule is termed a polypeptide, or protected polypeptide if it has been synthesized and is bound to the resin.

[3] I. Clark-Lewis, B. Moser, A. Walz, M. Baggiolini, G. J. Scott, and R. A. Aebersold, *Biochemistry* **30,** 3128 (1991).

[4] M. Loetscher, B. Gerber, P. Loetscher, S. A. Jones, L. Piali, I. Clark-Lewis, M. Baggiolini, and B. Moser, *J. Exp. Med.* **184,** 963 (1996).

[5] P. D. Ponath, S. Qin, D. J. Ringler, I. Clark-Lewis, J. Wang, N. Kassam, H. Smith, X. Shi, J.-A. Gonzalo, W. Newman, J.-C. Gutierrez-Ramos, and C. R. Mackay, *J. Clin. Invest.* **97,** 604 (1996).

[6] C. C. Bleul, M. Farzan, H. Choe, C. Parolin, I. Clark-Lewis, J. Sodroksi, and T. A. Springer, *Nature (London)* **382,** 829 (1996).

[7] E. Oberlin, A. Amara, F. Bachelerie, C. Bessia, J.-L. Virelizier, F. Arenzana-Seisdedos, O. Schwartz, J.-M. Heard, I. Clark-Lewis, D. F. Legler, M. Loetscher, M. Baggiolini, and B. Moser, *Nature (London)* **382,** 829 (1996).

[8] I. Clark-Lewis, C. Schumacher, M. Baggiolini, and B. Moser, *J. Biol. Chem.* **262,** 23128 (1991).

[9] I. Clark-Lewis, B. Dewald, T. Geiser, B. Moser, and M. Baggiolini, *Proc. Natl. Acad. Sci. U.S.A.* **90,** 3574 (1993).

[10] I. Clark-Lewis, B. Dewald, M. Loetscher, B. Moser, and M. Baggiolini, *J. Biol. Chem.* **269,** 16075 (1994).

[11] J.-H. Gong and I. Clark-Lewis, *J. Exp. Med.* **181,** 631 (1995).

[12] K. Rajarathnam, I. Clark-Lewis, and B. D. Sykes, *Biochemistry* **33,** 6623 (1994).

ies have demonstrated that chemical synthesis is a straightforward route to chemokines and novel analogs. In this chapter, we describe the rationale and procedures that we have developed based on our experience with the synthesis of chemokines.

Synthetic Approach to Obtaining Chemokines

The first and foremost requirement for the study of a particular chemokine is obtaining sufficient pure material for the experiments. Furthermore, the study of structure–activity relationships involves building conclusions from the results of multiple analogs. This requires the generation of large numbers of analogs, so the methods chosen must be capable of fulfilling this need. Natural sources are not useful because the amounts that can be derived are very low, and it is not possible to modify the sequence. Both the chemical and DNA approaches can provide native chemokines and analogs with specific changes. The two approaches are fundamentally different, with the synthesis method being less known, and therefore some comparisons are useful.

Although there has never been a reason to assume that synthetic chemokines are any different from ribosome-assembled chemokines, there is now ample experimental evidence that the two are equivalent. In situations where synthetic and cell-derived chemokines have been compared, no evidence has been found for any difference in their functional or receptor binding properties.[3,16] Furthermore, the three-dimensional structures of synthetic chemokines are indistinguishable from those generated by recombinant DNA methods.[12,13] Thus, synthetic chemokines have the same properties as natural and genetically engineered forms.

For structure and functional analysis the ability to rapidly test ideas with new analogs is critical. With the synthesis approach, the protein synthesis can be started immediately on selection of the sequence. With the assistance of a peptide synthesizer it takes about 10 days to generate a chemokine from the starting resin to the final folded and purified product (Table I). For the DNA expression route, DNA must be synthesized, an expression vector must be prepared, cells must be transformed and selected, and the expressed protein must be identified and purified. Costs for the

[13] K.-S. Kim, I. Clark-Lewis, and B. D. Sykes, *J. Biol. Chem.* **269,** 32909 (1994).

[14] K. Rajarathnam, I. Clark-Lewis, and B. D. Sykes, *Biochemistry* **34,** 12983 (1995).

[15] K.-S. Kim, K. Rajarathnam, I. Clark-Lewis, and B. D. Sykes, *FEBS Lett.* **395,** 277 (1996).

[16] I. Clark-Lewis, K.-S. Kim, K. Rajarathnam, J.-H. Gong, B. Dewald, M. Baggiolini, and B. D. Sykes, *J. Leukocyte Biol.* **57,** 703 (1995).

TABLE I
TOTAL YIELDS FOR AVERAGE CHEMOKINE SYNTHESIS

Stage	Weight	Moles[a]
Starting resin	0.8 g	500 μmol
Polypeptide–resin	4 g	250 μmol
Crude product	2 g	250 μmol
Preparative step 1 (reduced)	400 mg	50 μmol
Preparative step 2 (after folding)	175 mg	22 μmol
Semipreparative step (final)	100 mg	12.5 μmol

[a] Moles of total synthetic polypeptide including the desired product and by-products.

DNA approach are difficult to assess and are dependent on the particular methods used. The synthesis approach requires expensive chemicals. However, the longer time required to generate analogs by the DNA approach is a significant cost factor. Although not a difficulty for chemokines, the stepwise chemical synthesis method by its nature has size limitations. Novel technologies that combine stepwise synthesis with new methods for piecing synthetic fragments together (ligation) promise to extend the limits of the chemical approach.[17–19] The DNA approach, however, is not intrinsically limited by protein size. Overall, the major advantages of the synthetic approach are that it is rapid and provides sufficient material for detailed structural and functional analysis.

Another advantage of the chemical synthesis approach is that it is unlikely that the protein will be contaminated with molecules of biological origin. With material produced in a biological system there is always the possibility that biomolecules, for example, formylmethionylleucylphenylalanine (fMLP) and related peptides, endotoxins, and mitogens, could contaminate the preparation and interfere with function. In addition, with the synthetic route, the final protein has a defined covalent structure. Furthermore, chemokines can be proteolytically processed when made in biological systems. Even if the native protein is produced in the desired form, analogs of the molecule could be processed. Purification of the raw protein material requires approaches that differ depending on the nature of the contaminants. For example, whereas affinity chromatography is more useful for a recombinant material, more general chromatographic methods, such as high-performance liquid chromatography (HPLC), are more suitable for

[17] M. Schnolzer and S. B. H. Kent, *Science* **256,** 221 (1992).
[18] P. E. Dawson, T. W. Muir, I. Clark-Lewis, and S. B. H. Kent, *Science* **266,** 776 (1994).
[19] J. P. Tam, Y. A. Lu, C. F. Liu, and J. Shao, *Proc. Natl. Acad. Sci. U.S.A.* **92,** 12485 (1995).

synthetic products. If quantities greater than several hundred milligrams are needed, then an alternative synthetic method to that outlined here will be required. Recombinant DNA technology has been developed for production on an industrial scale.

Peptide synthesis has many advantages for single domain sized proteins, so it is reasonable to ask why peptide synthesis has not been more widely applied to protein engineering. One reason is that most researchers approach proteins from a biological rather than a chemical standpoint. Thus, the requirement for chemicals such as hydrogen fluoride (HF) is possibly a barrier to the widespread adoption of the technology among protein groups. The Fmoc (9-fluorenylmethoxycarbonyl) strategy, for which HF is not needed, could potentially overcome this. Another reason is the perception that peptide synthesis and protein synthesis are mutually exclusive, that is, that the former is done chemically and the latter is done by genetic engineering. Certainly the transition from peptide synthesis to protein synthesis has been slow. In the early days of solid-phase synthesis, the chemical methods had shortcomings and inefficiencies such that it was limited to short peptides. However, gradual optimization led to improvements in the chemistry and demonstrated that a range of different proteins could be synthesized. Thus the notion that the solid-phase method is intrinsically limited to short peptides has been proved incorrect.

In the future, nonnatural chemically engineered proteins will probably become the realm of the synthetic approach. Incorporating nonnatural residues or chemical groups into proteins can be as straightforward as the usual coded amino acids.[20] Despite some creative studies involving chemically engineered nonnatural proteins,[20–22] the value of nonnatural protein derivatives into protein chemistry has not yet been realized. Nevertheless, the ease with which nonnatural and chemically modified constructs can be generated is a clear advantage of the synthesis approach.

Methods for Synthesis, Purification, and Characterization

The steps in the synthetic procedure that are covered in this chapter are (1) designing the sequence, (2) synthesis of the resin-bound polypeptide, (3) cleavage of the polypeptide from resin, (4) analysis by reversed-phase

[20] K. Rajarathnam, B. D. Sykes, C. M. Kay, T. Geiser, B. Dewald, M. Baggiolini, and I. Clark-Lewis, *Science* **264**, 90 (1994).
[21] R. C. de L. Milton, S. C. F. Milton, and S. B. H. Kent, *Science* **256**, 1445 (1992).
[22] L. E. Canne, A. R. Ferré-D'Amaré, S. K. Burley, and S. B. H. Kent, *J. Am. Chem. Soc.* **117**, 2998 (1995).

HPLC, (5) purification, (6) folding, (7) evaluation of purity, and (8) verification of the covalent structure.

Designing the Sequence

In protein engineering, hypotheses for the role of particular structures in function are tested by designing a modified amino acid sequence and studying the resultant protein analog. Before designing analogs, it is essential to know the complete sequence of the fully active native protein. For all chemokines tested, the NH$_2$-terminal region is important for receptor interactions.[8,11,16] For example, changing just the NH$_2$-terminal residue of monocyte chemoattractant protein-1 is sufficient to change receptor selectivity and function.[23] The native protein must be first synthesized to ensure that it has the same chemical and functional properties as the molecule derived from natural sources. This native protein will provide the basis for comparison of the analogs.

Synthesis of Resin-Bound Polypeptide

Two different solid-phase synthesis methods are widely used. They are designated tBoc (tertiary-butoxycarbonyl) and Fmoc, according to the abbreviation for the $^\alpha$NH$_2$-protecting group that is used.[24] Suitable side-chain protecting groups for both strategies have been developed. The tBoc strategy is based on differential acid lability of the tBoc and the side-chain protecting groups, whereas the Fmoc strategy is dependent on the base lability of the Fmoc group compared to the acid lability of the side-chain protecting groups. The Fmoc chemistry has advanced considerably in recent years; however, synthesis of chemokines by this method has not been evaluated, and it is not discussed further.

The chemistry that we have used in our studies is based on the tBoc $^\alpha$NH$_2$-protection strategy. Observations made during the course of our studies have indicated that the more time that it takes to complete a synthesis, the lower the quality of the final product. There are two aspects to this: first, the more efficient the chemistry, the shorter is the time needed to add an amino acid, and therefore there is less opportunity for undesired side reactions; second, leaving the peptide–resin standing during synthesis leads to reduced yields. Thus, the synthesis is best done rapidly and continuously. Completed or intermediate fully protected peptide–resins, which

[23] M. Weber, M. Uguccioni, M. Baggiolini, I. Clark-Lewis, and C. A. Dahinden, *J. Exp. Med.* **183**, 681 (1996).
[24] S. B. H. Kent, *Annu. Rev. Biochem.* **57**, 957 (1988).

could have any number of residues already added, can be stored indefinitely at $-80°$ in dimethylformamide (DMF).

Chemokine synthesis can be performed manually using a fritted glass reaction vessel. However, it is simpler and more efficient to use a programmable peptide synthesizer, which is capable of performing the steps automatically with greater reliability, reproducibility, and rapidity than can be accomplished manually.[24-27]

A chemokine polypeptide chain–containing residue is started by linking the penultimate amino acid $(n - 1)$, to the COOH-terminal amino acid that is already on the resin. To do this, first the COOH group of the $n - 1$ amino acid must be activated to make it reactive. We have synthesized our chemokines using amino acids that are activated as symmetric anhydrides[3] or as active esters. Both methods have been found to result in good quality products, but active esters formed with HBTU (see below) reduce synthesis time about 3-fold.

The addition of each amino acid (see Fig. 1) involves removal of the tBoc protecting group with acid (deprotection step), addition of base to remove residual acid and ensure that the peptide $^\alpha NH_3^+$ is converted to $^\alpha NH_2$ (neutralization step), followed by addition of the activated amino acid (first coupling step) and then replacement of the reacted amino acid with a second solution of freshly activated amino acid (second coupling step). The second coupling step improves the average stepwise yield by 0.3–0.5% and thus significantly increases final overall yield. (Note that the second coupling does not result in double addition because the tBoc protecting group prevents reactivity.)

Yields. The term yield in peptide synthesis has little meaning, unless the context and the basis by which it is measured are specified. At the end of the experiment what matters most is the amount of pure material that is obtained, and whether it is enough to do the experiment. Nevertheless the final yield is the product of the yields that result from every side reaction that has occurred during the synthesis and cleavage. Side reactions have been reviewed elsewhere.[24,25] Most of the by-products are closely related to the molecule of interest and give the RP-HPLC profile a lumpy appearance (see *Analysis by Reversed Phase Chromatography* and Figs. 2A, 3A, and 4A). In chemical synthesis there are many possible ways to generate

[25] J. Schneider and S. B. H. Kent, *Cell* **54**, 363 (1988).

[26] I. Clark-Lewis and S. B. H. Kent, *in* "Receptor Biochemistry and Methodology: The Use of HPLC in Protein Purification and Characterization" (J. C. Ventor and L. C. Harrison, eds.), p. 43. Alan R. Liss, New York, 1989.

[27] M. Schnolzer, P. Alewood, A. Jones, D. Alewood, and S. B. H. Kent, *Int. J. Pept. Protein Res.* **40**, 180 (1992).

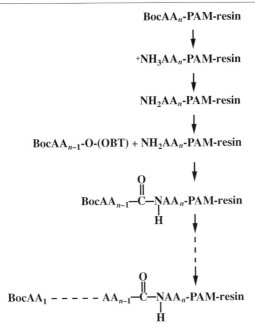

FIG. 1. Outline of the steps in tBoc solid-phase peptide synthesis.

by-products, and therefore yields and the factors that determine them must be considered.

The stepwise yields, which reflect the failure of the amino acid coupling step to go to completion, result in by-products that are shortened by one or more residues. These by-products are closely related to one another and to the final product and, therefore, are difficult to resolve; thus, they represent an important factor in determining the quality of the desired polypeptide. Moreover, the final yield due to incomplete coupling reactions is derived by multiplying the stepwise yields together [e.g., 71 steps for interleukin-8 (IL-8) which contains 72 amino acids]. The stepwise yields can be determined by assaying the level of completion of the coupling reaction during the synthesis[3,28] or by Edman sequencing methods.[3,29]

The yields that result from the various side reactions, if they can be determined, are informative because they provide information about the levels of the by-products that must be eliminated during the purification procedure. The yield in grams is the final arbiter of the success of the

[28] V. K. Sarin, S. B. H. Kent, J. P. Tam, and R. B. Merrifield, *Anal. Biochem.* **117**, 147 (1981).
[29] I. Clark-Lewis, R. A. Aebersold, H. Ziltener, J. W. Schrader, L. E. Hood, and S. B. H. Kent, *Science* **231**, 134 (1986).

synthesis. The absolute maximum possible yield of a chemokine with a molecular mass of 8000 Da that was synthesized starting with 0.5 mmol of resin is

$$0.5 \times 10^{-3} \text{ mol} \times 8000 \text{ Da} = 4 \text{ g}$$

However, before deprotection, the mass of the protected polypeptide and the resin itself is approximately 8 g—double the deprotected polypeptide. Because the resin swells as the peptide grows, the maximum amount of peptide–resin that can be accommodated in a standard 40-ml reaction vessel is about 4 g. In principle, the yield can be increased by increasing the size of the reaction vessel instead of removing resin; but the concentration of reagents must be maintained, so more reagents, including amino acid, will be required.

Resin Splitting. The technique of splitting the peptide–resin allows intermediates in the synthesis to be stored and restarted, thus allowing several analogs to be generated with a common COOH-terminal sequence. At the appropriate point the peptide–resin is simply divided, some removed and stored, and the synthesis continued. There is no problem with yield as the synthesis is started with an excess of resin, so there is enough to synthesize up to 10 full-length analogs (see above). This is an extremely efficient procedure, and is particularly useful for chemokines as most of the functional motifs are near the NH_2-terminal region. If analogs with different COOH-terminal residues were needed, then they would have to be synthesized separately.

Detailed Methods

Materials. For chemokine synthesis the aminoacyl(4-carboxyamidomethyl)benzyl ester (pam) resin is the best choice because after synthesis the polypeptide is released as the carboxylic acid,[30] which is the naturally occurring form of most proteins including chemokines. With this resin, the protected COOH-terminal amino acid is already covalently bound to the resin via a linker: poly(styrene–1% divinylbenzene). Thus, to cover the 20 naturally occurring amino acids that could be at the COOH-terminal position, 20 different resins are required. Protected amino acids that are used for the chain assembly are tBoc · Ala, tBoc · Asn(Xan), tBoc · Arg(Tos), tBoc · Asp(Bzl), tBoc · Cys(4MeBzl), tBoc · Glu(Bzl), tBoc · Gln, tBoc · Gly, tBoc · His(Bom), tBoc · Leu, tBoc · Ile, tBoc · Lys(2ClZ), tBoc · Met, tBoc · Pro, tBoc · Phe, tBoc · Ser(Bzl), tBoc · Tyr(2BrZ), tBoc · Thr(Bzl),

---

[30] A. R. Mitchell, S. B. H. Kent, M. Engelhard, and R. B. Merrifield, *J. Org. Chem.* **43,** 2845 (1978).

tBoc · Trp(CHO), and tBoc · Val [where tBoc is tertiary butyloxycarbonyl; Xan, xanthanyl; Tos, toluenesulfenyl; Bzl, benzyl; 4MeBzl, 4-methylbenzyl; Bom, benzyloxymethyl; 2ClZ, 2-chlorobenzoxycarbonyl; 2BrZ, 2-bromo-benzoxycarbonyl; and CHO, formyl].

Dimethylformamide (DMF) and dichloromethane (DCM) are from Burdick and Jackson (Muskeon, MI); trifluoroacetic acid is from Halocarbon (Riveredge, NJ) and is redistilled before use. Diisopropylethylamine (DIEA) is from Perkin-Elmer Applied Biosystems (Foster City, CA), and HBTU is from Richlieu (Montreal).

Synthesis Protocol. From active esters with 2-(1*H*-benzotriazol-1-yl)-1,1,3,3-tetramethyluronium hexafluorophosphate (HBTU)[31] using the following protocol, which has been adapted from Schnolzer *et al.*[27] Activate the amino acid by mixing 2 mmol of protected amino acid (dry powder), 1 ml DIEA, and 4 ml of 0.45 *M* HBTU (1.8 mmol) in DMF (HBTU/DMF) for 10 min. The HBTU/DMF should be made fresh daily.

Initiate the synthesis by adding the appropriate pam resin, which has been prewashed with DCM, to a 40-ml reaction vessel. Keep the resin suspended, except during delivery or draining steps. Add sufficient reagents to keep the resin in suspension. Remove the tBoc group by adding 65% trifluoroacetic acid (TFA) in DCM for 1 min and drain, and then add a second aliquot of 65% TFA in DCM for 10 min. Drain the reaction vessel, wash the resin thoroughly with DMF, and neutralize the resin with 5% DIEA in DMF. Activate the amino acid corresponding to residue $n - 1$, where n is the number of residues in the sequence (residue n, the COOH-terminal residue is already linked to the resin), as described above. Add the activated amino acid to the vessel and allow it to react for 15 min. Drain the vessel and then wash the resin with DMF. Add a second 2 mmol of the same amino acid, which has been freshly activated by repeating the activation protocol. After 15 min, drain and rinse the resin with DMF. The resin now has two linked residues attached. Repeat these steps for each amino acid until the entire protected polypeptide has been formed.

Cleavage of Polypeptide from Resin

We use a variation on the "low–high" hydrogen fluoride (HF) method.[32] Compared to the simpler "high" only HF procedure, this method reduces side reactions during cleavage. In the case of chemokines, the cysteines are particularly prone to side reactions. The procedure is normally carried out using a Teflon apparatus, and care is taken to eliminate any likelihood of exposure to HF liquid or vapors.

[31] R. Knorr, A. Trzeciak, W. Bannwarth, and D. Gillessen, *Tetrahedron Lett.* **30,** 1927 (1989).
[32] J. P. Tam, W. F. Heath, and R. B. Merrifield, *J. Am. Chem. Soc.* **105,** 6442 (1983).

Cleavage Protocol. Before the HF steps, first remove the CHO protecting group from tryptophan residues to prevent possible formylation of the peptide later during deprotection. To accomplish this add 5% (v/v) piperidine in DMF for 10 min at room temperature. Some chemokines contain no tryptophan, in which case this step should be eliminated. Second, remove the tBoc group from the NH_2-terminal residue, to prevent a possible tertiary butylation side reaction during the cleavage procedure. To do this use the deprotection and neutralization steps that have been described in the synthesis section. Wash the resin with DCM and then dry under vacuum.

For the "low HF" step, add up to 4 g resin to a Teflon vessel with 30 ml of a mixture of 65% (v/v) dimethyl sulfide (DMS) 5% (v/v) *p*-thiocresol, 5% (v/v) *p*-cresol, and 25% (v/v) anhydrous HF. After 1 hr at 0°, filter the peptide–resin from the mixture using a Teflon filter holder and a disposable Teflon filter cut from a sheet of porous Teflon (Chemplast, Wayne, NJ). For the "high HF" step, add 5% (v/v) *p*-thiocresol, 5% (v/v) *p*-cresol, and 90% (v/v) HF to the dried resin and allow to react for 1 hr at 0°. Draw off the HF into a 9 M KOH trap using a gentle N_2 stream, and then draw off any remaining HF into a calcium oxide trap using a high vacuum pump. An oily mixture should remain. This contains the polypeptide, resin, protecting groups, and the *p*-cresol/*p*-thiocresol. To precipitate the peptide and extract the low molecular weight by-products, add ethyl acetate and, after 5 min, filter on a Buchner funnel and wash the precipitate that remains with more ethyl acetate. The precipitate contains the free, deprotected polypeptide and the original resin. Dissolve the precipitated polypeptide in sufficient 6 M guanidine hydrochloride, 0.3 M Tris, pH 8.5, 10% (v/v) 2-mercaptoethanol, to make an approximately 30 mg/ml solution. Acidify this solution to pH 3.0 by addition of 20% (v/v) acetic acid and filter through a 0.2-μm disposable filter. This material is the crude product from the synthesis.

Analysis by Reversed-Phase Chromatography

Reversed-phase HPLC (RP-HPLC) and peptide/protein synthesis are well-matched methods. This is partly because polypeptides are compatible with the solvents that are used, and with the C_{18}–silica matrix. Another reason is that the nature of the by-products is such that the product can be resolved. Thus RP-HPLC is ideal for analysis of the crude synthetic product, as well as monitoring the various stages of folding and purification (Figs. 2, 3, and 4).

Analytical scale HPLC is carried out on a 4.6 × 250 mm C_{18} column (218TP54 Vydac, Hesperia, CA) using conventional acetonitrile/water/ 0.1% (v/v) TFA solvent system. A 60% (v/v) acetonitrile/water gradient over 60 min gives optimal separation. The elution is monitored by UV

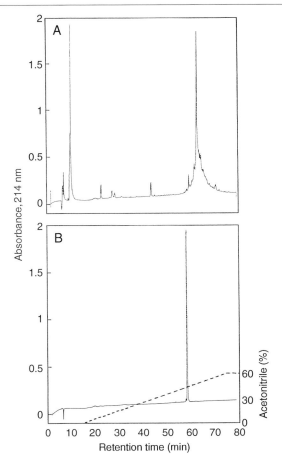

FIG. 2. RP-HPLC profiles of synthetic stromal cell–derived factor-1 (SDF-1). SDF-1 is a C-X-C chemokine of 67 residues. The profile of the crude SDF-1 product (A) shows a major sharp peak that contains the unfolded product, and minor peaks and shoulders are apparent; the accumulated effect of these is the broadening under the main peak. The final folded and purified SDF-1 is shown in B. Note the difference in retention time of the final protein compared to the reduced product.

absorption at 214 nm, 2.0 absorbance units full scale. Peptides have strong absorbance at this wavelength, and the loading should be adjusted so that the maximum peptide peak is 50 to 75% of full scale. We have found that it is best to keep the procedure constant as it allows us to compare profiles with past syntheses, whether they were done last week or 10 years ago.

Chromatograms of crude products of three different chemokine synthe-ses are shown in Figs. 2A, 4A, and 5A. The molecules do not elute as

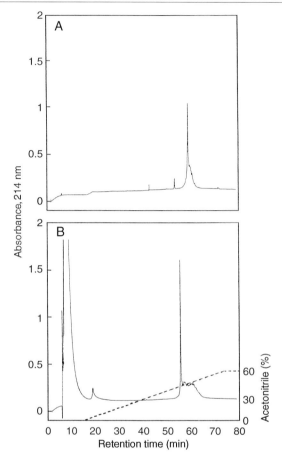

Fɪɢ. 3. RP-HPLC profiles of partially purified synthetic SDF-1, before (A) and after folding (B). The profile of the material after the first purification step is shown in A (compare Figs. 2A and 3A). After the folding step (B), the main peak is sharper and is eluting earlier. Moreover, the by-products now elute after rather than under the main peak.

discrete peaks but rather coalesce to give a broad, lumpy chromatogram. Nevertheless, the correct product is usually at the apex of the broad peak. When fractions are taken and reanalyzed, then discrete peaks can be observed indicating that the actual separation is better than it appears from the original chromatogram. Thus, even though the peaks are not fully resolved, the overall shape is indicative of the relative purity of the crude product.

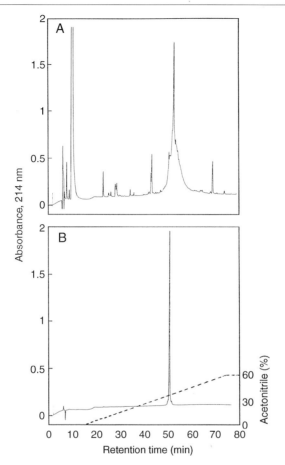

Fig. 4. RP-HPLC profiles of monocyte chemoattractant protein (MCP-1). MCP-1 is a C-C chemokine of 76 residues. The profiles are of the crude product (A) and the final folded and purified protein (B). The sharpness of the main peak and the breadth and height of the accumulated products either side of the main peak indicate that MCP-1 crude product is not as pure as the SDF-1 crude product (compare with Fig. 2A).

Purification

Not only is RP-HPLC useful as an analytical tool, but it can be readily scaled up for purification. Resolution is retained over a wide range of column sizes with different capacities and loading. Our routine analysis and purification is by RP-HPLC using C_{18} columns of appropriate sizes. We use a preparative scale column (Vydac 218TP1022, which is 22×250 mm, C_{18} silica, 10–15 μm, 300 Å pore size) and a semipreparative scale

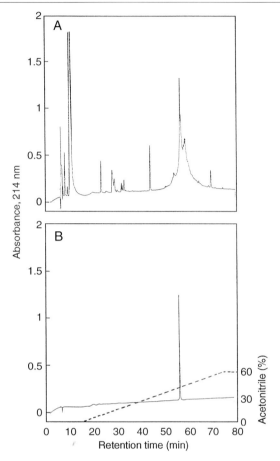

FIG. 5. RP-HPLC profiles of monokine induced by interferon-γ (MIG). MIG is an unusually large C-X-C chemokine of 104 residues; the increased size is due to a COOH-terminal extension of the typical chemokine fold. The profile of the crude product (A) shows a major peak and a close peak with slightly longer retention time. The major peak is not as sharp as with SDF-1 or MCP-1, indicating that there are more closely eluting by-products. Nevertheless, the final folded product is apparently pure (B).

column (Vydac 218TP510, which is 10×250 mm C_{18} silica, 5 μm, 300 Å pore size). The semipreparative column has the higher resolution of the two primarily because of the smaller particle size. Note that it is important to avoid excessive dilution, both on the column and during elution, to prevent losses. RP-HPLC columns have high capacities, and loading in the mid-capacity range of the column is best. A guide is 2 g for the preparative column, and 100 mg for the semipreparative column. Whereas low loading results in poor yields, overloading causes peak broadening due to solubility problems. The preparative columns are run at 15 ml/min and the semipreparative at 3 ml/min. A chromatography system that can handle flow rates up to 50 ml/min is required for the preparative scale, whereas the semipreparative column can usually be run with the same hardware as the analytical system.

For up to 2 g of crude polypeptide product the following strategy is used. (a) The crude product is applied to a preparative RP-HPLC column. (b) The pooled fractions are folded. (c) The folded protein is applied to the same preparative scale RP-HPLC column. (d) The selected fractions are applied to a semipreparative size column. Pure fractions are pooled and lyophilized to give the final protein.

Purification of the crude product before the folding step has two benefits: One is that it eliminates contaminants that interfere with folding, including reducing agent by-products that may precipitate; and the other is that it avoids the need to exchange the crude material into folding buffer as would be necessary if the folding was to be done first.

Purification Protocol. Load the crude reduced synthetic product, as obtained from the cleavage step, directly onto a preparative column through the solvent inlet of the A pump, and apply a 200-min gradient from 0% to 60% acetonitrile. Collect 1-min fractions. Note that for preparative HPLC it is not possible to effectively monitor the eluted polypeptide by UV absorbance because the concentrations are too high for any UV transparency. Therefore, analyze selected fractions by analytical scale HPLC and compare them to the chromatogram of the original crude material. An option is to analyze fractions by electrospray mass spectrometry to confirm the presence of the correct product. (Mass spectrometry of the original crude product is unlikely to give an unambiguous mass.) On the basis of the analytical profiles, select fractions, pool, and lyophilize. Note that it is essential that pooled fractions be lyophilized after each HPLC step, because good separation in the subsequent step will not be achieved unless the acetonitrile is completely eliminated and the sample is loaded as an aqueous solution.

In the case of chemokines the folded form elutes about 3 min earlier than the unfolded form, assuming a 1%/min acetonitrile gradient. Thus, the retention time of all the samples analyzed after folding must be compared with the folded material and not the original crude product. The uniformity in the retention time difference among chemokines probably reflects the similarity in their folded structures.[10] Of more practical significance is that on folding the forms that are able to fold elute earlier but not contaminants that are unable to fold. Thus the folding actually assists in the purification of the final chemokine product.

Acidify the folded material, filter, and run it on the preparative HPLC column as described for the crude product. To accomplish the final purification step, lyophilize the fractions selected from the preparative run, solubilize the resulting powder with 0.1% TFA, and load the sample onto the semipreparative column. With difficult samples, for which the peaks cannot be resolved by RP-HPLC, an alternative purification method such as ion-exchange HPLC may be used.

Storage. After purification store the lyophilized chemokine at $-80°$. For use, prepare a stock solution by dissolving a weighed sample of chemokine in water to make a 10 mg/ml solution and store aliquots at $-80°$. This can be diluted into medium or buffer, as required.

Folding

Chemokines fold readily, and the folded structure is stabilized by two disulfide bridges, which are essential for function.[10] Furthermore, analogs of chemokines generally form the folded form as well, although not always as readily as the native form. However, if an analog does not fold, this could be because of the changes made to the native sequence, rather than the synthesis or folding conditions.

As the two disulfides are formed after folding to stabilize the structure it follows that if the two disulfides have formed, then folding must have occurred. This can be inferred by the behavior of the protein on analytical RP-HPLC and mass spectrometry (as described above).

Two parameters must be considered for optimal *in vitro* folding: solubility and method of oxidation. First, the solvent must be selected so that the polypeptide is soluble in both its unfolded and folded form. The unfolded form is always less soluble than the folded form in aqueous solution; however, precipitation in the folding reaction will dramatically affect yields. Solvent variables include the type of denaturant and its concentration, pH, buffers, and ionic strength, among others. It is best to avoid using urea as a denaturant because of the potential for protein modification. Second, a mild oxidant is required that will convert the SH group to S^-, but not to higher oxidation states. It is essential that the reaction between the oxidant and the SH group be reversible. The oxidants that we prefer for chemokines are oxygen from air, which is dissolved by vigorous stirring, and dimethyl sulfoxide (DMSO).

From our experience with synthetic chemokine analogs of both the C-X-C and C-C classes, the chemokine fold accepts changes readily. However, not all analogs will fold under the same conditions as the corresponding native molecule. Changes in the primary structure affect the solubility and the kinetics and thermodynamics of folding. Partial folding is sometimes observed; therefore, it is best to check a number of different conditions on a small scale before committing the entire product to a particular method.

Establish folding conditions by preparing a series of 0.5-ml samples (1 mg/ml) of the chemokine polypeptide, using the following solvents and conditions: (a) 1 M guanidine hydrochloride, 0.1 M Tris, pH 8.5, 10% (v/v) DMSO in a sealed flask, in the dark without stirring; (b) 1 M guanidine hydrochloride, 0.1 M Tris, pH 8.5, stir in air; (c) water, stir vigorously in air (avoid frothing); (d) 10% (v/v) DMSO in a sealed flask, in the dark without stirring; and (e) phosphate-buffered saline (PBS), stir in air. After 24 hr, analyze each sample by RP-HPLC. Assess folding by comparing the chromatogram with that of the unfolded sample and look for conversion to a peak that elutes 3 min earlier, based on a 1% (v/v) acetonitrile/min gradient. The extent of conversion to this early eluting form reflects the extent of folding. Select the conditions that give optimal conversion for folding of the remaining material as a single large batch.

Evaluation of Purity

To establish purity the first step should be to run analytical RP-HPLC as has already been described. A single major peak should be present, but the presence of minor peaks or, more likely, of shoulders or a distorted peak shape are indicators that the material needs further purification. However, analysis by the same method that is used to purify the protein might not detect coeluting material. Thus, to confirm purity an orthogonal separation method should be used, for example, analytical ion-exchange HPLC, for which the separation is based on ionic rather than nonpolar interactions. This could also be used for further purification. Alternatively, analytical isoelectric focusing or electrophoresis can be used to detect impurities. Mass spectrometry is also capable of detecting impurities by the number of distinct species that are apparent. Nevertheless, background peaks are always observed, so minor contaminants may not be detected; also, an unambiguous measure of purity is not always possible by mass spectrometry.

Verification of Covalent Structure

It is important to take reasonable steps to show that the purified material has the covalent structure that was originally intended. This means that it has the correct sequence and chemical composition, including the disulfides. In reality it is neither practical nor necessary to determine this routinely. The analytical techniques available for establishing covalent structure have their limitations and do not provide absolute proof.

From a practical perspective electrospray mass spectrometry gives an accurate molecular mass.[3,10] If the mass is as predicted and reasonable care has been taken in the synthesis, it provides strong evidence (but does not prove) that the correct molecule has been made. However, if the mass detected is incorrect, then the wrong molecule has been synthesized and/ or isolated. A major advantage of this mass spectrometric method is that the whole protein is analyzed without fragmentation or chemical treatment. The electrospray method has the advantage of speed, high accuracy (for a chemokine with a mass of 8000 within 1 amu), and reliability.[33] RP-HPLC solvents are compatible with the analysis, and thus the material can be easily evaluated during purification.

Formation of two disulfides results in a mass 4 amu lower than the reduced form. Although this indicates that two disulfides have formed, it does not prove that the disulfide pairing is correct. To do this it is necessary to fragment the molecules and isolate each disulfide on two different fragments. This is difficult with chemokines because the disulfide pairs are extremely close. Nevertheless, nuclear magnetic resonance (NMR) structures of known chemokines, including synthetic chemokines and their analogs, show that cysteine pairing is correct, and the 1 to 3 and 2 to 4 arrangement is favored. Thus, even though mass spectroscopy is a verification rather than proof of structure, it is the most powerful method available for the analysis of synthetic peptides and proteins.

Acknowledgments

This work was supported by the Protein Engineering Network of Centres of Excellence (Canada), and the U.S. National Institutes of Health. I.C.-L. is the recipient of a Medical Research Council of Canada Scientist award. The authors acknowledge all the participants in the structure and function aspects of the synthetic chemokines.

[33] G. Siuzdak, *Proc. Natl. Acad. Sci. U.S.A.* **91**, 11290 (1994).

[17] Identification of Inflammatory Mediators by Screening for Glucocorticoid-Attenuated Response Genes

By Jeffrey B. Smith and Harvey R. Herschman

Introduction

The glucocorticoid hormones are important negative regulators of the inflammatory system. Even at the basal, physiological concentrations that circulate in unstressed animals, glucocorticoids exert an important restraining effect on inflammatory responses.[1,2] In response to systemic stress, adrenal secretion of glucocorticoids increases markedly. At the high concentrations found in stressed animals, glucocorticoids help to limit the injury potentially caused by uncontrolled activation of inflammatory mediators. Thus, adrenalectomy sensitizes animals to the lethal effects of endotoxin (lipopolysaccharide, LPS), interleukin-1β (IL-1β), and tumor necrosis factor-α (TNF-α), whereas pretreatment with pharmacological doses of glucocorticoids is protective.[3,4]

Although other mechanisms also contribute to their anti-inflammatory effects, a major part of this glucocorticoid activity is attributable to their ability to attenuate the induction of genes whose products play crucial roles in inflammation. An important example is the gene encoding the inducible form of prostaglandin synthase (PGS2). PGS2 encodes an inducible form of the cyclooxygenase that converts arachidonate to prostaglandin H_2, the common precursor of all the prostanoids—prostaglandins, prostacyclins, and thromboxanes.[5] Glucocorticoids inhibit the induction of PGS2 by LPS and other inflammatory stimuli in a wide variety of cell types, but they usually have no effect on expression of PGS1, the noninducible form of prostaglandin synthase. Glucocorticoids also inhibit the induction of the macrophage (inducible) form of nitric oxide synthase (iNOS) and numerous inflammatory cytokines including IL-1β, TNF-α, interleukin-8 (IL-8), and many others. We refer to glucocorticoid-attenuated inducible genes such as PGS2, iNOS, and IL-1 as glucocorticoid-attenuated response genes, or GARGs.[6]

As far as we have been able to determine, all of the previously known GARGs were identified either via assays of their biological activity or by screening procedures based on inducibility or sequence homology. In each case, glucocorticoid attenuation was investigated after the gene or its product has been identified. In contrast, our procedure for identifying novel inflammation-related genes used glucocorticoid attenuation of induction as the primary screening characteristic.

Glucocorticoid-Attenuated Response Gene Hypothesis

Our approach to screening for GARGs evolved from earlier studies of immediate early/primary response genes. Transcription of primary response

[1] A. Munck, P. M. Guyre, and N. J. Holbrook, *Endocr. Rev.* **5,** 25 (1984).

[2] A. Munck and A. Naray-Fejes-Toth, *Ann. N.Y. Acad. Sci.* **746,** 115 (1994).

[3] J. L. Masferrer, K. Seibert, B. Zweifel, and P. Needleman, *Proc. Natl. Acad. Sci. U.S.A.* **89,** 3917 (1992).

[4] R. Bertini, M. Bianchi, and P. Ghezzi, *J. Exp. Med.* **167,** 1708 (1988).

[5] H. R. Herschman, *Biochim. Biophys. Acta* **1299,** 125 (1996).

[6] J. B. Smith and H. R. Herschman, *J. Biol. Chem.* **270,** 16756 (1995).

genes in response to ligand stimulation is induced through signal transduction cascades that activate latent, preexisting transcription factors. Because synthesis of new proteins is not required, transcription rates of primary response genes increase in response to cellular stimulation even in the presence of a protein synthesis inhibitor such as cycloheximide. Studies by many investigators of immediate early genes induced by mitogens, growth factors, and cytokines have lead to the identification of genes encoding a variety of transcription factors such as *jun*, *fos*, *myc*, and *egr-1*, cytokines such as the murine chemokines JE/MCP-1 and KC/GRO, and enzymes such as the inducible prostaglandin synthase PGS2.[5,7] Many inflammatory mediators discovered in other ways (e.g., the inducible nitric oxide synthase, iNOS) have also proved to be primary response genes.[8] Despite the "immediate early" designation, induction of inflammation-related primary response genes is typically slower than that of transcription factors such as *fos* and *egr-1*.[9] In 3T3 cells, for example, message expression of PGS2, iNOS, JE/MCP-1, and KC/GRO peaks 2 to 4 hr after induction.

We observed that the induction of primary response genes encoding transcription factors involved in the G_0 to G_1 transition (e.g., *fos*, *egr-1/TIS8*, c-*myc*) is not generally suppressed by glucocorticoids. [It should be emphasized that we are referring to a lack of glucocorticoid effects on induction of transcription of these genes by mitogens. This is a phenomenon entirely different from glucocorticoid inhibition of AP-1 (*jun/fos*) activity.] In contrast, glucocorticoids markedly attenuate PGS2, iNOS, JE/MCP-1, and KC/GRO induction.[6] These examples suggested to us that glucocorticoid attenuation may distinguish a functional subclass of primary response genes involved in intercellular communication. We hypothesized (1) that PGS2, iNOS, JE/MCP-1, and KC/GRO are representative of a larger class of primary response genes (including many not yet described) whose induction by mitogens, growth factors, or inflammatory stimuli is attenuated by glucocorticoids. We also hypothesized (2) that GARGs predominantly encode proteins whose functions are extracellular or intercellular rather than intracellular.[6]

Glucocorticoid-Attenuated Response Gene Screening Strategy

A simplified outline of the GARG screening strategy is provided in Table I. GARGs are a subset of all the primary response genes in a particu-

[7] H. R. Herschman, *Annu. Rev. Biochem.* **60**, 281 (1991).

[8] R. S. Gilbert and H. R. Herschman, *J. Cell Physiol.* **157**, 128 (1993).

[9] R. R. Freter, J. C. Irminger, J. A. Porter, S. D. Jones, and C. D. Stiles, *Mol. Cell. Biol.* **12**, 5288 (1992).

TABLE I

OUTLINE OF GARG SCREENING STRATEGY, COMPARED WITH SCREENING FOR INDUCTION
OF PRIMARY RESPONSE GENES

	Screening for induction of primary response genes	Screening for glucocorticoid attenuation of induction
Cell treatments for		
Library	Inducer (plus cycloheximide)	Inducer (plus cycloheximide)
"Plus" probe	Inducer	Inducer
"Minus" probe	No inducer	Inducer plus glucocorticoid
Differentially expressed	All primary response genes	GARG subset of primary response genes

lar cell or tissue that could be identified, in theory, by screening for increased message expression following stimulation with a particular inducer (e.g., LPS, a cytokine). No existing methods are capable of identifying all differentially expressed messages, however. By targeting the specific subset of induced genes whose message expression is attenuated by glucocorticoids, we hoped to identify GARGs that may have escaped detection in previous screens. Although we originally anticipated that identifying novel genes would require more sophisticated techniques, we decided to test these ideas using a straightforward differential hybridization procedure. We synthesized a cDNA library in lambda (λ) phage, using RNA from Swiss 3T3 cells stimulated with LPS in the presence of cycloheximide. Cycloheximide was included so that only primary response genes would be induced. The library was then screened by differential hybridization, using a "plus" cDNA probe synthesized from 3T3 cells treated with LPS and a "minus" probe from cells treated with LPS and dexamethasone (DEX), a synthetic glucocorticoid (Table I). We selected as GARG candidates those clones showing reduced signal intensity on a filter hybridized with the minus probe, in comparison with a duplicate filter hybridized with the plus probe.

The methods used in the screening are described in more detail below. To simplify this discussion, we have omitted certain features of the screening we performed[6] that would not be relevant to implementation of the GARG strategy in other systems. In particular, we included transforming growth factor-$\beta 1$ (TGF-$\beta 1$) in the cell treatments for the library and the plus probe. This was done because TGF-$\beta 1$ enhances the expression of PGS2 and iNOS, our prototype GARGs, in Swiss 3T3 cells. In retrospect, the inclusion of TGF-$\beta 1$ proved to be unimportant for most of the GARGs we cloned. We also required candidate phage from the library to demonstrate increased hybridization with the plus probe versus a third (control) probe from uninduced cells. This was done to ensure that candidate clones were specifically

TABLE II

TWELVE GARG cDNAs IDENTIFIED BY DIFFERENTIAL HYBRIDIZATION OF 15,000 PHAGE[a]

Clone	Abundance	Identity	Properties/homologies
Known GARG cDNAs			
GARG-6	1/42	TSP1	Thrombospondin 1
GARG-10	4/42	crg2	Murine IP-10 (a C-X-C chemokine)
GARG-13	23/42	JE	Murine MCP-1 (a C-C chemokine)
GARG-17	5/42	MARC/fic	Murine MCP-3 (a C-C chemokine)
GARG-33	1/42	MCSF	Macrophage colony-stimulating factor
GARG-42	1/120	cyr61	Secreted growth regulator
GARG-49	1/120	IRG2	LPS and interferon-induced TPR-domain protein (IFIT-3)
Newly identified GARG cDNAs			
GARG-8	4/42	LIX	New CXC chemokine
GARG-16	1/42		LPS and interferon-induced TPR-domain protein (IFIT-1)
GARG-34	1/42		Unknown
GARG-39	1/120		LPS and interferon-induced TPR-domain protein (IFIT-2)
GARG-61	1/120		Unknown

[a] Abundance is the number of cross-hybridizing clones among the first 42 or 120 candidate phage. References for the identified genes are provided in Refs. 6 and 10.

induced by LPS. This step proved to be unnecessary, because only one potential candidate clone was eliminated by this criterion.

Testing the Glucocorticoid-Attenuated Response Gene Hypothesis

From an initial screening of 15,000 phage plaques, we identified 120 GARG candidates. After elimination of duplicates among these 120 candidates, and verification by Northern blot analysis, we obtained 12 independent GARG cDNAs. The 12 clones did not include our prototype GARGs (PGS2 and iNOS) because phage representing these genes were identified and eliminated in the screening process. Searches of the protein and nucleotide databases with partial sequences of the candidate clones showed that 7 of the 12 GARG cDNAs represented known murine genes, and 5 of the 12 were novel (Table II). Four of the 12 candidates were represented by single clones, which indicates that this screening is unlikely to have exhausted the pool of potential candidate GARGs. This observation, together with the high proportion of previously unknown murine sequences obtained from this initial screening of only 15,000 clones, provided support for the first part of our hypothesis, that GARGs are a large subclass of primary response genes, many of which have not been described. This result also shows that unknown genes can still be discovered using a relatively unsophisticated technique such as differential hybridization, if the screening strategy targets a selected subclass of genes. The entire class of GARGs expressed in different cell types and tissues may be quite large, because it should include glucocorticoid-attenuated genes that are inducible by other stimuli but not by LPS. Screening for GARGs induced by agents other than LPS should yield subsets of genes overlapping, but distinct from, the LPS-induced subset of GARGs expressed in particular cell types.

Of the seven known GARG cDNAs, six encode secreted proteins that function in intercellular communication (Table II). Among these are three chemokines: JE/MCP-1, MARC/fic/MCP-3, and crg2/IP-10. The seventh known GARG cDNA (GARG-49/IRG2) encodes a protein of unknown function (IRG2). Notably, absent from this group of GARGs are any genes that encode transcription factors, protein kinases, or protein phosphatases, categories of genes frequently identified in typical screens for ligand-induced primary response genes. Taken together, the characteristics of the seven known GARG cDNAs we cloned provides strong support for the second part of our hypothesis, that GARGs mainly encode products that function in extracellular rather than intracellular processes. It should be noted, again, that our hypothesis does not require the products of GARGs to be secreted proteins. For example, PGS2, the inducible prostaglandin synthase, and iNOS, the inducible nitric oxide synthase, are intracellular enzymes.[5]

One of the five novel GARG cDNAs we cloned encodes a new C-X-C chemokine, designated LIX, which is described below. GARG-16 and GARG-39 encode distinct proteins related to GARG-49/IRG2 and to two human interferon-inducible proteins of unknown function, IFI-56K and ISG-54K. Thus, 3 related but distinct members of this family are among the 12 GARG cDNAs we cloned. Analysis of the complete cDNA sequences of GARG-16, GARG-39, and GARG-49/IRG2, together with their homologs in other species, allowed us to identify multiple highly conserved tetratricopeptide repeat (TPR) domains in this family of proteins.[10] We have hypothesized that these LPS- and interferon-induced proteins are regulatory factors that participate in multicomponent assemblies. The remaining novel sequences, GARG-34 and GARG-61, have not yet been completely characterized.

It is interesting to note that, among the 12 GARGs identified in our screen, those whose message levels are most strongly attenuated by dexa-

---

[10] J. B. Smith and H. R. Herschman, *Arch. Biochem. Biophys.* **330**, 290 (1996).

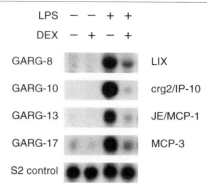

Fɪɢ. 1. GARG message expression in Swiss 3T3 fibroblasts. Attenuation by dexamethasone of LPS-induced message expression is shown for the four chemokines cloned in our screen. Swiss 3T3 fibroblasts were pretreated with 2 μM dexamethasone (DEX) for 3 hr before addition of 10 ng/ml LPS and harvested after 4 hr more. Northern analysis was performed on replicates from the same preparation of total cellular RNA (10 μg per lane). Each filter was reprobed with the constitutive control S2 to verify uniform loading.

methasone are the four chemokines (Fig. 1). Dexamethasone reduces the LPS-induced expression of GARG-10/crg-2 by more than 95%, and that of GARG-8/LIX, GARG-13/JE, and GARG-17/MARC by 75–80%. The effect of dexamethasone on LPS-induced expression of the other GARG genes is more modest. We also observed that, for several GARGs, dexamethasone attenuation varies with the inducer. For example, dexamethasone attenuates basal expression of GARG-6/thrombospondin 1 by 87%, attenuates TGF-β1 or LPS induction by 40–50%, but has a minimal effect (<10%) on serum induction.[6] This observation emphasizes that the ability of glucocorticoids to attenuate expression of a particular primary response gene may be inducer-specific, and need not be a universal property of the gene.

LIX: A New C-X-C Chemokine

The GARG-8 clone encodes a novel member of the chemokine family, which we designated LIX, for LPS-induced C-X-C chemokine.[6] The predicted LIX protein has a 40-amino acid signal sequence and a 92-amino acid mature peptide. The COOH-terminal region of LIX has a distinctive length and sequence not present in any known chemokine. An alignment of the LIX mature peptide with its four closest relatives among the C-X-C chemokines is shown in Fig. 2. These four peptides, porcine alveolar macro-

FIG. 2. Alignment of the LIX protein with its four closest relatives among the C-X-C chemokines. Amino acids identical to those in the corresponding position in LIX are shown in reverse type. The four conserved cysteines of the C-X-C chemokine family are indicated by asterisks (*). Porcine (por), bovine (bov), human (hu), and murine (mu) sequences are shown.

phage chemotactic factor-II (AMCF-II),[11] bovine and human granulocyte chemotactic peptide-2 (GCP-2),[12] and human epithelial neutrophil activating peptide-78 (ENA-78),[13] are more closely related to one another than to LIX. In the alignment shown in Fig. 2, all four of these chemokines differ from LIX at 18 sites, in addition to the unique C-terminal region of LIX. At 8 of the 18 sites where they all differ from LIX, those four chemokines, from three species, are identical to one another. At the remaining 10 sites, two or more of the four chemokines have an identical residue. The Q-51 residue in LIX is unique: all other known C-X-C peptides have a G at the corresponding position, except for D in crg2. These amino acid differences, plus the distinctive COOH-terminal region, lead us to suggest that the LIX is a novel C-X-C chemokine, and not the murine ortholog of any known chemokine in other species.[6]

On the basis of the presence in LIX of the ELR sequence preceding the first cysteine, and other structural features shown by mutation analysis to be important in IL-8, we predicted that the LIX product would be a neutrophil chemoattractant.[6] Amino- and carboxyl-terminal processed forms of LIX were subsequently purified from fibroblast cell lines by Wuyts et al.[14] and shown to have potent neutrophil chemoattractant activity. These

[11] R. B. Goodman, D. C. Foster, S. L. Mathewes, S. G. Osborn, J. L. Kuijper, J. W. Forstrom, and T. R. Martin, *Biochemistry* **31,** 10483 (1992).

[12] P. Proost, A. Wuyts, R. Conings, J. P. Lenaerts, A. Billiau, G. Opdenakker, and J. Van Damme, *Biochemistry* **32,** 10170 (1993).

[13] A. Walz, R. Burgener, B. Car, M. Baggiolini, S. L. Kunkel, and R. M. Strieter, *J. Exp. Med.* **174,** 1355 (1991).

[14] A. Wuyts, A. Haelens, P. Proost, J. P. Lenaerts, R. Conings, G. Opdenakker, and J. Van Damme, *J. Immunol.* **157,** 1736 (1996).

authors referred to LIX as murine GCP-2. However, we have cloned the human GCP-2 gene as well as the murine LIX gene (J. B. Smith and H. R. Herschman, unpublished, 1997) and analyzed the nucleotide as well as protein sequence relationships of LIX, human GCP-2, and human ENA-78. This analysis does not support the idea that LIX is the specific murine counterpart of human GCP-2. In particular, we find that LIX is no more closely related to human GCP-2 than it is to human ENA-78. LIX could be the murine counterpart of this closely related pair of human genes. Alternatively, a true human ortholog of LIX may remain to be discovered.

The LIX chemokine is induced by LPS in Swiss 3T3 cells, a fibroblastlike cell line, at concentrations as low as 0.1 ng/ml. Expression of LIX peaks at 2 to 4 hr, but it remains well above basal levels for at least 24 hr. Serum produced no detectable induction of LIX. In contrast, the three other chemokines we cloned (GARG-10/crg2, GARG-13/JE, and GARG-17/ MARC) are induced by serum in 3T3 cells. We also showed that LIX message expression is induced by LPS in early-passage mouse embryo fibroblasts, and we suggested that LIX may participate in the recruitment of inflammatory cells by injured or infected tissue. We have subsequently demonstrated substantial induction of LIX message in lung, heart, spleen, and other organs of mice injected with LPS *in vivo* (J. B. Smith and H. R. Herschman, unpublished, 1997). However, we were unable to detect LIX expression in two macrophage cell lines or in peritoneal macrophages stimulated *in vitro* or *in vivo* with LPS. In contrast, porcine AMCF-II, the closest structural relative of LIX, is expressed in alveolar macrophages.[11] Further investigation of the cell-specific regulation of LIX in comparison with related chemokines in other species will help to determine if LIX has a unique biological role.

The 3'-untranslated region of the LIX message contains a 125-bp segment with strong similarity to sequences in the 3'-untranslated regions of three other C-X-C chemokines: porcine AMCF-II, human ENA-78, and human GCP-2[6] (J. B. Smith and H. R. Herschman, unpublished, 1997). (We expect that the 3'-untranslated region of bovine GCP-2, which has not yet been described, will prove to have a similar sequence.) Remarkably, the nucleotide sequence identities in this 3'-untranslated segment are much greater than in the protein coding regions of these chemokines. This segment, conserved in a subset of chemokines from at least three species, is likely to serve an important regulatory function.

Glucocorticoid-Attenuated Response Gene Screening Methods

Although we cloned LIX and the other GARGs listed in Table II by differential hybridization screening, the basic idea of the GARG cloning strategy is not limited to this technique. A variety of more powerful methods for cloning differentially expressed genes could be used to search for GARGs in future studies. These methods include differential display,[15] representational difference analysis,[16–18] serial analysis of gene expression (SAGE),[19] and a number of approaches for large-scale automated analysis of gene expression that are currently under development.[20,21]

General Considerations

The most important determinants of the outcome of a search for GARGs may be the choice of cell type and inducer to be used. Although some GARGs may be expressed in a wide range of cell types, others will have a more restricted range of expression, which will vary with the inducer. Thus, a search for GARGs induced in Swiss 3T3 cells by agents other than LPS should yield subsets of genes overlapping, but distinct from, the LPS-induced subset of GARGs.

It is helpful to identify as a positive control at least one glucocorticoid-attenuated gene known to be expressed in the system chosen for the search. Our positive controls were PGS2 and iNOS. It is also helpful to identify a constitutive control, that is, a gene whose message expression is unaffected by the inducer or by glucocorticoids. Because glucocorticoid attenuation of induction is generally less than 100%, the magnitude of the difference in expression between induced and glucocorticoid-attenuated is less than the difference between induced and uninduced (Table I and Fig. 1). A well-chosen constitutive control will improve the sensitivity of screening for expression of genes with modest levels of attenuation (e.g., 30–50%) and minimize the number of false positives. We used a cDNA for the murine ribosomal S2 protein for this purpose. The positive and constitutive controls may be used in preliminary experiments to optimize induction and glucocorticoid attenuation of the positive controls. The mechanism of gene induction and glucocorticoid attenuation, and therefore the optimal dose and timing of the glucocorticoid and the inducer, will vary for different genes and different inducers. Of course, the conditions that will prove optimal for the unknown genes to be cloned will not be known in advance.

[15] P. Liang and A. B. Pardee, *Science* **257,** 967 (1992).

[16] B. S. Braun, R. Frieden, S. L. Lessnick, W. A. May, and C. T. Denny, *Mol. Cell. Biol.* **15,** 4623 (1995).

[17] N. Lisitsyn, N. Lisitsyn, and M. Wigler, *Science* **259,** 946 (1993).

[18] M. Hubank and D. G. Schatz, *Nucleic Acids Res.* **22,** 5640 (1994).

[19] V. E. Velculescu, L. Zhang, B. Vogelstein, and K. W. Kinzler, *Science* **270,** 484 (1995).

[20] M. Schena, D. Shalon, R. W. Davis, and P. O. Brown, *Science* **270,** 467 (1995).

[21] E. S. Lander, *Science* **274,** 536 (1996).

Probes and Library Construction

RNA for the probes should be prepared from a single batch of cells, treated identically except for the addition of glucocorticoid for the minus probe cells. In our experiments, we added 10 ng/ml of LPS to the plus cell medium 2 hr before harvesting the cells for RNA isolation. For the minus probe, we added dexamethasone to final concentration of 2 μM 3 hr before addition of LPS. These conditions were determined by preliminary experiments using PGS2 and iNOS as positive controls. Optionally, uninduced cells may be harvested for preparation of a control, uninduced, cDNA probe. The library to be screened can be made from the same RNA preparation as the plus cells, or from a separate batch of cells treated simultaneously with the inducer and with cycloheximide, so that only primary response genes are induced. Alternatively, subtraction techniques could be used to increase the representation of differentially expressed genes in the library.

Before proceeding further, one should assess the integrity of the RNA preparations and verify the differential expression of the known GARG controls. This may be done by Northern blot analysis of the plus, minus, and library RNA preparations, using the positive and constitutive control cDNAs as probes.

After construction of the library, use the known GARG and constitutive control cDNAs as probes in a preliminary screening to identify clones containing those inserts. Plaque-purify the control phage for use as internal controls in the differential screening, and suspend the phage from a single plaque in 1 ml of sterile SM buffer[22] with 50 μl of chloroform. The frequency of representation of the known GARG control phage in the library provides a useful estimate of the number of phage that will need to be screened to identify candidate phage expressed at similar levels.

Differential Screening

Detailed descriptions of general λ phage techniques are available elsewhere[22] and are not repeated here. We provide an abbreviated account of the procedures we used, emphasizing specific techniques that may increase the sensitivity and reliability of screening for GARGs. We used Lambda Zap II phage with XL-1 Blue MRF' bacteria (Stratagene, La Jolla, CA).

1. Using a freshly prepared stock of bacteria, titer the cDNA library and then plate appropriate dilutions of the library to obtain a density of approximately 2000 clones per 150-mm petri plate. We found it convenient

---

[22] J. Sambrook, E. F. Fritsch, and T. Maniatis, "Molecular Cloning: A Laboratory Manual," 2nd Ed. Cold Spring Harbor Laboratory, Cold Spring Harbor, New York, 1989.

to work with eight plates at one time, because autoradiographs of eight filters can be obtained simultaneously on a single 11- by 17-inch sheet of film.

2. After the top agarose solidifies at room temperature, mark each plate to define its orientation on a template that identifies the locations for inoculating control phage. It is helpful to inoculate each plate with the constitutive control phage at two or three different locations, and with each of the known GARG control phage in one location. These phage may be inoculated by dipping the end of a sterile toothpick into an aliquot of the plaque-purified phage in SM buffer, then lightly touching the agarose with the toothpick at the specific locations indicated by the template. Incubate the plates overnight at 37°. The control phage spots may be several times the size of the individual plaques of the library phage.

3. Make replicate transfers of the phage to nitrocellulose filters. To obtain reproducible transfers, use fresh nitrocellulose filters (132 mm diameter, 0.45 μm pore size; Schleicher & Schuell, Keene, NH, or equivalent) from the same lot. Using gloves, and touching the filters only at the edge, label each filter using a waterproof marking pen. Carefully place a filter on the surface of each plate, avoiding wrinkles or air bubbles. Use a sterile needle to make multiple punctures through the filter and agar in an asymmetric pattern around the edge. Subsequent filters on the same plate must be punctured in identical positions (using the holes in the agar) to allow precise orientation of the filters with one another. To transfer equal amounts of phage DNA, successive filters must remain in contact with the phage for progressively longer periods of time. We obtained good results for quadruplicate transfers with contact times of 1.5, 3, 6, and 12 min for the first through fourth filters, respectively, which we labeled A–D. The first two transfers (to filters A and B) are the most reliable, and they should be used for the differential hybridization. The third and fourth, which are optional, may be used for additional controls. After transfer, carefully place each filter (phage side up) for 3 min on Whatman (Clifton, NJ) No. 1 filter paper soaked with a solution of 0.5 M NaOH and 1.5 M NaCl. Neutralize the filter on paper soaked with 3 M NaCl plus 0.5 M Tris-HCl, pH 7.45, for 3 min, and then on 2× SSPE for 3 min. After air drying, vacuum-bake the filters at 80° for 1 hr.

4. Carefully stack all of the A filters (first transfer), rotating the filters so that the labeling marks are not directly aligned. (There will be some bleeding of the marking ink onto adjacent filters). Do the same for the other groups of filters. Preincubate each stack in separate sealed plastic bags for 2–4 hr at 42° in 50% formamide, 6× Denhardt's reagent, 6× SSC, 60 mM sodium phosphate (pH 6.8), 1 mM sodium pyrophosphate, 145 μg/ml ATP, and 60 μg/ml sonicated nondenatured salmon sperm DNA.

5. Use equal quantities of the plus, minus, and (optionally) uninduced control RNA to prepare ^{32}P-labeled cDNA probes using Moloney murine leukemia virus (MMLV) reverse transcriptase and suitable primers [we used a mixture of both oligo(dT)[12-18] and random hexamer primers]. After alkaline denaturation of the RNA and removal of unincorporated nucleotides, check for equal incorporation of label in the probes. Add equal counts of the ^{32}P-labeled plus and minus cDNA probes to bags containing the stacks of first and second filter lifts, respectively, and hybridize at 42° for 4 days. The third lifts may be hybridized with the control cDNA probe (from uninduced cells). Alternatively, both the third and fourth lifts may be hybridized with specific cDNA probes for the known positive control genes. Wash the filters repeatedly in 2× SSC + 0.5% sodium dodecyl sulfate (SDS) at room temperature and then at 60°. Continue washing at each temperature until the radioactivity detectable in the wash buffer with a handheld counter falls to background levels. Perform a final high-stringency wash in 0.1× SSC + 0.1% SDS at 65°.

6. Expose the filters to X-ray film, ensuring that the filters are flat, with no air bubbles under the plastic film used to cover the filters. A range of exposure times, from a few hours to several days, will be needed to obtain optimal differential signals for different clones, depending on the abundance of their messenger RNAs. Adjust the lengths of exposures of the plus, minus, and control filters, if necessary, to ensure that the constitutive control phage spots on lifts from the same plate have equal signal intensity, then verify that the positive control phage spots show a greater hybridization signal on the plus compared with the minus filters. Select candidate plaques that show greater hybridization signal on the plus filter as compared with both the minus and control filters on at least two sets of autoradiograms exposed for different times. Remove the candidate plaques using a sterile Pasteur pipette and suspend in 1 ml sterile SM buffer with 50 μl chloroform.

7. Plaque purify the candidate phage by repeating the differential hybridization procedure for the individual phage picks. Replate each candidate at a density of 50–100 plaques on 100-mm petri dishes and spot the constitutive control phage at two or three locations on each plage. Perform duplicate transfers to 82-mm nitrocellulose filters, and hybridize stacks of filters (we have used as many as 20 filters per stack) with freshly prepared plus and minus cDNA probes. Select isolated plaques of the rescreened phage that demonstrate reproducible differential hybridization relative to the constitutive control. (There should be multiple examples of each phage on the plate, which makes this evaluation more reliable than the primary screen.) Eliminate those candidates that fail to show differential hybridization on rescreening.

Amplification of Phage Inserts by Polymerase Chain Reaction

To obtain probes for Northern blot analysis, the inserts in the candidate phage can be amplified by polymerase chain reaction (PCR) with standard primers for sequences flanking the polycloning region. If the PCR products are to be used as hybridization probes on grids of the λ phage, as described below, it is advantageous to minimize the amount of vector sequence included in the region amplified. We used primers (5'-CGGGCTGCAG-GAATTC-3' and 5'-CCCCTCGAGTTTTTTTTTT-3') designed to hybridize just outside the 5' end of the insert at the *Eco*RI cloning site, and at the overlap of the poly(A) tail of the insert with the *Xho*I cloning site of Lambda Zap II.[23]

1. To obtain phage DNA free of the salts in the SM buffer, replate the plaque-purified phage and grow overnight. Pick a single plaque, and place in 0.5 ml distilled water with 25 μl chloroform. Freeze and thaw twice, and store frozen at −20°.

2. Heat a 25-μl aliquot of the phage to 95° for 5 min. Amplify the insert using 2.5 units of Ampli-Taq polymerase and buffer (Perkin-Elmer, Norwalk, CT), dNTP concentrations of 200 μM, and 1 μM each of the above primers in a total volume of 50 μl. After an initial denaturation at 94° for 3 min, cycle the reactions 30 times (15 sec at 94°, 15 sec at 48°, 1 min at 72°), with a final extension at 72° for 15 min.

3. Analyze the PCR products by electrophoresis. If PCR for particular candidate phage is unsuccessful or results in multiple bands, the candidate inserts may be obtained by converting the Lambda ZAP II phage to Bluescript II phagemids by *in vivo* excision[24] and then removing the insert with suitable restriction enzymes.

Eliminate Phage Duplicates by Cross-Hybridization

Differentially expressed genes whose messages are abundant are likely to be represented by multiple candidates. Duplicate phage can be efficiently identified as follows.

1. Spot the plaque-purified candidates in a grid pattern on a lawn of bacteria on several replicate 150-mm plates. After overnight growth of the phage, prepare multiple lifts from each plate as described for the library screening.

[23] L. Vician, I. K. Lim, G. Ferguson, G. Tocco, M. Baudry, and H. R. Herschman, *Proc. Natl. Acad. Sci. U.S.A.* **92,** 2164 (1995).
[24] J. M. Short and J. A. Sorge, *Methods Enzymol.* **216,** 495 (1992).

2. Use PCR to amplify inserts directly from the candidate phage, as described above. Select a number of PCR products or inserts of different sizes, and prepare ^{32}P-labeled cDNA probes from them.

3. Hybridize each of the selected probes with one of the replicate filters containing the grid of candidates, wash under stringent conditions, and expose to film. Plaques hybridizing with the same signal intensity as the plaque corresponding to the probe are likely to be duplicate isolates, and can be eliminated from further consideration. (However, this procedure might eliminate a distinct gene that has extremely high sequence similarity to another candidate. If this is a concern, each candidate may be sequenced.)

4. Repeat the cross-hybridization with additional insert probes until all multiples have been identified, or the number of candidates has been reduced sufficiently so that the labor involved in plasmid preparation and sequencing is not excessive.

Verification and Analysis of Candidates

Although the use of the constitutive control phage will significantly improve the reliability of the screening, it is essential to verify by Northern blot analysis that the candidates truly represent inducible genes whose induction is attenuated by glucocorticoids. After verification, sequence the ends of the candidate inserts and search for database matches.

Summary

We describe an approach for identifying novel inflammatory mediators, based on screening for immediate early/primary response genes whose induction by an inflammatory stimulus is attenuated by glucocorticoids. This procedure can be applied to a wide range of cell types and tissues, using a variety of inducers. In an initial test of this idea, we identified cDNAs for 12 LPS-induced, glucocorticoid-attenuated response genes (GARGs) by differential hybridization screening of a λ phage cDNA library from murine 3T3 fibroblasts. Seven of the GARGs were known genes, including the chemokines JE/MCP-1, fic/MARC/MCP-3, and crg2/IP-10. One of the novel cDNAs was a new C-X-C chemokine that we designated LIX, for LPS-induced C-X-C chemokine. Because the 12 GARG cDNAs were identified in a single screening of only 15,000 phage, and four were found as single isolates, these results suggest that there are many GARGs not yet described. Furthermore, six of the seven known GARGs encode proteins that modulate intercellular communication. These results support our hypothesis that GARGs predominantly encode products that function in paracrine cell communication. Here we provide an overview of the GARG strategy and the differential hybridization procedures used in our initial screening. A variety of other methods for identifying differentially expressed genes may be used in future searches for novel GARGs.

Acknowledgments

We thank Linda Vician for contributions to the development of our λ phage screening techniques. This work was supported in part by the American Cancer Society Institutional Grant IN/IRG-131N to the Jonsson Comprehensive Cancer Center at the University of California, Los Angeles, and by the National Institutes of Health Grant GM24797.

[18] Chemokine-Induced Human Lymphocyte Infiltration and Engraftment in huPBL-SCID Mice

By Dennis D. Taub, Michael L. Key, Dan L. Longo, and William J. Murphy

Introduction

The extravasation of leukocytes from the circulation into inflammatory or lymphoid tissues requires a series of soluble and cell-bound signals between the responding leukocyte and the vascular endothelial barrier. Through a coordinated series of signals generated within a tissue lesion, the local vascular endothelial barrier becomes "sticky," permitting responding leukocytes to roll along this primed vasculature and slowing their transit through the circulation. On interacting with leukocyte-specific as well as nonspecific proadhesive molecules, these rolling leukocytes are signaled to tightly adhere to and spread out along the endothelial layer through the activation of cell surface integrin molecules. These primary adhesion events are prerequisite for the successful trafficking of circulating leukocytes into extravascular tissues.

The subsequent steps leading to leukocyte transendothelial migration and their directed movement through the extravascular space toward the signaling tissue are also dependent on the presence of additional adhesive interactions involving integrin molecules as well as the presence of leukocyte-specific chemotactic molecules.[1-4] Similar interactions are believed to

---

[1] M. L. Dustin and T. J. Springer, *Annu. Rev. Immunol.* **9,** 27 (1991).
[2] Y. Shimizu, W. Newman, Y. Tanaka, and S. Shaw, *Immunol. Today* **13,** 106 (1992).
[3] T. A. Springer, *Cell* **76,** 301 (1994).
[4] E. C. Butcher and L. J. Picker, *Science* **272,** 60 (1996).

occur for lymphocyte trafficking into lymphoic or thymic tissues.[1–4] On encountering a chemotactic molecule, responding leukocytes begin to migrate directionally from regions of low ligand concentrations toward the sites of chemoattractant production, which typically possess more substantial levels of soluble chemotactic factors (a process called chemotaxis). Many chemoattractants have also been shown to directly bind to the cell surface of vascular endothelial cells and various adhesive substrates within the tissue, facilitating the directed migration of leukocytes while still maintaining a bound concentration gradient (a process known as haptotaxis).[5–10] Using this mechanism, only those leukocytes that have already established their first adhesive interaction with the endothelium should be permitted exposure to chemotactic agents. On entering the region of chemoattractant production, this directed leukocyte movement is shut down, permitting the infiltrating cells to localize within the signaling tissue.

Activated serum components (e.g., C5a), platelet-activating factor (e.g., PAF), eicosinoids [e.g., prostaglandin E_2 (PGE_2), leukotriene B_4 (LTB_4)], bacterial-derived peptides (e.g., f-Met-Leu-Phe, fMLP), endotoxin, growth factors [e.g., human growth factor (HGF)], cytokines [e.g., interleukin-1 (IL-1), tumor necrosis factor-α (TNF-α), interferon-γ (IFN-γ)], and neuroendocrine hormones [e.g., growth hormone (GH), prolactin (PRL), insulin-like growth factor-1 (IGF-1), opiates] have all been shown to induce leukocyte chemotaxis and/or adhesion to endothelial cells and purified adhesive ligands.[10] The subtype of leukocytes that appear in inflammatory infiltrates can differ markedly depending on the identity of the inflammatory stimuli as well as the duration of irritation. The chemotactic factors responsible for these differences are likely to be cell type-specific (selective) chemoattractants and/or chemoattractant receptors.

Although most of the chemotactic and proadhesive mediators listed above facilitate leukocyte trafficking, their lack of leukocyte subset specificity has brought their relevance in selective leukocyte recruitment into question. Since the late-1980s, a rapidly growing superfamily of small, soluble, structurally related molecules called chemokines have been identified and shown to selectively promote the rapid adhesion and chemotaxis of a variety of leukocyte subtypes both *in vivo* and *in vitro*.[5–10] Chemokines are produced by almost every cell type in the body in response to a number of inflammatory signals, in particular those that activate leukocyte–endothelial

[5] M. Baggiolini, B. Dewald, and B. Moser, *Adv. Immunol.* **55,** 97 (1994).

[6] D. D. Taub and J. J. Oppenheim, *Ther. Immunol.* **1,** 229 (1994).

[7] K. B. Bacon and T. J. Schall, *Int. Arch. Allergy Immunol.* **109,** 97 (1996).

[8] T. J. Schall and K. B. Bacon, *Curr. Opin. Immunol.* **6,** 865 (1994).

[9] R. M. Strieter, T. J. Standiford, G. B. Huffnagle, L. M. Colletti, N. W. Lukacs, and S. L. Kunkel, *J. Immunol.* **156,** 3583 (1996).

[10] D. D. Taub, *Cytokine Growth Factor Rev.* **7**(4), 355 (1996).

cell interactions. On the basis of the presence or absence of an intervening amino acid residue located between the first two of the four conserved cysteine residues or the absence of one of the cysteine residues at the amino terminus, the chemokine superfamily can be separated into distinct subfamilies called the C-X-C, C-C, and C subfamilies, respectively.[5-10] Chemokines also appear to play important roles in cellular activation and leukocyte effector functions.[5-10] These activities appear to be relevant as chemokines appear to play a role in a variety of disease states, including rheumatoid arthritis, sepsis, atherosclerosis, asthma, psoriasis, ischemia/reperfusion injury, a variety of pulmonary disease states, and acquired immunodeficiency syndrome (AIDS).[5-10] As the previous chapters have described, many of the characteristics and properties of chemokines, this chapter focuses on the effects of chemokines on human T-cell trafficking.

The effect of chemokines on lymphocyte migration has been a hot subject of investigation since the early 1990s. Many of the C-C chemokines, including macrophage inflammatory protein-1α (MIP-1α), MIP-1β, RANTES, macrophage chemotactic protein-1 (MCP-1), MCP-2, MCP-3, MCP-4, and C10, as well as the C-X-C chemokines interleukin-8 (IL-8), growth related oncogene α (GROα), interferon-inducible protein-10 (IP-10), stromal cell derived factor 1 (SDF-1), and MIG, have been shown to induce significant human T-lymphocyte and/or natural killer (NK) cell migration *in vitro*.[11-22] The initial studies by Larsen *et al.*[12] and Schall *et al.*[11] were the first to demonstrate the ability of IL-8 and RANTES,

[11] T. J. Schall, K. Bacon, K. I. Toy, and D. V. Goedell, *Nature (London)* **347,** 669 (1990).

[12] C. G. Larsen, A. O. Anderson, E. Appella, J. J. Oppenheim, and K. Matsushima, *Science* **243,** 1464 (1989).

[13] Y. Tanaka, D. H. Adams, S. Hubscher, H. Hirano, U. Siebenlist, and S. Shaw, *Nature (London)* **361,** 81 (1993).

[14] D. D. Taub, K. Conlon, A. R. Lloyd, J. J. Oppenheim, and D. J. Kelvin, *Science* **260,** 355 (1993).

[15] T. J. Schall, K. Bacon, R. D. R. Camp, J. W. Kaspari, and D. V. Goeddel, *J. Exp. Med.* **177,** 1821 (1993).

[16] D. D. Taub, A. R. Lloyd, K. Conlon, J. M. Wang, J. R. Ortaldo, A. Harada, K. Matsushima, D. J. Kelvin, and J. J. Oppenheim, *J. Exp. Med.* **177,** 1809 (1993).

[17] M. W. Carr, S. J. Roth, E. Luther, S. S. Rose, and T. A. Springer, *Proc. Natl. Acad. Sci. U.S.A.* **91,** 3652 (1994).

[18] D. D. Taub, P. Proost, W. J. Murphy, M. Anver, D. L. Longo, J. Van Damme, and J. J. Oppenheim, *J. Clin. Invest.* **95,** 1370 (1995).

[19] P. Loetscher, M. Seitz, I. Clark-Lewis, M. Baggiolini, and B. Moser, *FASEB J.* **8,** 1055 (1994).

[20] D. D. Taub, T. Sayers, C. Carter, and J. R. Ortaldo, *J. Immunol.* **155,** 3877 (1995).

[21] G. S. Kelner, J. Kennedy, K. Bacon, S. Kleyensteuber, D. A. Largaespada, N. A. Jenkins, N. G. Copeland, J. F. Bazan, K. W. Moore, T. J. Schall, and A. Zlotnik, *Science* **266,** 1395 (1994).

[22] C. C. Bleul, R. C. Fuhlbrigge, J. M. Casasnovas, A. Aiuti, and T. A. Springer, *J. Exp. Med.* **184,** 1101 (1996).

respectively, to facilitate human T-lymphocyte migration *in vitro*. Depending on the laboratory, subsequent studies have demonstrated selective $CD4^+$, $CD8^+$, $CD45RA^+$, $CD45RO^+$, $CD26^+$, or $CD26^-$ T-cell subset migration in response to C-X-C and C-C chemokines and, in some cases, depending on the chemokine concentration utilized in the assay.[5-10] The C chemokine lymphotactin and the C-X-C chemokine SDF-1 have been shown to be potent chemoattractants for human and murine lymphocytes or thymocytes with little to no significant biological activity on other leukocyte subsets.[21,22] A number of factors appear to affect T-cell migration including the T-cell activation state, the memory–naive status, the presence of adhesive ligands, the differences in levels and signaling through both promiscuous and specific chemokine receptors on the lymphocyte cell surface, variations in the responsiveness of individual T cells and their subsets, the T-cell donor, and (perhaps analogous to basophils and eosinophils) a need for cytokine priming. The ability of C-C chemokines to selectively promote human T-cell subset migration may play a role in sorting out various lymphocytes and lymphocyte subpopulations into different lymphoid compartments.

The most common *in vivo* study performed to assess chemokine activity is the ability of a given chemokine to recruit leukocytic subpopulations to sites of challenge. *In vivo* injection of both C-X-C and C-C chemokines into rodent and rabbit models have been shown to induce a number of proinflammatory effects.[5-10] Subcutaneous and intradermal injections of C-X-C chemokines such as IL-8, $GRO\alpha$, MIP-2, and neutrophil-activating peptide-2 (NAP-2) into rabbits, rats, mice, and humans have been reported to induce predominantly neutrophil infiltration, beginning as soon as 3 hr postinjection.[5-10] Very few monocytes or lymphocytes have been observed in these studies; however, only short-term histological analyses (<24 hr) were performed by these investigators.

Our laboratories, in collaboration with Dr. Hugh Perry,[23] has shown that intracranial injections of IL-8 and MIP-2 injections into the murine hippocampus using specialized micropipettes revealed a dramatic polymorphonuclear (PMN) leukocyte recruitment across the brain parenchyma of the central nervous system (CNS). This parenchymal accumulation of neutrophils was associated with a breaching of the blood–brain barrier that was particularly severe after MIP-2. Similarly, injections of some of the C-C chemokines, such as MIP-1α, MIP-1β, RANTES, MCP-1, and I-309/ TCA-3, have been reported to cause an edema and a mild early neutrophil accumulation followed by a more prominent monocytic infiltration.[5-10] Hippocampal injections of C-C chemokines into murine brains have also

[23] M. Bell, D. D. Taub, and V. H. Perry, *Neuroscience (Oxford)* **74,** 283 (1996).

revealed that MCP-1 induces significant accumulation of mononuclear cells across the brain parenchyma.[23] However, no blood–brain barrier breach was observed using MIP-1α, RANTES, or MCP-1. Although all of these studies and others (described in Refs. 5–10) suggest that C-X-C and C-C chemokines have profound chemotactic effects on murine neutrophils and mononuclear cells (particular monocytes), little is known about the *in vivo* effects of chemokines on human cell migration and homing.

The use of animal models has greatly increased our understanding of the role of chemokines during the initial and intermediate stages of an inflammatory response. Many laboratories have examined the production, cellular localization, and effects of *in vivo* neutralization of specific chemokines in various animal models of disease. In a number of inflammatory and noninflammatory disease states, C-X-C and C-C chemokines appear to play a key role in the accumulation of leukocytes at the site of an inflammatory lesion.[5–10] Although little is known about the role of chemokines in the progression of disease in humans, it is believed that chemokine production is a prerequisite for the initiation of leukocyte recruitment into the areas of tissue damage.

Several laboratories have established animal models to explore the various roles of chemokines in inflammation and disease progression. Our laboratories have chosen to examine human lymphocyte migration and trafficking using a human/mouse chimeric model in mice possessing the gene for severe combined immunodeficiency (SCID). SCID mice have a defect in their DNA repair system that does not permit productive rearrangement of their immune receptor genes.[24] Thus, these mice lack mature T and B cells and are incapable of rejecting solid tissue allografts. However, these mice do possess normal NK cell function and can reject bone marrow allografts after lethal irradiation. Mosier and colleagues[25] were the first to describe the ability of human (hu) peripheral blood lymphocytes (PBLs) to engraft into SCID mice. The huPBLs were injected intraperitoneally, and human T and B cells persisted in these mice for months in the peritoneum and peripheral lymphoid organs with no overt pathology being observed. Subsequent studies by our laboratories using this model have demonstrated significant effects of neurohormones on human T-cell functions and engraftment into murine lymphoid tissues.[26–28]

[24] W. J. Murphy, D. D. Taub, and D. L. Longo, *Semin. Immunol.* **8,** 233 (1996).
[25] D. E. Mosier, R. J. Gulizia, S. M. Baird, and D. B. Wilson, *Nature (London)* **335,** 256 (1988).
[26] W. J. Murphy, S. K. Durum, and D. L. Longo, *J. Exp. Med.* **178,** 231 (1993).
[27] W. J. Murphy, S. K. Durum, and D. L. Longo, *J. Immunol.* **149,** 3851 (1992).
[28] W. J. Murphy, S. K. Durum, and D. L. Longo, *Proc. Natl. Acad. Sci. U.S.A.* **89,** 4481 (1992).

The huPBL-SCID mouse has been used extensively by many laboratories as a model for the examination of human immune function as well as for the study of human disease states. However, questions concerning the effects of placing immunocompetent human cells into an immunocompromised xenogeneic host still remain an issue.[24] Despite the obvious limitations of such a chimeric model, the huPBL-SCID model remains the most accessible and easiest *in vivo* system in which to examine human lymphocyte function and activity.

On the basis of the *in vitro* effects of C-C and C-X-C chemokines on T-cell migration and adhesion, we examined the effects of chemokines on human T-cell trafficking in huPBL-SCID mice. We have found that human CD3$^+$ T cells migrate within 4 hr into injection sites in response to recombinant preparations of human MIP-1α, MIP-1β, RANTES, and MCP-1 and within 72 hr in IL-8 and IP-10 injection sites.[29-33] Furthermore, huPBL-SCID mice treated with multiple injections of the chemokines MIP-1β or RANTES or the human neurohormone GH, but not recombinant human (rh) IL-2, macrophage colony-stimulating factor (M-CSF), or platelet factor 4 (PF4), demonstrated significant human T-cell trafficking to the thymus and various peripheral lymphoid tissues.[34] As with the infiltration studies, chemokine-mediated human T-cell engraftment was found to be chemokine-specific and required the presence of active integrin molecules. This chapter describes several of our chemokine studies in huPBL-SCID mice as well as the benefits and pitfalls of this chimeric system.

Materials and Methods

Mice

BALB/c and C.B-17 *scid/scid* (SCID) mice are obtained from the Animal Production Area at the U.S. National Cancer Institute–Frederick Cancer Research and Development Center (NCI-FCRDC, Frederick, MD).

[29] W. J. Murphy, D. D. Taub, M. Anver, J. J. Oppenheim, D. J. Kelvin, and D. L. Longo, *Eur. J. Immunol.* **24,** 1823 (1994).

[30] D. D. Taub, M. Anver, J. J. Oppenheim, D. L. Longo, and W. J. Murphy, *J. Clin. Invest.* **97,** 1931 (1996).

[31] D. D. Taub, D. L. Longo, and W. J. Murphy, *Blood* **87,** 1423 (1996).

[32] D. D. Taub, S. M. Turcovski-Corrales, M. L. Key, D. L. Longo, and W. J. Murphy, *J. Immunol.* **156,** 2095 (1996).

[33] W. J. Murphy, Z.-G. Tian, O. Asai, S. Funakoshi, P. Rotter, M. Henry, R. M. Strieter, S. L. Kunkel, D. L. Longo, and D. D. Taub, *J. Immunol.* **156,** 2104 (1996).

[34] D. D. Taub, G. Tsarfaty, A. Lloyd, S. Durum, D. Longo, and W. Murphy, *J. Clin. Invest.* **94,** 293 (1994).

Mice are used at 8–12 weeks of age and are kept under specific pathogen-free conditions. SCID mice are housed in microisolator cages, and all water, food, and bedding are autoclaved before use. SCID mice receive 40 mg trimethoprim and 200 mg sulfamethoxazole per 320 ml of drinking water.

Transfer of Human Cells into SCID Mice and Treatment with Chemokines or Recombinant Human Growth Hormone

Human peripheral blood lymphocytes (huPBL) are obtained from healthy normal donors who have provided informed consent. The huPBL are purified by countercurrent elutriation. The lymphocyte fractions containing >90% lymphocytes are used for transplantation into SCID recipients. In studies using purified human T cells, the peripheral blood mononuclear cells (PBMCs) are passaged over Ficoll–Hypaque to remove erythrocytes, granulocytes, and cellular debris. T cells and T-cell subsets are then purified according to a modified procedure using R&D T cell and T-cell subset enrichment columns (R&D Systems, Minneapolis, MN). Viable cells isolated from the Ficoll–Hypaque interface (1×10^7 cells/ml) are subsequently incubated at 4° with 10 μg/ml anti-CD16 antibody for 30 min and then passaged over an R&D Systems T-cell enrichment column. Preincubation of PBMCs with anti-CD16 antibody facilitates the removal of NK cells from the T-cell preparations. This isolation procedure typically yields >95% motile $CD3^+$ $CD14^-$ $CD19^-$ $CD16^-$ $CD56^-$ lymphocytes.

The huPBL or purified human T cells (1×10^8) are injected intraperitoneally into recipient SCID mice. All mice receive 20 μl of anti-asialo-G_{M1} (anti-ASG_{M1}, Wako Chemicals, Dallas, TX) intravenously 1 day before huPBL injection, which has previously been shown to improve human cell engraftment in SCID mice with rhGH and anti-CD3 monoclonal antibody (MAb).[26–28] Untreated SCID and huPBL-SCID recipients receive either 1 μg rhIP-10, rhIL-8, rhRANTES, rhMIP-1α, rhMIP-1β (Peprotech, Rocky Hills, NJ), rhPF4 (Sigma, St. Louis, MO), rhIL-2 (Hoffman-LaRoche, Nutley, NJ), or rhM-CSF (Chiron, Emeryville, CA) in 100–200 μl of Hanks' balanced saline solution (HBSS), or HBSS alone subcutaneously. Endotoxin levels are determined for all chemokine preparations by a *Limulus* lysate assay and are found to contain less than 0.1 ng/ml. Mice are assayed either 4 or 72 hr after the injection. Experiments contain three to six mice per group.

In certain experiments, PBLs are incubated with rhMIP-1β, rhRANTES (both at 10 ng/ml), or medium overnight. This concentration is chosen on the basis of the ability of 10 ng/ml to optimally induce T-cell migration and costimulation.[32,35] Overnight treatment is utilized as shorter time periods fail

[35] D. D. Taub, D. Clark, and S. M. Turcovski-Corrales, *J. Immunol. Methods* **184**, 187 (1995).

to induce significant trafficking (data not shown). After incubation, the PBLs are extensively washed [3× with 50 fold phosphate-buffered saline (PBS)] and injected intraperitoneally into SCID mice. After 72 hr, the thymuses of these mice are harvested and tested by flow cytometry for the presence of human T cells. As a positive control for augmentation of human T-cell localization to the murine thymus, purified human T cells are incubated on anti-CD3 MAb-coated plates for 18–24 hr at 37°.[32] The cells are subsequently harvested and injected into SCID mice intraperitoneally.

In addition, certain SCID recipients receive 10 μg of either rhGH (Genentech, south San Francisco, CA) or ovine GH (ovGH; provided by the National Institute of Diabetes and Digestive and Kidney Diseases, the Center for Population Research of the National Institute of Child Health and Human Development, and the Agricultural Research Service of the U.S. Department of Agriculture, as well as University of Maryland School of Medicine, Baltimore) in 200 μl of HBSS, or HBSS alone, intraperitoneally every other day until time of assay 6–10 weeks later. In certain experiments, PBLs are incubated with rhGH (10 ng/ml) overnight in the presence or absence of anti-human CD18 (IOT18) and anti-human CD29 (IOT29, Amac, Westbrook, ME) antibodies (each at 10 μg/ml). After incubation, the PBLs are extensively washed and injected intraperitoneally into SCID mice. After 72 hr, the thymuses of these mice are harvested and tested by flow cytometry for the presence of human T cells.

Immunohistochemistry

Skin injection sites and the underlying body wall are embedded in OCT compound tissue medium (Baxter Health Care, Charlotte, NC), snap-frozen on dry ice, and stored at −70° until sectioning. Tissues are sectioned on a cryostat at 5 μm. Before staining, tissue sections are warmed to room temperature, fixed for 10 min in acetone, and rinsed in PBS. The following primary mouse MAb are used: DAKO-CD3, T3-4B5 (normal human blood T lymphocytes); DAKO-CD4, MT310 (helper/inducer T lymphocytes, CD4); DAKO-CD8, and DK25 (suppressor/cytotoxic T lymphocytes, CD8) (Dako, Carpinteria, CA). Dilutions of antibody are 1 : 150, 1 : 40, and 1 : 30, respectively; incubation time is 30 min. The avidin–biotin complex method is used with the mouse avidin–biotin complex Vectastain Elite Kit (Vector Laboratories, Burlingame, CA) as the secondary antibody and 3,3'-diaminobenzidine as the chromogen. Mayer's hematoxylin (HE) is used for counterstaining. An injection site negative control (mouse Elite kit and no primary antibody) and a positive control (pelleted normal human lymphocytes) are included for each sample run. Injection sites are evaluated microscopically without knowledge of the experimental treatment. Immunostaining is graded from minimal (1+) to extensive (4+) based on the intensity and distribution of peroxidase-positive cells. Background staining for immunohistology is assessed in normal adjacent murine tissues (data not shown).

Histology

Injuection sites from one to three mice per experimental group are mixed in 10% neutral buffered formalin, embedded in paraffin, sectioned at 5 μm, and stained with hematoxylin and eosin (H&E). Slides are evaluated microscopically without knowledge of the experimental treatment.

Flow Cytometric Analysis

The following anti-human MAb are used. HLA-ABC fluorescein isothiocyanate is purchased from Olympus (Lake Success, NY). CD4 (Leu3a-fluorescein isothiocyanate), CD8 (Leu2a-biotinylated), and CD3 (Leu4-biotinylated) are purchased from Becton Dickinson, San Jose, CA. Antibodies are used to label single-cell suspensions of thymocytes. Labeling is done in the presence of 2% (v/v) human AB serum (GIBCO-BRL, Grand Island, NY) to saturate human and mouse Fc receptors. After incubation with primary antibody, cells are washed, fixed with 1% (v/v) paraformaldehyde, and analyzed using an EPICS flow cytometer (Coulter Electronics, Hialeah, FL). The amount of human cells present in the thymus is determined by multiplying the total cellularity by the percentage of positive cells as determined by flow cytometric analysis. Statistics are done by comparing the occurrence and extent of engraftment using a Wilcox random sum analysis and Student's t test.

Results

Subcutaneous Injections of Recombinant Chemokines into Murine Skin Inducing Significant Leukocyte Accumulation

Initial studies performed to determine the effects of rhMIP-1α, rhMIP-1β, rhRANTES, rhMCP-1, rhIP-10, rhIL-8, rhPF4, or rhM-CSF administration in unreconstituted SCID or BALB/c mice (Table I) demonstrated that all of the endotoxin-free recombinant murine and human C-X-C and C-C chemokines induced significant neutrophil and/or mononuclear cell infiltration into the subcutaneous injection sites within 4 hr. In all of these studies, mice received 1 μg injections of either recombinant chemokines or PBS and were subsequently examined for cellular infiltrates at various time points. Almost all of the mononuclear cells observed in C-C chemokine

TABLE I
Chemokine-Mediated Leukocyte Infiltration in huPBL-SCID Mice[a]

Chemokine	4 hr postinjection			72 hr postinjection		
	PMN	MNC	Human T cells	PMN	MNC	Human T cells
IL-8	4+	1+/−	−	3+	1+/−	4+
MCP-1	−	3+	1+	−	2+	2+
MIP-1α	1+/−	1+	1+	1+	1+	1+
MIP-1β	1+/−	1+	1+	1+	1+	1+
RANTES	1+/−	1+	2+	1+	1+	3+
IP-10	1+/−	1+/−	1+/−	1+/−	1+	1+
PF-4	4+	1+	−	ND	ND	ND
PBS	−	−	−	1+/−	1+/−	−

[a] Skin injection sites and the underlying body wall were embedded and examined either histologically or immunohistologically poststaining with anti-human CD3 MAb as described in Materials and Methods. The results are summarized in histological and immunohistological gradings with − (none), 1+ (minimal), 2+ (moderate), 3+ (high), or 4+ (extensive) infiltration or as ND (not determined).

injection sites were found to be monocytes and macrophages with very few lymphocytes (Table I). MIP-1α, MIP-1β, and RANTES but not MCP-1 also incuded a mild polymorphonuclear cell infiltration into these injection areas. In contrast to *in vitro* chemotaxis studies,[16,31] the C-X-C chemokine rhIP-10 induced a mild infiltration in both SCID and BALB/c mice of murine mononuclear cells and polymorphonuclear cells into the subcutaneous tissues at both the 4- and 72-hr time points (Table I). The extent and type of IP-10-induced mononuclear cell infiltration were similar to that observed with the chemokine RANTES.[29,31]

In contrast, injection of rhIL-8 into nonreconstituted SCID mice revealed significant murine neutrophil but not mononuclear cell accumulation within 4 hr of injection (Table I, Fig. 1). By 72 hr postinjection, neutrophils were still the predominant cell type within IL-8 injection sites; however, a mild monocytic cell infiltrate was also observed. Surprisingly, rhPF4 also induced significant murine neutrophil accumulation, quite similar to that seen with IL-8, when examined 4 hr postinjection. These studies are in direct contrast to previous reports demonstrating that PF4 exhibits no chemotactic activity on human or murine neutrophils[5–10] and suggest that rhPF4 may indirectly induce murine neutrophil accumulation through the activation or alteration of the extravascular environment.[30] It should be noted that injections of PBS, HBSS, or M-CSF all fail to induce any significant accumulation of murine leukocytes when examined histologically or immunohistologically at either 4, 24, 48, or 72 hr.

These studies provide important information regarding the bioactivity of chemokines in a subcutaneous tissue environment in mice. As *in vitro* chemotaxis and transmigration studies can be quite variable, it is important to confirm *in vitro* findings in an *in vivo* setting. However, *in vivo* injection studies are not without their obstacles. To ensure that chemokine injection studies are correctly interpreted, careful and consistent injection procedures into a designated injection site without excessive trauma to the tissues must be utilized to avoid significant nonspecific cell migration (background). Excessive trauma to the hindflank tissue may induce the release of a multitude of mediators and cytokines that may facilitate cellular infiltration and vascular permeability. The use of small-gauge needles prevents excessive tissue trauma. Furthermore, animals should be examined over various times to determine if the chemokines are mediating specific and reproducible effects.

Our observations demonstrating differences in the incident, extent, and type of cellular infiltration in response to a given chemokine suggest that each chemokine is mediating a distinct signal—whether direct or indirect—promoting leukocyte recruitment. The specificity of our model is strongly supported by the ability of chemokine-specific antibodies to block chemokine-mediated leukocyte recruitment. It should be noted that the injection procedure itself may prime the treated tissue to be more responsive to a chemokine stimuli. Typically, one would not expect leukocyte recruitment in response to a chemokine alone. Priming of the vascular endothelial barrier by IL-1, TNF-α, or other inflammatory stimuli seems to be prerequisite for optimal leukocyte extravasation. Thus, our chemokine infiltration data most likely reflect multiple activation mechanisms including the release of inflammatory mediators within the tissue injection site. Fortunately for our studies, this injection trauma alone (in the majority of our studies) does not provide all the necessary signals for cellular recruitment.

Human T-Cell Infiltration into Skin of huPBL-SCID Mice Injected with C-C and C-X-C Chemokines

To assess the *in vivo* effects of recombinant chemokines on human T-cell migratory capabilities, SCID mice were given an intraperitoneal injection of 1×10^8 huPBLs (that had been elutriated to remove monocytes) or purified human T cells, immediately followed by a subcutaneous injection of recombinant human chemokines (1 μg) or HBSS alone. Injection sites were examined, at 4 or 72 hr after injection, by histological and immunohistochemical analysis. Cellular infiltration could be detected in the dermis, muscle, and fat at the site of chemokine injection in huPBL-SCID

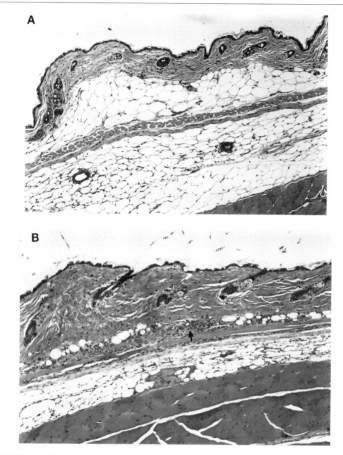

FIG. 1. Histological analysis for murine leukocytes infiltrating the dermis of mice after recombinant human IL-8 and MIP-1β administration. (A) Control PBS injection. Magnification: ×25. No cellular infiltrate was observed. (B) Biopsy of injection site in mice receiving rhMIP-1β. Magnification: ×25. The tissue was examined 4 hr after injection. A significant mononuclear cell infiltrate was observed at 4 hr with minimal PMN recruitment. (C) Biopsy of injection site in mice receiving rhIL-8. Magnification: ×25. A significant neutrophil infiltrate with no mononuclear cells was observed at 4 hr. (D) Biopsy of injection site in mice receiving rhIL-8. Magnification: ×80. A significant mononuclear and PMN infiltrate was observed at 72 hr.

mice.[18,29-33] Mononuclear cells were consistently the predominant cell type to accumulate at the sites of rhMIP-1α, rhMIP-1β, rhRANTES, and rhMCP-1 administration.[18,29,33] Because SCID mice lack their own T and B cells[24] and treatment with anti-ASG$_{M1}$ depleted their NK cells, the mononuclear cells could either be SCID monocytes and/or human lymphocytes.

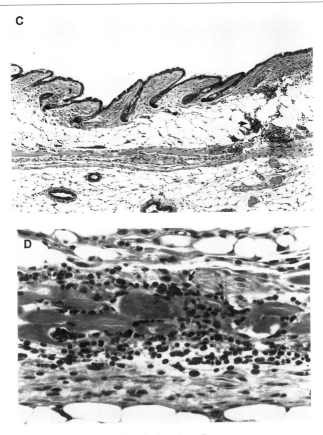

Fig. 1. (*continued*)

Immunohistological examination of the skin using human-specific anti-CD3, -CD4, and -CD8 MAb indicated that these chemokines are inducing significant human T-cell migration into the mouse tissues, as demonstrated by the numerous human CD3+ cells present in the dermis and underlying tissues after 4 and 72 hr. These cells were either CD4+ or CD8+ T cells, with no subset preference being observed in response to any of the chemokines.[18,29–31,33] The extent of human T-cell infiltrate varied depending on the huPBL donor and the chemokine tested, with rhRANTES exhibiting the most potent effects at 4 hr (Table I). In addition, although significant human T-cell infiltration occurred within 4 hr of rhMIP-1α, rhMIP-1β, rhRANTES, or rhMCP-1 injection, a more extensive (and consistent) human T-cell accumulation was observed 48–72 hr postinjection. This rapid C-C chemokine-induced human T-cell infiltration supports a direct effect

TABLE II
EFFECTS OF ANTI-MIP-1β ANTISERA ON HUMAN CD3⁺ CELL INFILTRATION
INTO rhMIP-1β INJECTION SITES[a]

Animal	PBS	MIP-1β	MIP-1β plus anti-MIP-1β
1	NS	NS	NS
2	NS	NS	NS
3	NS	Minimal, MF	Minimal, MF
4	NS	Minimal, MF	NS
5	NS	NS	NS
6	NS	Minimal, MF	NS
7	NS	Moderate, MF	NS
8	NS	Moderate, MF	NS

[a] SCID mice received 1×10^8 huPBLs intraperitoneally followed by daily subcutaneous injections of rhMIP-1β or PBS (1 μg) in the presence or absence of anti-human MIP-1β antisera for 3 days. Skin injection sites and the underlying body wall were embedded and stained with anti-human CD3 MAb as described in Materials and Methods. Injection sites were evaluated microscopically without knowledge of the experimental treatment. Immunostaining was graded from minimal to extensive based on intensity and distribution of positive cells. MF, Multifocal; NS, not significant.

on T cells and supports the previous findings that C-C chemokines are potent human T-cell chemoattractants.[11–19] In multiple experiments, the extent of human T-cell infiltration was variable from mouse to mouse, and occasionally high background levels of infiltration with certain human donors complicated interpretation. It should also be noted that murine or human leukocyte infiltration was rarely detected in SCID mice receiving a single injection or multiple injections of PBS, HBSS, human immuno-globulin G (IgG), or the cytokine rhM-CSF (data not shown).

Similar to the C-C chemokines, rhIP-10 also induced significant human T-cell accumulation in the huPBL-SCID mouse model.[31] However, only mild human CD3⁺ T cells were observed within the mononuclear cell infiltrates 4 hr after rhIP-10 injection. By 72 hr, a greater and more repro-ducible level of human T-cell accumulation was observed.

In all of these studies, the chemokine-induced T-cell infiltrates were found to be chemokine-specific and not due to a contaminant within the chemokine preparation or the injection procedure itself. This can be more clearly seen in Table II; rhMIP-1β coinjected with MIP-1β-specific antisera into the hindflanks of huPBL-SCID mice completely blocked human and murine cellular infiltrates into the subcutaneous injection sites. However, PBS alone or PBS coinjected with anti-MIP-1β antibody failed to elicit any significant T-cell infiltration (data ot shown). Overall, these results suggest that the chemoattractant effects of chemokines in this human–mouse chimeric model can be rapid and selective as well as chemokine-specific.

Similar to the studies described above, injection of rhIL-8 into these chimeric SCID mice also led to an early (4 hr) infiltration by murine leukocytes; however, IL-8 injections induced a predominant neutrophil infiltration, with no mononuclear cell infiltration being observed (Table I, Figs. 1 and 2). Interestingly, on examination of these IL-8 injection sites 72 hr after administration, a significant number of human CD3[+] T lymphocytes infiltrated the IL-8-treated subcutaneous tissues (Fig. 2). Interestingly, the extent of human CD3[+] T-cell infiltration was even greater than that observed with any of the C-C chemokines, which are potent *in vitro* T-cell chemoattractants.[30] In the IL-8-treated mice, both CD4[+] and CD8[+] T cells were found to be present in comparable amounts.[30] This lymphocyte migration appears to occur after the induction of murine neutrophil infiltration in the injection site, suggesting that rhIL-8 may indirectly induce human T-cell infiltration in this huPBL-SCID model through the possible release of substances by infiltrating neutrophils. The prerequisite for neutrophil infiltration prior to human T-cell accumulation in this chimeric model was clearly demonstrated by systemically depleting murine neutrophils using the granulocyte-specific 8C5 antibody prior to the rhIL-8 injections.[30]

Overall, these studies suggested that neutrophil infiltration may be an important prerequisite for subsequent human T-cell migration into inflammatory areas. As IL-8 has previously been shown to induce human and murine neutrophil degranulation *in vitro*,[5–10,30] on murine neutrophil entry into sites of IL-8 production or injection, the neutrophils may be induced to degranulate, releasing additional proadhesive and promigratory mediators present in cytoplasmic endosomes and granules. Subsequent to these studies, we have demonstrated that purified neutrophil granules derived from either human or murine neutrophils contain potent human T-cell and monocyte chemoattractants.[30] In collaboration with the laboratory of Dr. Joost Oppenheim,[36] we have demonstrated that biochemically purified neutrophil granule-derived defensins and azurocidin/CAP-37 are potent human T-cell and monocyte chemotactic agents both *in vitro* and *in vivo*. Despite these neutrophil granule findings, the biological relevance of these and other neutrophil-derived molecules to cellular recruitment in both acute and chronic inflammatory disease states remains to be determined.

[36] O. Chertov, D. F. Michiel, L. Xu, J. M. Wang, H. A. Pereira, D. D. Taub, and J. J. Oppenheim, *J. Biol. Chem.* **271**, 2935 (1996).

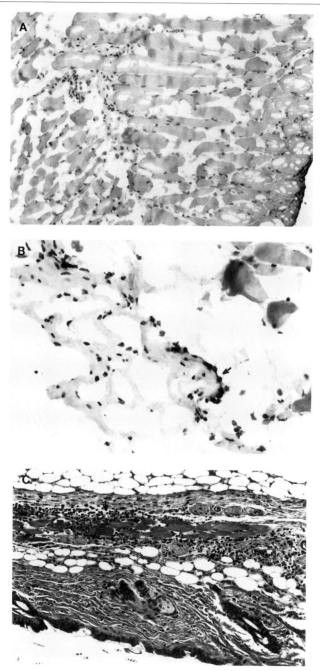

*Treatment of Human T Cells with Recombinant Human MIP-1β and
RANTES as Well as Recombinant Human Growth Hormone
Permitting Human Lymphocyte Engraftment in Lymphoid Tissues of
SCID Mice*

Our previous studies with rhGH have demonstrated that continuous injections of rhGH promotes the engraftment of human T cells in the lymphoid organs of SCID mice.[24,26-29] An example of this trafficking activity can be seen in Table III. In these studies, continuous injection of SCID mice with rhGH (10-μg injections every other day starting the day of huPBL transfer) induces significant human T-cell engraftment in the lymphoid organs of these animals, particularly in the thymus. Treatment with rhGH significantly increased both the incidence and extent of human cell engraftment in the murine thymus.[34] The predominant cell type to engraft were CD3+, HLA+ lymphocytes.[34] This rhGH-mediated human cell engraftment appears to be mediated through its direct interaction with GH receptors on the human T cells, as only rhGH but not ovGH induced human T-cell engraftment in the murine thymus. Interestingly, ovGH, which is not capable of binding human T cells through their GH receptors but can affect murine cells through their GH receptors, failed to promote significant human CD3+ T-cell engraftment in either the murine thymus or other peripheral lymphoid organs (Table III). Thus, it seems likely that rhGH is directly stimulating the injected human T cells, promoting their peripheral engraftment in the SCID thymus. Furthermore, human PBLs pretreated overnight with 10 ng/ml of rhGH also demonstrated a significant increase in both the incidence and extent of human T-cell engraftment in the murine thymus.

These results suggest that rhGH promotes human T-cell engraftment in SCID mice by directly altering the adhesive capacity of T cells via human adhesion molecules on the surface of the transferred lymphocytes. Although our studies demonstrating the ability of rhGH to induce human T-cell migration and adhesion *in vitro*[34] strongly support an rhGH effect on T-cell adhesion and homing *in vivo,* they fail to provide direct evidence that rhGH is directly altering the adhesive or homing capacity of the human T cells within this chimeric model. We have reported that overnight pretreatment

---

FIG. 2. Immunohistological analysis for human CD3+ T cells within huPBL-SCID mice injected subcutaneously with recombinant human MIP-1β or IL-8. (A) Control PBS injection site stained with anti-human CD3 peroxidase-labeled antibody. Magnification: ×20. No significant staining or infiltrate was observed. (B) Presence of CD3+ human lymphocytes within subcutaneous fat examined 4 hr postinjection. Magnification ×20. (C) Presence of significant CD3+ human lymphocytes only 72 hr but not 4 hr after rhIL-8 injection. Magnification: ×40.

TABLE III

RECOMBINANT HUMAN GROWTH HORMONE BUT NOT OVINE GROWTH HORMONE
ADMINISTRATION TO INCREASE HUMAN T-CELL ENGRAFTMENT IN THYMUS OF
SCID MICE VIA HUMAN INTEGRIN MOLECULES[a]

Experiment and animal	Treatment	Cell number ($\times 10^6$)	Percent of thymocytes staining for HLA-ABC$^+$/CD3$^+$
A1	—	1.6	0.0
A2	—	1.9	8.8
A3	—	1.2	5.6
A4	—	1.7	0.0
A5	—	1.3	0.0
A6	rhGH	1.2	6.9
A7	rhGH	0.9	12.7
A8	rhGH	1.8	1.0
A9	rhGH	1.4	8.2
A10	rhGH	1.7	5.6
A11	ovGH	0.7	1.1
A12	ovGH	0.9	0.0
A13	ovGH	0.9	0.0
A14	ovGH	1.7	9.4
A15	ovGH	0.7	0.0
B1	—	1.5	0.0
B2	—	1.0	0.0
B3	—	1.4	0.0
B4	—	1.4	1.1
B5	rhGH	1.7	1.5
B6	rhGH	1.5	1.1
B7	rhGH	1.1	2.3
B8	rhGH	1.2	0.0

[a] In Experiment A, SCID mice were given 10-μg injections of rhGH or ovGH every other day starting the day of human peripheral blood lymphocyte transfer. Mice were then analyzed 6 weeks after transfer by flow cytometric analysis of the cells in the thymus of these treated mice. In Experiment B, resting T cells were preincubated with 10 ng/ml of rhGH or PBS overnight after which the cells were harvested and injected into the peritoneal cavity of SCID mice. Mice were then analyzed 72 hr after transfer by flow cytometric analysis of the human cells of the thymus of these treated mice. In both experiments, the results are expressed as the percentage of CD3$^+$ HLA-ABC$^+$ cells engrafted in the murine thymus. The results are representative of two to three experiments.

of human T cells with rhGH in the presence of anti-human CD18 and anti-human CD29 antibody (specific for the $\beta2$ and $\beta1$ integrin chains) inhibited human T lymphocyte engraftment in SCID thymic tissue, whereas T cells pretreated with rhGH alone exhibited significant engraftment.[34] The data that rhGH pretreatment promotes T-cell engraftment and that anti-human integrin antibodies block GH-mediated trafficking suggest that this hormone directly alters T-cell adhesion on interacting with cell surface hormone receptors.

On the basis of these rhGH studies as well as our previous data demonstrating the ability of C-C chemokines to promote human T-cell adhesion and migration *in vitro*,[14,16,18,20,29,30,32,37] it was of interest to determine whether certain C-C chemokines could also facilitate T-cell trafficking to these murine lymphoid organs. Thus, mice were given three daily 1 μg subcutaneous injections of rhMIP-1β, rhRANTES, rhPF4, rhIL-2, or PBS daily starting the day of huPBL transfer, after which the thymocytes from these mice were analyzed by flow cytometric analysis for the presence of human HLA$^+$/CD3$^+$ T cells. As with rhGH, MIP-1β treatment significantly increased the extent of human cell localization within the murine thymus (Fig. 3A). The predominant cell type to localize were CD3$^+$, HLA$^+$ lymphocytes. Treatment with MIP-1β also induced modest increases in the number of human T cells in murine lymph nodes and spleen (data not shown). In addition, injections of RANTES also promoted human T-cell engraftment in the SCID thymus (Fig. 3A), whereas rhPF4 and rhIL-2 failed to induce any significant T-cell trafficking in this human/mouse model.

Differences in the background numbers of human CD3$^+$ T cells in the thymus appear to be human T-cell donor-dependent.[24,34] Different human lymphocyte donors generally gave rise to different effects on the extent of engraftment. As seen with *in vitro* chemotaxis studies using different T-cell donors, variable chemokine responsiveness can often be observed. This variation may be due to differences in the activation state, phenotype, or adhesive ability of the donor T cells.

In addition, the use of this xenogeneic model may also contribute to the variable engraftment responses observed, as the presence of xenoantigen-specific T cells may induce a graft-versus-host disease (GVHD)-like effect leading to systemic alterations in cell trafficking.[24,38] However, owing to the limited time periods in which we are examining chemokine-mediated T-cell engraftment, the role of xenogeneic T cells seems unlikely. In most of our studies using huPBL-SCID mice, long-term engraftment experiments in SCID mice using various huPBL donors do reveal significant GVHD effects.[38] In addition, the presence of human B cells from Epstein-Barr virus (EBV)-positive huPBL donors in this model have been shown to spontaneously transform into human B-cell lymphomas.[24]

In an effort to determine whether the chemokine trafficking effects

[37] A. R. Lloyd, J. J. Oppenheim, D. J. Kelvin, and D. D. Taub, *J. Immunol.* **156,** 932 (1996).
[38] W. J. Murphy, M. Bennett, M. R. Anver, M. Baseler, and D. L. Longo, *Eur. J. Immunol.* **22,** 1421 (1992).

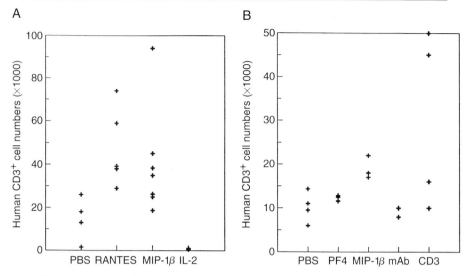

FIG. 3. Recombinant human MIP-1β and RANTES administration increased human T-cell localization in the thymus of SCID mice. (A) SCID mice received 1×10^8 huPBLs intraperitoneally followed by daily subcutaneous injections of rhMIP-1β, rhRANTES (1 μg), rhIL-2, or PBS for 3 days. Thymic localization was assayed for the number of human CD3⁺ T cells by flow cytometric analysis as described in Materials and Methods. The significance of the differences between the various treatments is as follows: PBS versus rhRANTES, $p < 0.05$; PBS versus rhMIP-1β, $p < 0.05$; PBS and rhMIP-1β versus IL-2. NS, not significant. (B) SCID mice were given 1-μg injections of rhMIP-1β in the presence or absence of anti-rhMIP-1β antisera or PBS daily for 3 days starting the day of huPBL transfer. The significance of the differences between the various treatments is as follows: PBS versus rhPF4, not significant; PBS and PF4 versus rhMIP-1β, $p < 0.05$; PBS and rhPF4 versus rhMIP-1β and anti-rhMIP-1β, not significant; PBS and rhPF4 versus anti-CD3 MAb, $p < 0.05$. The results in both experiments are expressed as the total number of human CD3⁺ T cells within the SCID thymus 72 hr postinjection with chemokine.

described above were specific, an anti-human MIP-1β antibody was coinjected with rhMIP-1β or PBS and subsequently examined for the level of lymphocyte thymic engraftment. The results in Table II and Fig. 3B clearly demonstrate that the combination of anti-MIP-1β antisera with MIP-1β abrogated the increased infiltration and trafficking of human CD3⁺ T cells not only in the murine skin but also into the murine thymus, whereas administration of MIP-1β alone induced significant human lymphocyte infiltration into this organ.[33] A comparison between of the ability of chemokine-treated mice and anti-CD3 MAb-treated human T-cells to traffic to the SCID thymus was also performed. We have previously shown that human T cells stimulated with anti-CD3 MAb *in vitro* promoted peripheral lymphocyte trafficking of human T cells in SCID mice.[39] This trafficking response was found to be activation-dependent and persisted for well more than 2 months. The results presented here demonstrate that anti-CD3 MAb-treated T cells had a greater ability to localize to the thymus than MIP-1β-treated T cells (Fig. 3B).

These studies strongly suggest that the activation state of the T cells *in vivo* can greatly influence the extent and incidence of human T-cell thymic engraftment in SCID mice. On the basis of our studies showing that many of the C-C chemokines including MIP-1β and RANTES are costimulators of T-cell activation *in vitro*,[32,33] we believe that cellular activation may be one of the mechanisms mediating these effects. PF4, a C-X-C chemokine that we have previously shown not to activate or chemoattract human T cells,[14,16,18,19,32] failed to induce any significant human T-cell accumulation in the thymic tissues. Injections of PBS in the presence or absence of antibody also failed to induce an increase in human lymphocyte accumulation in the thymus. These results strongly support the findings shown in Table II demonstrating that MIP-1β and RANTES can induce human CD3$^+$ T-cell trafficking to the murine thymus in a chemokine-specific fashion.

Although these results demonstrate that certain C-C chemokines directly stimulate human T-cell infiltration into chemokine injection sites as well as trafficking to the SCID thymus, it fails to provide direct evidence that these chemokines are directly signaling human T cells to traffic to the periphery. In an effort to determine if these chemokines directly promote human T-cell thymic homing through interactions with cell surface chemokine receptors on T cells, we pretreated human PBLs with 10 ng/ml of rhMIP-1β and rhRANTES or medium alone *in vitro* overnight, after which the cells were extensively washed and tested for their ability to engraft in the SCID thymus. Thus, in these experiments, no additional chemokines were injected in the skin or coinjected intraperitoneally with the cells. Analysis of the thymocytes 72 hr after PBL transfer in these mice clearly demonstrated (Fig. 4) that rhMIP-1β and rhRANTES pretreatment significantly increased the extent of human cell engraftment within the murine thymus, suggesting that the C-C chemokines may be directly altering the T-cell adhesive or homing capacity after overnight incubation. Thus, when placed in these SCID mice, the chemokine-primed T cells may have an altered adhesive capacity, permitting their migration into the peripheral lymphoid tissues; alternatively, analogous to anti-CD3 activation of T cells,[39] perhaps the chemokines are costimulating xenoreactive T cells and promoting their trafficking to the thymus.

---

[39] W. J. Murphy, K. C. Conlon, T. J. Sayers, R. H. Wiltrout, T. C. Back, J. R. Ortaldo, and D. L. Longo, *J. Immunol.* **150**, 3634 (1993).

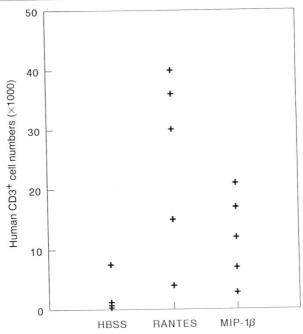

FIG. 4. Pretreatment of human T cells with recombinant human MIP-1β and RANTES permits lymphocyte engraftment in the thymus of SCID mice. Purified human lymphocytes were preincubated with 10 ng/ml of rhMIP-1β, rhRANTES, or PBS overnight, after which the cells were harvested and injected into the peritoneal cavity of SCID mice. Mice were then analyzed 72 hr after transfer by flow cytometric analysis for the number of human CD3$^+$ T cells in the thymuses of the treated mice. The significance of the differences between the various treatments is as follows: HBSS versus rhRANTES, $p < 0.05$; PBS versus rhMIP-1β, $p < 0.05$.

Discussion

The huPBL-SCID chimera model allows the examination of the effects of various chemokines using human cells and reagents in an *in vivo* setting. However, there are significant limitations to this model that are due to the uncharacterized effects of placing human lymphocytes in an xenogeneic environment. The original finding of Mosier and colleagues[25] indicated that huPBLs engrafted only after intraperitoneal but not subcutaneous or intravenous injection into SCID recipients. These results have been confirmed by other laboratories.[38] In addition, the number of huPBLs injected was critical for engraftment as well as the time of huPBL transfer, and the techniques used to assess human cell engraftment also add to the complexities of the model and of interpreting data. Many investigators use more than one technique to measure the extent of human cell engraftment, including polymerase chain reaction (PCR) detection of human-specific mRNA sequences, flow cytometric and immunohistological analysis using antibodies directed against human-specific markers (i.e., HLA-A, B, C CD3, or CD45), and quantitation of human immunoglobulin in mouse serum.[24] Although PCR is the most sensitive method to qualitatively detect the presence of human cells, it does not permit the precise quantitation of human cell engraftment or trafficking. We prefer to utilize flow cytometric and immunohistological analysis of injection sites and organ suspensions as it permits both a phenotypic and quantitative analysis of the trafficking human cells.

In addition, for still undetermined reasons, there appears to be considerable mouse-to-mouse variation in the extent of human cell trafficking despite the use of the same huPBL donor. This variation necessitates the use of several huPBL donors in multiple experiments as well as the use of multiple mice (at least five mice per experimental group). One possible explanation for this variability may be the presence of murine NK cells within the SCID mice. Studies have demonstrated that SCID mice can resist human cells by way of their NK cells.[24] Removing these NK cells with the administration of anti-asialo-G_{M1} (ASG_{M1}) markedly enhances human cell engraftment and reduces mouse-to-mouse variations. In our current model, all mice are treated with anti-asialo-G_{M1} antibody intravenously 24 hr prior to huPBL transfer, a technique we have previously shown to systemically eliminate murine NK cells and improve human cell engraftment posttreatment with anti-CD3 MAb and rhGH.[24,28] However, this does not abrogate the variability within the model.

Another potential problem with the injection of human lymphocytes into xenogeneic SCID mice is the occurrence of GVHD. Although the initial report on the huPBL-SCID model did not demonstrate any overt pathology associated with the injection of huPBL,[25] xenogeneic GVHD can occur after huPBL transfer, resulting in splenomegaly, liver pathology, and, in some instances, death. The incidence and extent of GVHD within the huPBL-SCID model appear to be quite variable and huPBL donor-dependent.[38] As stated earlier, this GVHD-like effect may predispose xenogeneic T cells or the mouse host to the effects of the chemokines being evaluated.

Despite the various methods used to facilitate engraftment, including administration of various activating agents such as anti-CD3 MAb, superantigens, and neurohormones in these animals, there still remains considerable laboratory-to-laboratory, donor-to-donor, experiment-to-experiment, and mouse-to-mouse variability using this model due to mechanisms that are still poorly understood.

Few *in vivo* studies have directly demonstrated T-cell infiltration in response to chemokine injection. Early studies by Larsen and colleagues[12] demonstrated that intradermal injection of IL-8 in rats causes an accumulation of both neutrophils and lymphocytes in the connective tissue. In addition, injections of IL-8 into the lymphatic drainage areas of lymph nodes of these rats resulted in an accelerated emigration of only lymphocytes in high endothelial venules of the draining lymph nodes.[12] However, in many additional studies examining C-C chemokines in rodents, only sparse lymphocytic infiltration has been observed with chemokine injection sites. Our studies demonstrated that human CD3[+] T cells migrate into chemokine injection sites in response to recombinant preparations of human IL-8 and IP-10 in huPBL-SCID mice.[30,31] The injection of human IL-8 into these chimeric SCID mice also led to an early (4 hr) infiltration by murine neutrophils. Interestingly, 72 hr after IL-8 injection, a high number of human T lymphocytes infiltrated the injection site.

These IL-8/huPBL-SCID studies suggest that neutrophil infiltration may be an important prerequisite for lymphocyte entry into inflammatory sites. Systemic depletion of neutrophils within the huPBL-SCID prior to chemokine administration blocked human T-cell recruitment to the IL-8 injection site, supporting the hypothesis that the chemotactic effects of IL-8 on human T cells were indirect.[30] Similarly, subcutaneous injections of rhIP-10 into these huPBL-SCID mice induced significant murine leukocyte infiltration, particularly of monocytes, as well as human CD3[+] T-cell infiltration within 24–48 hr of IP-10 injection.[31] Our current hypothesis is that IP-10 can both directly and indirectly induce T-cell and monocyte migration. *In vivo,* this probably occurs through the development of a gradient (both free and bound IP-10) reaching and endothelium and leaking into the circulation. Circulating T cells infiltrate the injection sites following the IP-10 gradient. This would account for the highest number of cells being located on the needle path. As for its indirect effects, we believe that IP-10 can operate on cells within the tissues (such as endothelial cells)[5–10] and may alter the vascularization, adhesion molecule expression, and basement membrane integrity. In addition, the C-C chemokines, MIP-1α, MIP-1β, MCP-1, and RANTES, have been shown to induce significant infiltration of murine monocytes within 4 hr of injection in huPBL-SCID mice.[18,29–31,33] Over a 72-hr period, the human CD3[+] T-cell infiltrates observed in this model increased; however, at least with the C-C chemokines, the subpopulations of leukocytes entering the injection site do not appear to alter.

This model supports the *in vitro* chemotaxis results of several laboratories examining human and mouse T-cell migration in response to chemokines.[11–19] In addition, this model also demonstrates that there is adequate similarity between the adhesion molecules of mouse and human to allow trafficking of human cells into the peripheral tissues of mice.

We are examining the adhesion molecule expression in the tissues after chemokine injection, particularly in brain tissues.[23] Previous studies demonstrated that anti-CD3 MAb-primed T cells (primed for 4–6 hr but not 18–24 hr *in vitro*) respond more effectively to certain chemokines, including the C-X-C chemokines IP-10 and MIG, *in vitro*[16] (data not shown), but other studies have revealed that freshly isolated quiescent T cells also efficiently migrate in response to C-C chemokines.[35] Contrary to one report,[40] we have found that treatment of human T cells for extended times with high doses of rhIL-2 failed to promote human T-cell migration *in vitro*.[35] However, treatment of human T cells with very low doses of IL-2, IL-4, or IGF-1 for short periods at 37° appears to potentiate human T-cell responsiveness to chemokines (data not shown). This augmentation may be due to an increase in cell surface chemokine receptor expression on peripheral T cells. *In vivo,* IL-2 alone fails to promote human T-cell engraftment in the SCID mice, suggesting that an activating cytokine like IL-2 is not directly capable of signaling T-cell trafficking. Perhaps the addition of a second signal such as a chemokine or an additional proadhesive factor(s) may potentiate cell engraftment and peripheral trafficking. Species differences in adhesion molecules may also affect human T-cell trafficking in these chimeras. However, we have shown that human T cells are capable of binding murine intercellular adhesion molecule-1 (ICAM-1), vascular cell adhesion molecule-1 (VCAM-1), and various extracellular matrix proteins.[34,37] Despite this fact, it is quite possible that human T cells from certain donors require additional adhesion molecules necessary for optimal trafficking in this chimeric model.

As described above, we have demonstrated[33] the ability of rhMIP-1β and rhRANTES to directly affect human T-cell engraftment in SCID mice. These chemokines may induce the expression or activation of certain adhesion molecules or homing receptors on the surface of the T cell, permitting lymphocyte movement and entry into lymphoid tissues. Although the molecular mechanisms involved in chemokine priming of T cells for enhanced trafficking are not yet defined, the fact that MIP-1β and RANTES treatment of purified primary human T cells *in vitro* results in no increase in the cell surface expression of known adhesion molecules suggests that these chemokines may alter the avidity of adhesion receptors for their respective ligands rather than upregulate adhesion molecule expression.[37] However, it should be noted that certain human T-cell clones derived by our labora-

[40] M. Loetscher, B. Gerber, P. Loetscher, S. A. Jones, L. Piali, I. Clark-Lewis, M. Baggiolini, and B. Moser, *J. Exp. Med.* **184,** 963 (1996).

tory are induced by C-C chemokines to express increased levels of cell surface $\beta 1$ and $\beta 2$ integrin molecules (data not shown). However, on a bulk T-cell level, it seems more likely that the "avidity alteration" hypothesis is operating within this system.

Human T cells have been shown to exhibit enhanced VCAM-1, ICAM-1, and fibronectin binding in the presence of C-C chemokines within a 15- to 30-min period *in vitro*.[37] Our current hypothesis is that MIP-1β and RANTES can both directly and indirectly induce T-cell migration and homing. *In vivo*, this probably occurs through the development of a gradient (both free and bound chemokine) reaching the endothelium and subsequently leaking into the circulation. Circulating T cells infiltrate the injection sites following the gradient as well as being activated to "home" to various lymphoid tissues in the periphery. As for indirect effects, we believe that these chemokines can operate on cells within the tissues (such as endothelial cells, monocytes/macrophages, and tissue leukocytes)[5–10] and may alter the vascularization, adhesion molecule expression, and basement membrane integrity. Multiple subcutaneous injections of MIP-1β and RANTES over a 3-day period may result in systemic levels of chemokine, perhaps signaling the circulating human T cells to adhere to certain adhesive substrates such as ICAM-1 or VCAM-1. As lymphoid tissues constitutively express higher levels of ICAM and VCAM than normal endothelial cell populations within the microvasculature, these "triggered" cells may more freely enter the murine thymus and lymph nodes. It is also possible that the cells that traffic into the chemokine injection sites are permitted to enter the draining lymph and traffic to the lymph nodes and subsequently to the thymus. The pretreatment studies in Fig. 4 strongly support a direct effect of rhMIP-1β and RANTES on human T cells promoting lymphocyte trafficking *in vivo*, as no additional chemokine was injected into the SCID mice.

It is quite difficult to prove or disprove each of the possibilities in this model; however, it seems likely that several mechanisms may be at work here, particularly a direct effect on T-cell adhesion. Engraftment studies using several additional C-C chemokines including MIP-1α and MCP-1 have yielded positive results in several huPBL donors; however, the results were quite jumpy and inconsistent. In all cases, IP-10 and IL-8 failed to demonstrate any T-cell thymic trafficking in any of the mice examined. Thus, although all of these chemokines permitted cell entry into the subcutaneous injection sites, only MIP-1β and RANTES (and possibly other C-C chemokines) permitted lymphoid engraftment. These results suggest that the signals required for human T-cell engraftment in this SCID model may be quite different than those required for T-cell infiltration into challenged sites. We believe that these chemokines are capable of inducing human T-cell migration both *in vitro* and *in vivo* and may have a potential for clinical use to promote T-cell development and recirculation *in vivo,* particularly in instances of T-cell deficiencies.

In addition to lymphocyte trafficking into compromised tissues, lymphocytes also emigrate into normal lymphoid and nonlymphoid organs to fulfill their role in immunosurveillance. Despite the fact that chemokines appear, in some laboratories,[5–10] to induce the adhesion and migration of different lymphocyte subsets, their contribution to the process of lymphocyte homing into lymphoid tissues remains circumstantial. However, the constitutive production of chemokines in normal healthy tissues such as skin, thymus, and lymph nodes strongly suggests that these molecules may play a role in leukocyte homing and recirculation *in vivo.*[10] Under normal physiological conditions, lymphocytes are the predominant infiltrating cell type within these various tissues. Furthermore, the presence of MIP-1β on the surface of murine lymphoid high endothelial venules (HEV) and the presence of IP-10 and lymphotactin in murine thymic tissue suggest a possible role in populating the lymph nodes and thymus.[10,13,41] A role for chemokines in thymocyte homing is also supported by the ability of pertussis toxin to block access of murine thymocytes as well as T and B lymphocytes to peripheral lymphoid compartments.[42,43] These studies suggest that lymphocyte adhesion to HEV involves a G-protein-signaling pathway, most probably chemokines, which facilitate firm lymphocyte attachment. Overall, these findings, as well as our studies describing the ability of various C-C chemokines to mediate lymphocyte adhesion to various adhesive substrates as well as facilitate human T-cell trafficking in huPBL-SCID mice, strongly support a role for chemokines in lymphocyte trafficking into normal lymphoid organs.

Acknowledgment

The contents of this publication do not necessarily reflect the views and policies of the Department of Health and Human Services, nor does mention of trade names, commercial products, or organizations imply endorsement by the U.S. Government.

[41] C. R. Gattass, L. B. King, A. D. Luster, and J. D. Ashwell, *J. Exp. Med.* **179,** 1373 (1994).
[42] K. E. Chaffin and R. M. Perlmutter, *Eur. J. Immunol.* **21,** 2565 (1991).
[43] R. F. Bargatze and E. C. Butcher, *J. Exp. Med.* **178,** 367 (1993).

[19] High Throughput Screening for Identification of RANTES Chemokine Expression Inhibitors

By Debra A. Barnes, Steven W. Jones, and H. Daniel Perez

Introduction

The immune system is a complicated network of interreactive cells whose major role is to detect and eliminate pathogens from the body. In most cases, the immune system works as expected, but sometimes it can react inappropriately either by overresponding, in the case of allergy, or by suddenly recognizing the body's own components as foreign, in the case of autoimmune disorders. Autoimmune diseases are characterized by huge cellular infiltrates, the specificity and location of which are unique to each particular disease. It has become increasingly clear that chemokines are one of the major sets of signals that direct immune cells to their cellular targets,[1] and as such they have become candidates for therapeutic intervention.

The chemokine RANTES is a member of the C-C family, and its name is an acronym derived from "regulated on activation, normal T-cell expressed and secreted."[2] The mature protein is 8 kDa and is processed from a larger molecule. RANTES functions as an attractant for several different cell types including eosinophils, basophils, monocytes, and T lymphocytes.[3–5] The ability to attract such a diverse number of cells suggests that this chemokine plays an important role in a number of inflammatory diseases including multiple sclerosis (MS) and rheumatoid arthritis.

Multiple sclerosis is a disease that can take two forms; one is chronic and characterized by a continual degenerative process, and the other is cyclical in nature, consisting of periods of relapse and remission. An animal model for this disease, called experimental allergic encephalomyelitis (EAE), has been extensively used to define the pathophysiology. Results indicate that chemokines may play a part in the pathology of this disease. Indirect data in the form of *in situ* hybridization experiments showed that certain chemokine transcripts [RANTES, interferon-inducible protein-10

[1] J. J. Oppenheim, C. O. C. Zachariae, N. Mukaida, and K. Matsushima, *Annu. Rev. Immunol.* **9**, 617 (1991).
[2] T. J. Schall, J. Jongstra, and B. J. Bradley, *J. Immunol.* **141**, 1018 (1988).
[3] T. J. Schall, K. Bacon, K. J. Toy, and D. V. Goeddel, *Nature* (*London*) **347**, 669 (1990).
[4] R. Alam, S. Stafford, and P. A. Forsythe, *J. Immunol.* **150**, 3442 (1993).
[5] S. C. Bischoff, M. Krieger, and T. Brunner, *Eur. J. Immunol.* **23**, 761 (1993).

(IP-10), and macrophage chemotactic protein 1 (MCP-1)][6,7] appeared just prior to the symptoms of the disease and disappeared during remission. More direct evidence was obtained when it was shown that an antibody to a single chemokine, macrophage inflammatory protein-1α (MIP-1α), could ameliorate or prevent the onset of EAE.[8]

Another disease characterized by T-lymphocyte and monocyte infiltration is rheumatoid arthritis (RA). The inflammatory process in RA is also believed to be mediated by chemotactic factors released by the inflamed tissues. It is known that passive immunization of mice with antibodies directed against MIP-1α or MIP-2 decreases the severity of adjuvant-induced arthritis.[9] It was found that rheumatoid synovial fibroblasts up-regulate RANTES mRNA in response to interleukin-1β (IL-1β), tumor necrosis factor-α (TNF-α), and interferon-γ (IFN-γ). In addition, Rathanaswami et al. demonstrated by Northern blot and enzyme-linked immunosorbent assay (ELISA), that cultured synovial fibroblasts isolated from rheumatoid patients were capable of expressing and producing RANTES and other chemokines in response to IL-1β.[10] Snowden et al. have used reverse transcriptase (RT) PCR (polymerase chain reaction) to detect RANTES mRNA in four of seven synovial tissue samples from RA patients.[11] In contrast, osteoarthritis tissue does not express RANTES mRNA.[11] Furthermore, it appears that RANTES expression plays an important role in RA because, in an adjuvant arthritis model of RA in the rat, the pathology of the disease can be greatly reduced by pretreatment with polyclonal antibody raised to RANTES (J. Tse, D. A. Barnes, J. Hesselgesser, R. Horuk, and H. D. Perez, unpublished results).

At present little has been published on the expression of RANTES in allergic diseases, but what has been reported is provocative. Several investigations suggest that expression of RANTES may be important in the induction of the T-cell infiltrates and eosinophilia characteristic of allergy. Stellato et al. have demonstrated in vivo RANTES expression by pulmonary epithelial cells.[12] In addition, RANTES has been found in the

[6] R. Godiska, D. Chantry, G. N. Dietsch, and P. W. Gray, J. Neuroimmunol. **58,** 167 (1995).
[7] R. M. Ransohoff, T. A. Hamilton, and M. Tani, FASEB J. **7,** 592 (1993).
[8] W. J. Karpus, N. W. Lukacs, B. L. McRae, R. M. Strieter, S. L. Kunkel, and S. D. Miller, J. Immunol. **155,** 5003 (1995).
[9] T. Kasama, R. M. Strieter, N. W. Lukacs, P. M. Lincoln, M. D. Burdick, and S. L. Kunkel, J. Clin. Invest. **95,** 2868 (1995).
[10] P. Rathanaswami, M. Hachicha, M. Sadick, T. J. Schall, and S. R. McColl, J. Biol. Chem. **268,** 5834 (1993).
[11] N. Snowden, A. Hajeer, W. Thomson, and B. Ollier, Lancet **343,** 547 (1994).
[12] C. Stellato, L. A. Beck, T. J. Schall, D. Proud, S. J. Ono, and R. P. Schleimer, FASEB J. **8,** A225 (1994).

bronchoalveolar lavage fluid of asthmatic individuals at levels higher than that of normal subjects.[13] Beck and co-workers demonstrated RANTES expression in the nasal epithelium of patients suffering from nasal polyps, a disease characterized by an eosinophilic infiltrate.[14] Basophils and mast cells are also an important component in the allergic response.[15] As described above, RANTES is chemotactic for basophils and causes them to release histamine.[16]

One possible course of intervention in the inflammatory disease process is to block the action of a particular chemokine. This may be achieved at several levels by interfering with binding to specific cell surface receptors, by modifying postreceptor signaling pathways, or by interfering with the transcription of the specific chemokine mRNA and by inhibiting protein processing. We illustrate these approaches for RANTES by discussing methods that could be used to identify possible inhibitors of expression.

The primary screen to identify inhibitors of chemokine production should be set up to identify compounds that are efficacious at micromolar concentrations. These lead templates can then be chemically modified to identify compounds that are active in the nanomolar range. As noted above, inhibitors of RANTES expression could be useful in the treatment of a number of inflammatory diseases including MS, RA, and atopic dermatitis. Therefore, we have established a high throughput screening assay to develop new lead templates for optimization.

Guidelines for Assays Amenable to High Throughput Screening

High throughput screening (HTS) is a process by which a large number of molecules (small chemical compounds or peptides) can be tested for their effect on a biochemical or biological reaction. This process can be automated and is easily adapted to a 96-well microtiter format. Compounds for screening may come from existing chemical libraries, from combinatorial chemistry, or from natural sources. Traditionally, the most extensive chemical libraries are the private domain of large pharmaceutical companies. Access to these libraries may be limited or restricted. Combinatorial chemistry utilizes methods to make synthetic analogs for screening. This approach has been used with phage-display peptide libraries.[17] Because peptides

[13] R. Alam, P. Forsythe, and S. Stafford, *J. Allergy Clin. Immun.* **93,** 183 (1994).

[14] L. A. Beck, T. J. Schall, and L. D. Beall, *J. Allergy Clin. Immunol.* (*Abstr.*) **93,** 234 (1994).

[15] K. J. Isselbacher, E. Braunwald, J. D. Wilson, J. B. Martin, A. S. Fauci, and D. L. Kasper, eds., "Harrison's Principles of Internal Medicine." McGraw-Hill, New York, 1994.

[16] P. Kuna, S. R. Reddigari, T. J. Schall, D. Rucinski, M. Y. Viksman, and A. P. Kaplan, *J. Immunol.* **149,** 636 (1992).

[17] H. M. Geysen, R. H. Meloen, and S. J. Barteling, *Proc. Natl. Acad. Sci. U.S.A.* **81,** 3998 (1984).

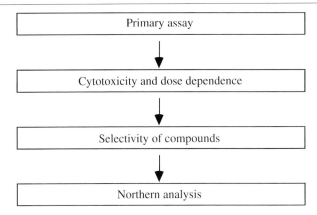

FIG. 1. Flow diagram leading to identification and characterization of lead compounds that inhibit RANTES expression.

make poor oral candidates, however, this is not a favored method. Combinatorial chemistry is likely to be most useful after the lead compound has already been identified. Finally, many novel and as yet undiscovered compounds with a diversity of activities may be mined from natural sources. For a review on the general principles of high throughput screening, see Bevan et al.[18] Screening large diverse compound libraries even with the advent of automatable systems is still time-consuming, and many laboratories now make use of matrix pooling strategies[19] that were first applied to screening cDNA libraries.[20]

The biochemical reaction to be assayed may include anything from a purified enzyme with known quantifiable activity, to a membrane preparation in which the assay may measure specific binding activity of a ligand, to a whole cell assay where measurable results might include cell death or quantitation of a particular cellular protein. The most important points regarding the design of the primary assay are that the end point be quantifiable and reproducible and that there not be any significant variation between preparations of enzyme, membrane, cell line passage, etc.

Although the parameters of the primary assay are important, it is also essential to design a screening tree or flow chart where each step introduces a new criterion for selectivity, thus allowing the selection of chemical structures exhibiting specific properties. An example of a simplified flow diagram utilized in the search for RANTES expression inhibitors is given in Fig. 1.

[18] P. Bevan, H. Ryder, and I. Shaw, *Trends Biotechnol.* **13**, 115 (1995).
[19] J. J. Devlin, A. Liang, and C. Trinh, *Drug Dev. Res.* **37**, 80 (1996).
[20] R. Devos, G. Plaetinck, and H. Cheroutre, *Nucleic Acids Res.* **11**, 4307 (1983).

Primary Assay

To devise a strategy for the screening of the compound library we made use of the astrocytoma cell line CH235. These cells are adherent and attach to the bottom of flat-well plates, they can be induced to produce RANTES, and they were found to exhibit responses similar to primary human astrocytes when exposed to TNF-α, IL-1β, and IFN-γ.[21] It was previously determined that exposure of CH235 cells to IL-β (40 U/ml) and IFN-γ (10 U/ml) for a period of 2 hr resulted in the induction of RANTES transcription, and the RANTES protein product was observed in the supernatant by 5 hr poststimulation (determined by ELISA).[21] The induction continues in a linear fashion for at least 12 hr.[21] These data were used to establish a reliable 96-well format assay.

1. CH235 cells are plated at a concentration of 10,000 cells/well in 96-well plates (Corning, Corning, NY) in Dulbecco's modified Eagle's medium (DMEM)/F12 medium.[21]

2. The cells are allowed to settle and adhere to the bottom of the well for 72 h in a humidified 37° incubator in 5% (v/v) CO_2.

3. The cells are then stimulated with IL-1β (Genzyme, Cambridge, MA) and IFN-γ (Sigma, St. Louis, MO) for 2 hr after which compounds dissolved in dimethyl sulfoxide (DMSO) are added to a final concentration of 1.0 μM. To maintain consistency, DMSO is added to the positive control wells (stimulated with IL-1β and IFN-γ alone, no compounds added) and the negative control wells (not stimulated with cytokines, no compounds added) at the same final concentration as was achieved with the addition of compounds. Incubation is continued for an additional 4 hr.

4. At the end of the total 6-hr incubation period, supernatants are removed and assayed for RANTES using a 96-well ELISA (R&D Systems, Minneapolis, MN).

As the cells are fully induced after 2 hr of stimulation with IL-1β and IFN-γ, the rationale behind this approach is to detect compounds that would bypass inhibition of ligand binding and signal transduction.

Once the parameters of the assay have been decided, reasonable cutoffs are set that determine whether the compound or peptide continues to the next step in flow diagram. For example, in an enzyme inhibitor screen a number of parameters, including the percentage of inhibition required compared to a positive control and the maximal concentration of inhibitor allowed to induce this level of inhibition, are determined. In a sense, such parameters are somewhat arbitrarily set. For the purposes of our study we

[21] D. A. Barnes, M. Huston, and R. Holmes, *J. Neuroimmunol.* **71,** 207 (1996).

chose compounds that were (1) active at concentrations at or below 1.0 μM, (2) inhibited RANTES expression by at least 40% in the 6-hr assay, and (3) were composed of tractable structures so that chemical optimization would be possible.

After screening approximately 60,000 compounds according to the protocols described above, we identified 24 compounds that fell into 12 different structural classes that fulfilled the above criteria. A structural class is defined as a group of related compounds that share a common chemical foundation.

Cytotoxicity Testing

Although the primary assay identified 12 structural classes as inhibitors of RANTES expression, some could almost certainly be predicted to be cytotoxic. Because cytotoxicity also will result in a perceived inhibition of RANTES expression, the compounds are tested at the same concentration (1.0 μM) in two or more 24-hr cytotoxicity tests. All tests include microscopic examination to determine whether these adherent roughly star-shaped cells retain both shape and adherence when incubated with inhibitor. The microscopic examination, however, while quick and simple, is not sensitive. Cells may still be adherent and retain their shape but incapable of growth. Nevertheless, this assay is useful in eliminating compounds that are grossly cytotoxic.

There are more sensitive methods for determining levels of cell viability. The easiest of these utilize cellular staining, the stain denoting either viability or nonviability depending on the method employed. The trypan blue exclusion test stains nonviable cells; viable cells with intact membranes are capable of excluding the dye and remain unstained. Relative levels of cytotoxicity can be determined by counting the percentage of stained cells using a microscope and dividing by the total number of cells counted. Another cellular staining method that is even more sensitive uses the tetrazolium salt 3-(4,5-dimethylthiazol-2-yl)-2,5-diphenyltetrazolium bromide (MTT), which is cleaved by metabolically active mitochondria to produce a formazan product. This assay is easily adapted to the 96-well format, and the results can be obtained by a microtiter plate reader. The MTT assay (Promega, Madison, WI) compares favorably with [³H]thymidine incorporation, which measures DNA synthesis, in its sensitivity.[22]

Of the 12 structural classes of compounds identified in the primary assay, 8 were eliminated primarily because of problems with cytotoxicity. Although all of the compounds identified above withstood the requirements

---

[22] H. Tada, O. Shiho, K. Kuroshima, M. Koyama, and K. Tsukamoto, *J. Immunol. Methods* **93,** 157 (1986).

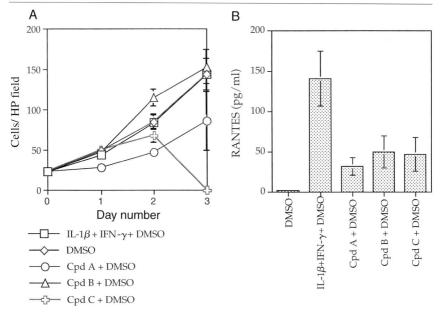

Fig. 2. Seventy-two-hour cytotoxicity assay and RANTES ELISA. (A) Cell count versus time; (B) Amount of RANTES in the supernatant of each culture at the 72-hr time point.

of the 24-hr cytotoxicity screen they still had to be tested for long-term cytotoxicity. In other words, whereas the 24-hr cytotoxicity screen would exclude obvious noxious compounds, it would not remove those that might still inhibit cell growth. Therefore, we set up an assay that would detect inhibitors of cell growth.

1. Cells are plated in 24-well plates at very low density (~400 cells/ well). The cells are allowed to settle for 24 hr.

2. After this time three wells are given DMSO to 1 μl/ml (these are the control cells, as all of the inhibitors are dissolved in DMSO) and three high-power microscope fields are counted/well.

3. Another three wells are given 40 U/ml of IL-1β and 10 U/ml of IFN-γ (plus DMSO at 1 μl/ml) or IL-1β and IFN-γ as well as compound A (1.0 μM) all at the same time and the cells are counted as above.

The compounds B and C are tested in a similar manner also at a concentration of 1.0 μM. Cell numbers are counted as described above at 24, 48, and 72 hr (Fig. 2A) Because the doubling time of the CH235 cells is about 36–48 hr, it is possible to determine whether the inhibitors have any effect on cell growth. At the final cell count (72 hr), the supernatants

TABLE I
IC$_{50}$ VALUES OF COMPOUNDS A, B, AND D

Source	Compound A	Compound B	Compound D
RANTES	1.9 ± 0.36	4.4 ± 1.16	0.3 ± 0.15
IL-6	1.24 ± 0.15	2.27 ± 0.63	0.1 ± 0.01
IL-8	24.21 ± 10.94	≫30	0.4 ± 0.20

are removed so that RANTES expression could also be assayed by ELISA (Fig. 2B).

Compound B clearly has no effect on the growth of CH235 cells but is a very effective inhibitor of RANTES expression. Compound A slightly inhibits the growth of CH235 cells but also reduces RANTES expression. Surprisingly, compound C, found to be nontoxic in the 24-hr assay, is found to be highly toxic in the 72-hr growth assay (Fig. 2A). For this reason compound C is eliminated from further study. It should also be noted that when the CH235 cells are given the inhibitor and the cytokines at the same time the inhibitory effect is much greater (~70% inhibition) and that the effect is long lasting as cells did not receive fresh inhibitor during the 3-day incubation period.

From our primary assay that identified 24 compounds and 12 different structural classes, three structural classes remained.

Dose Dependence and Inhibitory Ability of Compounds A, B, and D

Compounds A, B, and D are retested at a dosage range of 0.1–5.0 μM to determine their relative efficacy of RANTES inhibition. Effects on IL-6 and IL-8 expression are also determined. We choose to examine IL-8 and IL-6 expression because both of these are induced in CH235 cells by the combined IL-1β and IFN-γ treatment.

CH235 cells are grown as discussed previously and stimulated for 2 hr with IL-1β and IFN-γ followed by the addition of inhibitor at various concentrations for an additional 4 hr. The IC$_{50}$ values for RANTES, IL-6, and IL-8 are obtained from three different experiments. Results are given as micromolar concentrations (Table I). All compounds are found to inhibit both RANTES and IL-6. Compounds A and B exhibit at least some selectivity in that they do not inhibit IL-8. Compound D, however, is completely nonselective as it inhibits RANTES, IL-6, and IL-8 equally well. At this point, we eliminate compound D from the group of three structural classes, leaving two.

Scope of Activity of Compound A and Compound B

The astrocyte is an important, major cell type of the central nervous system (CNS). Although the precise role of the astrocyte has yet to be defined, evidence suggests that it plays an important part in the pathogenesis of various disease states, but that it also may be important in repairing tissues,[23] maintaining the integrity of the blood–brain barrier, and providing neurotrophic support.[24] Any compound that inhibits RANTES should be as specific as possible because compromising the ability of astrocytes to operate smoothly might seriously cripple the normal function of the CNS.

Although both compounds A and B have been found to inhibit RANTES and IL-6 but not IL-8 expression, the scope of their activities had yet to be determined. Therefore, CH235 cells are stimulated for 2 hr with IL-1β and IFN-γ prior to the addition of either compound. The cells are incubated overnight (15 hr), and the supernatants are removed and assayed for the following cytokines/chemokines: RANTES, IL-8, IL-6, MCP-1, TNF-α, transforming growth factor-β_2 (TGF-β_2), granulocyte colony-stimulating factor (G-CSF), granulocyte–macrophage colony-stimulating factor (GM-CSF), and macrophage colony-stimulating factor (M-CSF). These chemokine/cytokines are chosen because 96-well ELISA kits are available (R&D Systems) to measure their levels in the supernatant. It is important to determine whether the inhibitors affect the levels of other cytokines/chemokines because these molecules are important in cell-to-cell communication among astrocytes and other cells in the CNS, and among astrocytes and circulating T lymphocytes and monocytes. Activated astrocytes, however, are known to produce a large array of different proteins that may not be secreted into the medium (for a review, see Eddleston and Mucke[25]). The results from a typical screen are shown as percentage inhibition in Table II. Although both compounds inhibit RANTES and do not affect IL-8 expression (as required for selectivity), both are also found to inhibit IL-6. Compound B does not affect the expression of any other cytokine/chemokine assayed; compound A, however, inhibits MCP-1 and TNF-α expression but has no effect on TGF-β.

It is interesting to note that the lead compounds are able to inhibit both RANTES and IL-6 expression. Although at this time it is not known whether modification of the chemical structure will result in a RANTES-specific compound, it may be beneficial to have a single compound that inhibits both RANTES and IL-6. IL-6 is a multifunctional cytokine that

[23] A. Logan, S. A. Frautschy, A.-M. Gonzalez, M. B. Sporn, and A. Baird, *Brain Res.* **587**, 216 (1992).
[24] G. A. Banker, *Science* **209**, 809 (1980).
[25] M. Eddleston and L. Mucke, *Neuroscience* (*Oxford*) **54**, 15 (1993).

TABLE II
ACTIVITY OF COMPOUND A AND COMPOUND B[a]

| Chemokine/cytokine | Activity | |
	Compound A	Compound B
RANTES	60 ± 5.0	55 ± 10
IL-8	NDI[b]	NDI
IL-6	39 ± 4.7^c	35 ± 5.0^c
MCP-1	45 ± 10	NDI
TNF-α	60 ± 3.0	NDI
TGF-β	NDI	NDI
GM-CSF	Not tested	NDI
G-CSF	Not tested	NDI
M-CSF	Not tested	NDI

[a] Results are expressed as percent inhibition $\pm$ SD and represent the average of one to four experiments performed in triplicate.
[b] NDI, No detectable inhibition.
[c] These results were obtained from a 6-hr incubation (as in the primary assay). This experiment has not yet been performed for an overnight (15-hr) incubation.

plays a major role as a mediator of acute phase inflammatory responses. These include inflammatory cell and lymphocyte activation and hepatocellular stimulation of acute phase protein synthesis. It is known that serum levels of IL-6 increase with age, and it has been proposed that many of the diseases associated with aging, for example, osteoporosis and Alzheimer's disease, may be due to dysregulation of IL-6 gene expression.[26]

Interleukin-6 is produced by osteoblasts and is known to induce bone resorption *in vitro*[27] Estrogen is known to have the opposite effect and is found to inhibit bone resorption. It has been argued that estrogen may either inhibit IL-6 gene expression or interfere in some way with its function.[28] It is possible that bone loss occurring during menopause may be due to the lack of estrogen, an inhibitor of IL-6 function. Furthermore, IL-6 has been implicated in Alzheimer's disease, the pathogenesis of which is due to the deposit of β-amyloid fragments. The β-amyloid proteins are produced due to misprocessing of a larger precursor.[29] Several groups have found that α_2-macroglobulin acts as a protease inhibitor within human neuronal cells

[26] W. B. Ershler, *J. Am. Geriatr. Soc.* **41,** 176 (1993).
[27] Y. Ishimi, C. Miyaura, and C. H. Jin, *J. Immunol.* **145,** 3297 (1990).
[28] R. L. Jilka, G. Hangoc, and G. Girasole, *Science* **257,** 88 (1992).
[29] J. A. Hardy and G. A Higgins, *Science* **256,** 184 (1992).

and leads to the misprocessing of the β-amyloid precursor. The α_2-macro-globulin inhibitor could be produced in neuronal cells but only after they were stimulated with IL-6. In addition there was strong immunohistochemical staining for IL-6 and α_2-macroglobulin within senile plaques in brain sections from Alzheimer's disease patients.[30,31]

Northern Blot Analysis

To determine whether the remaining compounds identified had an effect on RANTES mRNA levels, a Northern blot analysis has been performed.

1. The negative control cells are unstimulated and untreated. The positive control cells are stimulated with IL-1β and IFN-γ.

2. Cells given inhibitor are first treated for 2 hr with cytokines prior to the addition of the inhibitor (final concentration 1.0 μM).

3. Cells are harvested at the end of the 6-hr incubation, and mRNA is purified using a Qiagen (Chatsworth, CA) mRNA preparation kit.

4. Blots are probed with cDNA probes for β-actin (RNA loading), RANTES, and IL-8.

The compounds tested in this assay are A, B, and D. Compound D has been shown to greatly inhibit RANTES expression, but although it is noncytotoxic, it has been eliminated from further consideration because it also greatly inhibits both IL-6 and IL-8 (see Table I).

These data indicate that both compounds A and B appear to lower the steady-state levels of RANTES transcript. At this point we do not know whether they affect initiation of transcription or whether they are affecting the stability of the RANTES message. Neither of these two inhibitors has any effect on the level of IL-8 transcription. Compound D, which inhibits both RANTES and IL-8 protein expression (as determined by ELISA), has no effect on the transcript levels of these two chemokines. This inhibitor serves as an internal control and may be working at a level that is posttranscriptional. The data obtained from compounds A and B (Fig. 3) have been quantitated as percentage inhibition of the relative amounts of RANTES and IL-8 mRNA (Fig. 4).

It should be noted that at the time the mRNA is isolated (6 hr) RANTES mRNA has reached its highest steady-state levels within the CH235 cells, whereas RANTES protein secretion into the medium does not achieve maximum levels until at least 15 hr poststimulation.[21] Thus, the inhibitors show a marked decrease in the RANTES mRNA levels at the end of the

[30] U. Ganter, S. Strauss, and U. Jonas, *FEBS Lett.* **282,** 335 (1991).
[31] S. Strauss, J. Bauer, and U. Ganter, *Lab. Invest.* **66,** 223 (1992).

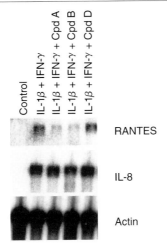

FIG. 3. Northern blot analysis of cells treated with inhibitors. CH235 cells were plated in 162-mm^2 culture flasks at 5×10^6 cells/dish, and mRNA was isolated for hybridization analysis. Unstimulated CH235 cells were the negative control (lane marked Control); lane IL-1β + IFN-γ shows results from cells treated with cytokines for 6 hr (positive control). The next three lanes contain mRNA from cells treated with cytokines for 2 hr and then given one of the RANTES inhibitors (compounds A, B, or D) at a final concentration of 1.0 μM for a further 4 hrs. The same Northern filter was hybridized to a RANTES PCR product,[21] an IL-8 PCR product, and a β-actin PCR product (Clontech, Palo Alto, CA).

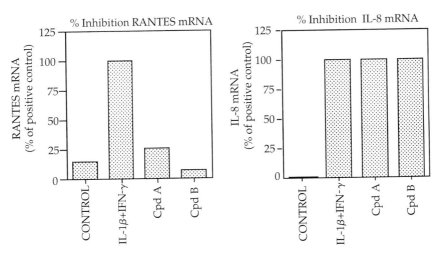

FIG. 4. Quantitation of mRNA. After hybridization, Northern blots from Fig. 3 were exposed to a PhosphorImager screen. Intensity of signal was quantified, and RANTES or IL-8 mRNA data were expressed as percentages of the positive control (IL-1β + IFN-γ treated cells).

6-hr incubation period even though there is not such a drastic decrease in the levels of RANTES in the medium (i.e., ~40%). One would predict that if the assay were allowed to proceed for 24 hr, or longer, inhibition of RANTES expression will be greater. This has indeed proved to be the case (data not shown).

Conclusion

Using this screening protocol we have identified a potential lead compound that inhibits RANTES expression in the astrocytoma cell line CH235. In our assay and experimental design we utilized a whole cell based assay with RANTES protein levels secreted into the medium determined by ELISA. Such methods do not allow for identification of mechanism of action for these compounds, which has yet to be determined. It is interesting to note, however, that the mechanism must be fairly universal because compound B has been found to inhibit RANTES expression in at least four different primary cell lines including synovial fibroblasts and peripheral blood lymphocytes (data not shown). Further work will now be necessary both to optimize the effectiveness of this compound and to elucidate its mechanism of action.

Acknowledgments

The authors thank Dr. Richard Horuk for a critical reading of the manuscript.

[20] Transgenic Methods to Study Chemokine Function in Lung and Central Nervous System

By Sergio A. Lira, M. Elena Fuentes, Robert M. Strieter, and Stephen K. Durham

Introduction

Trafficking of leukocytes throughout specific tissues is regulated by mechanisms that are not completely understood. Current models postulate that a series of complex events, initiated by local changes in vascular flow, favor the migration of leukocytes from the vascular compartment to the extravascular space.[1] Despite significant advances in the identification of factors promoting these changes and in the elucidation of mechanisms triggered by their expression, there is still uncertainty as to whether many of these molecules do regulate leukocyte trafficking *in vivo*. An approach to address this problem is to introduce specific mutations into the genes encoding these putative factors and study leukocyte trafficking in the resulting mutants. These genetic modifications can be transmitted through the germ line to a large number of animals, which can be used in a variety of experimental settings.

Among the many putative molecules regulating leukocyte trafficking are the chemokines. Chemokines are able to promote chemotaxis of leukocytes *in vitro*, but it has been only relatively recently that genetic experiments have been used to test their role *in vivo*.[2,3] Programmed expression of chemokines in tissues such as the thymus, stratified epithelium, and the central nervous system (CNS) can trigger the recruitment of leukocytes *in vivo*.[4–6] Conversely, deletion of specific chemokines or their receptors can lead to disturbances in hematopoiesis and leukocyte trafficking, leading to inappropriate immune responses.[7–9] Taken together, these observations indicate that chemokines can initiate the cascade of events leading to vascular extravasation and tissue infiltration by leukocytes, a far more important role in the regulation of leukocyte trafficking than previously anticipated.

In this chapter, we describe genetic methods to study chemokine biology, focusing on methods to express chemokines in specific tissues (lung and brain) of transgenic mice. We also describe methods to analyze the phenotypes of the resulting transgenic mice. The specific methodology to generate transgenic mice has been extensively covered in a number of excellent

[1] T. A. Springer, *Cell (Cambridge, Mass.)* **76,** 301 (1994).

[2] S. A. Lira, *J. Leukocyte Biol.* **59,** 45 (1996).

[3] D. Cook and S. A. Lira, *in* "Leukocyte Recruitment in Inflammatory Disease" (G. Peltz, ed.) p. 259. R. G. Landes, New York, 1996.

[4] S. A. Lira, P. Zalamea, J. N. Heinrich, M. E. Fuentes, D. Carrasco, D. S. Barton, S. K. Durham, and R. Bravo, *J. Exp. Med.* **180,** 2039 (1994).

[5] M. E. Fuentes, S. K. Durham, M. R. Swerdel, A. C. Lewin, D. S. Barton, J. R. Megill, R. Bravo, and S. A. Lira, *J. Immunol.* **155,** 5769 (1995).

[6] M. Tani, M. E. Fuentes, J. W. Peterson, B. D. Trapp, S. K. Durham, J. K. Loy, R. Bravo, R. Ransohoff, and S. A. Lira, *J. Clin. Invest.* **98,** 529 (1996).

[7] D. N. Cook, M. A. Beck, T. M. Coffman, S. L. Kirby, J. F. Sheridan, and I. B. Pragnell, *Science* **269,** 1583 (1995).

[8] T. Nagasawa, S. Hirota, K. Tachibana, N. Takakura, S. Nishikawa, Y. Kitamura, N. Yoshida, H. Kikutani, and T. Kishimoto, *Nature (London)* **382,** 635 (1996).

[9] G. Cacalano, J. Lee, K. Kikly, A. M. Ryan, S. Pitts-Meek, B. Hultgren, W. I. Wood, and M. W. Moore, *Science* **265,** 682 (1994).

sources[10,11] and is not repeated here. Other good sources of information on chemokines[12] and mouse genetics[13-15] can now be found on-line.

Expression of Chemokine KC in Lung

The lung, as with the skin and intestinal tract, represents a large surface for interaction with the environment. In the lung, the interactions between pathogens and pollutants happen at the level of the upper and lower respiratory tree, leading to a variety of inflammatory conditions. The expression of chemokines under many of these inflammatory conditions has been amply documented, but it remains unclear whether they can actually dictate the recruitment of inflammatory cells to the lung. One of the chemokines implicated in lung inflammation is the C-X-C chemokine KC. Murine KC, also known as N51, is an immediate early gene that encodes a secretory protein of approximately 8000 Da.[16,17] Induction of KC gene expression has been documented in mitogen-stimulated fibroblasts,[17,18] in lipopolysaccharide (LPS)-stimulated peritoneal macrophages,[19] and in endothelial[20] and vascular smooth muscle cells,[21] but not in a variety of unstimulated mouse tissues by Northern blot analysis.[17]

The chemokine KC is induced in rat lung cells during inflammatory responses *in vitro* and *in vivo*. Rat KC (rKC) is rapidly induced in the lung as consequence of chemical irritants such as vanadium[22] and sulfur dioxide.[23] Lipopolysaccharide treatment of rat alveolar macrophages *in vitro* causes

[10] B. Hogan, R. Beddington, F. Costantini, and E. Lacy, "Manipulating the Mouse: A Laboratory Manual." Cold Spring Harbor Laboratory, Cold Spring Harbor, New York, 1994.

[11] P. M. Wassarman and M. L. DePamphilis, eds., *Methods Enzymol.* **225** (1993).

[12] http://cytokine.medic.kumamoto-u.ac.jp

[13] http://www.cco.caltech.edu/~mercer/htmls/rodent_page.html

[14] http://www.jax.org

[15] http://www.hgmp.mrc.ac.uk/Public/rodent-gen-db.html

[16] P. Oquendo, J. Alberta, D. Z. Wen, J. L. Graycar, R. Derynck, and C. D. Stiles, *J. Biol. Chem.* **264,** 4133 (1989).

[17] R. P. Ryseck, H. Macdonald-Bravo, M. G. Mattei, and R. Bravo, *Exp. Cell. Res.* **180,** 266 (1989).

[18] B. H. Cochran, A. C. Reffel, and C. D. Stiles, *Cell (Cambridge, Mass.)* **33,** 939 (1983).

[19] M. Introna, R. C. Bast, Jr., C. S. Tannenbaum, T. A. Hamilton, and D. O. Adams, *J. Immunol.* **138,** 3891 (1987).

[20] X. Y. Shen, T. A. Hamilton, and P. E. DiCorleto, *J. Cell. Physiol.* **140,** 44 (1989).

[21] J. D. Marmur, M. Poon, M. Rossikhina, and M. B. Taubman, *Circulation* **86,** 53 (1992).

[22] L. M. Pierce, F. Alessandrini, J. J. Godleski, and J. D. Paulauskis, *Toxicol. Appl. Pharmacol.* **138,** 1 (1996).

[23] A. Farone, S. Huang, J. Paulauskis, and L. Kobzik, *Am. J. Respir. Cell. Mol. Biol.* **12,** 345 (1995).

rapid and marked increases in mRNA for rKC.[24] *In vivo,* rKC mRNA levels rise significantly within both BAL (bronchoalveolar lavage) cells and trachea homogenates after LPS instillation.[24] The neutrophilic infiltrate that results from LPS instillation can be markedly inhibited by a neutralizing anti-KC antibody.[25] In addition, intratracheal instillation of rKC induces a dose-dependent polymorphonucleocytic (PMN) influx into air spaces.[25] The rapid expression of rKC mRNA in BAL cells, prior to the observed neutrophilic infiltration, implicates KC as an important factor in the initiation of inflammation.

To further test the hypothesis that KC mediates the recruitment of neutrophils to the lung, we have genetically programmed its expression to a particular set of cells in the lungs of transgenic mice. By mimicking, in a noninvasive way, the pattern of expression of KC during lung inflammation, we expect to learn about the specific inflammatory properties of KC *in vivo.* To target the expression of mouse KC (mKC) to the lung, we have used regulatory elements from the CC10 gene. The CC10 gene encodes a 10-kDa protein of unknown function produced by nonciliated bronchiolar epithelial cells (Clara cells). Clara cells are tall, dome-shaped epithelial cells that are present in large numbers in the bronchioles of mammals, where they serve as progenitors for bronchiolar ciliated cells and other Clara cells. In some species, such as rodents, Clara cells are also found in the trachea and bronchi. Clara cells synthesize, store, and secrete protein components of the bronchiolar extracellular lining. In rodents, but not primates, Clara cells contain abundant smooth endoplasmic reticulum and play a major role in electrolyte balance and xenobiotic metabolism in the lung.

Segments of the rat CC10 gene have previously been used to target the expression of transgenes to Clara cells in mice.[26,27] The expression of transgenes under the regulation of a rat 2.3-kb CC10 promoter recapitulates the tissue- and cell-specific expression of the CC10 gene.[27,28] To clone this promoter element, we use a modification of the polymerase chain reaction (PCR) method described by Stripp *et al.*[27]

The primer set

[24] S. Huang, J. D. Paulauskis, J. J. Godleski, and L. Kobzik, *Am. J. Pathol.* **141,** 981 (1992).

[25] C. W. Frevert, S. Huang, H. Danaee, J. D. Paulauskis, and L. Kobzik, *J. Immunol.* **154,** 335 (1995).

[26] B. P. Hackett and J. D. Gitlin, *Proc. Natl. Acad. Sci. U.S.A.* **89,** 9079 (1992).

[27] B. R. Stripp, P. L. Sawaya, D. S. Luse, K. A. Wikenheiser, S. E. Wert, J. A. Huffman, D. L. Lattier, G. Singh, and S. L. Katyal, *J. Biol. Chem.* **267,** 14703 (1992).

[28] L. R. Margraf, M. J. Finegold, L. A. Stanley, A. Major, H. K. Hawkins, and F. J. DeMayo, *Am. J. Respir. Cell. Mol. Biol.* **9,** 231 (1993).

Upper primer:
    5'-ACGCGGCCGC<u>GGATCC</u>AATAACTTAAGCCCTGTAGCA-3'
Lower primer: 5'-CC<u>GGATCC</u>GGGCTGTCTGTAGATGTGGGCT-3'

is used to amplify a 2.3-kb segment from the 5' flanking region of the CC10 gene, using rat genomic DNA as a template. The underlined sequences encode a restriction site for *Bam*HI. The conditions used are 95° for 30 sec, 60° for 30 sec, and 72° for 3 min for 40 cycles. After amplification the PCR products are isolated using a PCR kit (Qiagen, Chatsworth, CA), restricted with *Bam*HI, and subcloned into plasmid pGEMN51, to yield the transgene CC51 (Fig. 1A) containing the CC10 promoter, the KC cDNA, and segments of the human growth hormone (hGH) gene. The CC51 transgene is excised from the vector using *Not*I and *Sfi*I and gel purified using a Qiagen kit. After gel purification the transgene is subjected to three consecutive washes through Microcon filters (Amicon, Beverly, MA) using microinjection buffer. Prior to injection the samples are quantitated, and an aliquot is used to check DNA integrity by gel electrophoresis.

The transgene is resuspended to a final concentration of 1–5 ng/ml and microinjected into fertilized [C57BL/6J × DBA/2] F_2 eggs, according to standard methodology.[10] Microinjected eggs are transferred into oviducts of ICR (Sprague–Dawley) foster mothers according to published proce-

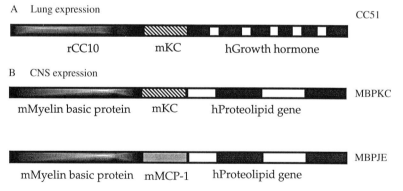

FIG. 1. Diagram of the transgenes used to target the expression of chemokines to lung (A) and CNS (B). The rat CC10 promoter (2.3 kb) drives the expression of the murine KC cDNA fused to human growth hormone genomic sequences (CC51 transgene). The myelin basic protein promoter (1.9 kb) drives the expression of murine KC and MCP-1 cDNAs fused to genomic sequences of the human proteolipid gene (MBPKC and MBPJE transgenes, respectively). AU-rich sequences present 3' to the translational stop signal in the KC and MCP-1 cDNAs were removed to increase mRNA stability. Genomic sequences from the human growth hormone and proteolipid genes were added to provide introns, exons, and polyadenylation sequences. Translation of the hybrid mRNAs terminates at the stop signal of the chemokine.

dures.[10] By 10 days of life, a piece of tail from the resulting animals is clipped and digested in 100 μl of 50 mM Tris-HCl (pH 8.3), 100 mM NaCl, 5 mM EDTA, 1% sodium dodecyl sulfate (SDS), 300 μg proteinase K for 5 hr at 55°. Samples are extracted with phenol/chloroform/isoamyl alcohol (25:24:1, by volume), and nucleic acid is precipitated with ethanol and resuspended in 150 μl of water. Identification of transgenic founders is carried out by PCR analysis. Two oligonucleotide primers recognizing hGH sequences are used for the screening of transgene incorporation (5'-TTTGGGGTTCTGAATGTGAG-3' and 5'-AGGCACTGCCCTCTT-GAAGC-3'), and, as an internal control, we use the endogenous β-TSH gene (5'-GTAACTCACTCATGCAAAGT-3 and 5'-TCCTCAAAGAT-GCTCATTAG-3'). These primers amplify 218- and 366-bp segments of the hGH and b-TSH genes, respectively. Two microliters of tail DNA are added to 18 μl of PCR mixture [50 mM KCl, 10 mM Tris-HCl (pH 8.4), 1.5 mM MgCl$_2$, 20 μg/ml gelatin, 0.2 mM NTPs (Pharmacia, Piscataway, NJ), 0.6 units *Thermus aquaticus* DNA polymerase (*Taq* polymerase, Per-kin-Elmer/Cetus), 200 ng of each oligonucleotide primer]. Samples are overlaid with 35 μl of mineral oil and reactions run in a thermal cycler (Perkin-Elmer). PCR conditions used are initial denaturation for 2 min at 94° and amplification for 30 cycles of 90 sec at 55°, 1 min at 72°, and 1 min at 94°. A 10-μl sample is subjected to electrophoresis on a 2% agarose gel, containing 100 ng/ml ethidium bromide, and DNA bands are visualized under UV light.

In our studies, 260 embryos were injected with the CC51 transgene and transferred to the oviducts of pseudopregnant animals. Of the resulting 77 mice, 8 were identified by PCR analysis of tail DNA as positive for the transgene. From these transgenic founders, five transgenic lines were estab-lished. The resulting transgenic animals are being kept under pathogen-free conditions.

The CC51 transgenic mice are healthy and fertile. To test for transgene expression, the lungs and other organs are dissected, and total RNA is extracted. Analysis of these samples demonstrated that the transgene is specifically expressed in the lungs in all transgenic lines (not shown). Be-cause transgenes are integrated at random in the genome, we usually select three or more of the expressing transgenic lines for phenotypic analysis. By doing so, we increase the likelihood that the observed phenotype results from the expression of the chemokine, rather than from a mutation induced by the integration of the transgene into the genome.

To investigate whether the expression of KC in the lung leads to in-flammatory infiltrates, a number of histological procedures are performed. Lungs from transgenic (three different lines) as well as control mice are evaluated by light and transmission electron microscopy following fixation by tracheal infusion at a constant pressure of 30 cm of water with McDowell–Trumps fixative with 0.1 M phosphate buffer, pH 7.4. After overnight fixation, two 1-mm-thick transverse sections are cut for embedding in epoxy resin. The first section is taken immediately caudal to the hilus; the second section is taken midway between the hilus and the most caudal aspect of the lung. The remainder of the left lung is cut transversally into sections 2–3 mm thick, embedded in paraffin by routine methods, sectioned at 4–6 μm, and stained with hematoxylin and eosin. Epoxy resin-embedded lung sections are cut 1.0 μm thick and stained with toluidine blue to delineate areas of interest. Areas of interest (mainstem bronchi, terminal bronchioles, parenchyma without airways) are sectioned at approximately 85 nm thick, mounted on 150-mesh copper grids, stained with lead citrate and uranyl acetate, and examined with a transmission electron microscope.

To examine whether the infiltrate is spatially coincident with the known distribution of Clara cells, we enumerate neutrophils infiltrating airways based on unit length as the denominator for quantification. Specifically for bronchi and bronchioles, we count the number of neutrophils per linear millimeter of airway basement membrane using an ocular reticle. To examine whether the infiltrate is spatially coincident with alveolar parenchyma, we count the number of neutrophils based on unit area of 1 mm^2 using an ocular reticle. A neutrophil is identified based on morphological criteria: round cell with a diameter of 10–12 μm, multilobulated nucleus, and indistinct cytoplasmic granules.

Results obtained by these measurements have indicated that the expression of the transgene in lung results in recruitment of neutrophils in at least a 10-fold increase to topographical areas rich in Clara cells (bronchioles) as compared to parenchyma within the same animal or a 20-fold increase as compared to the same anatomic zone in an age-matched control (Fig. 2A and B). Despite the significant influx of neutrophils to these areas, no morphological change indicative of tissue injury is observed. Because no morphological change is observed by light microscopy, we examine neutrophil ultrastructure. As shown in Fig. 2C, neutrophils still contain vast stores of cytoplasmic granules, indicating that they have not undergone degranulation, thus explaining the absence of tissue injury.

In previous experiments using the *lck* promoter, we have reported that the neutrophilic infiltrate induced by KC in the thymus is attenuated over time.[4] We decided to investigate whether the continued expression of KC in the lung induced a similar phenomenon. To this end, we enumerate neutrophils in specific topographical areas of the lung during selected time points, namely, days 18, 90, and 180. We perform the same methodology for the temporal relationship that we applied to the topographical study. The results of these measurements are shown in Fig. 3. The number of infiltrating neutrophils decrease significantly within the first 3 months of

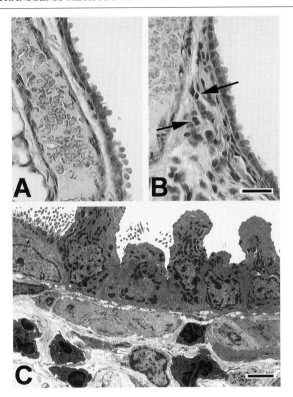

Fig. 2. Light photomicrographs of the bronchiole and adjacent vasculature from (A) control or (B) transgenic mice. Clusters of neutrophils (arrows) are subjacent to Clara cells in the transgenic animal and not the age-matched control. Sections were stained with hematoxylin and eosin. Bar: 20 μm. (C) Transmission electron micrograph of neutrophil ultrastructure from a transgenic animal, in which these cells still contained vast stores of cytoplasmic granules. Clara cells are unaffected. Sections were stained with lead citrate and uranyl acetate. Bar: 5 μm.

life. By 90 days, there are 5-fold fewer neutrophils in the bronchioles of transgenic mice than by 18 days of age.

To investigate whether the reduced number of cells could be attributed to changes in the level of expression of the transgene, we measure the levels of KC protein in plasma and lung extracts from control and transgenic animals using a modification of a double ligand method. At time of sacrifice, anticoagulated (50 units heparin per 500 μl of blood) ocular venous plexus blood is collected and centrifuged. The plasma is stored at −70° for later analysis. The left lung is snap frozen for subsequent homogenization and sonication in antiprotease buffer [1× PBS (phosphate-buffered saline) with 2 mM phenylmethylsulfonyl fluoride (PMSF) and 1 μg/ml each of antipain,

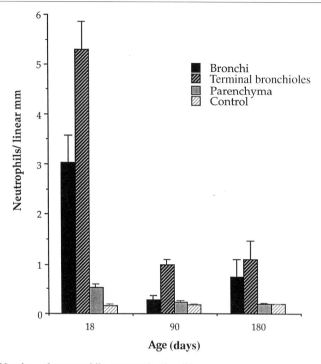

FIG. 3. Number of neutrophils present in the different areas of the lung of control and CC51 transgenic mice, at different ages. Notice a significant reduction in the number of neutrophils between 18 and 90 days of age. Control numbers were derived from measurements in terminal bronchioles of nontransgenic animals. Data are means ± SEM.

aprotinin, leupeptin, and pepstatin A], followed by filtration through 0.45-μm filters (Acrodiscs, Gelman Sciences, Ann Arbor, MI). The filtrate is stored at −70° for later analysis of KC by specific enzyme-linked immunosorbent assay (ELISA).

Flat-bottomed 96-well microtiter plates (Nunc Immuno-Plate I 96-F) are coated with 50 μl/well of the polyclonal anti-KC antibody (1 ng/μl in 0.6 M NaCl, 0.26 M H_3BO_4, and 0.08 N NaOH, pH 9.6) for 24 hr at 4° and then washed with PBS, pH 7.5, 0.05% (w/v) Tween 20 (wash buffer). Microtiter plate nonspecific binding sites are blocked with 2% (w/v) bovine serum albumin (BSA) in PBS and incubated for 60 min at 37°. Plates are rinsed three times. Fifty microliters of sample (neat and 1:10) is added, followed by incubation for 1 hr at 37°. Plates are washed three times, 50 μl/well of biotinylated polyclonal rabbit anti-KC antibody [3.5 ng/μl in PBS, pH 7.5, 0.05% Tween 20, and 2% fetal calf serum (FCS)] is added, and plates are incubated for 45 min at 37°. Plates are washed three times, streptavidin–peroxidase conjugate (Bio-Rad Laboratories, Richmond, CA) added, and the plates incubated for 30 min at 37°. Plates are washed again and chromogen substrate (Bio-Rad Laboratories) added. The plates are incubated at room temperature to the desired extinction, and the reaction is terminated with 50 μl/well of 3 M H$_2$SO$_4$ solution. Plates are read at 490 nm in an automated microplate reader (Bio-Tek Instruments, Winooski, VT). Standards are dilutions of recombinant KC from 100 ng/ml to 1 pg/ml (50 μl/well). This method has consistently detected KC concentrations greater than 50 pg/ml in a linear fashion.

As shown in Fig. 4, KC can be readily detected in the lungs of transgenic but not in control mice. Despite being detected in the lung, no immunoreactive KC can be detected in the plasma of the transgenic mice, indicating that the protein is not being secreted into the bloodstream. The levels of KC protein, as measured by a sensitive ELISA, do not change appreciably during the first 6 months of life, but, as observed above, a marked reduction in the number of infiltrating neutrophils occurs during this time frame. These findings underscore the need for systematic analysis of the phenotype(s) elicited by chemokine expression during different phases of life,

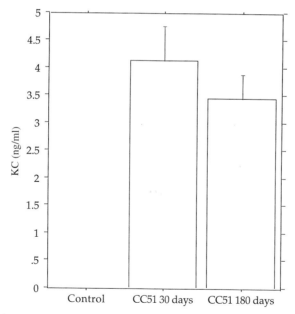

Fig. 4. Levels of KC protein, as determined by ELISA, in the lungs of control and transgenic animals at 30 and 180 days. Notice that the levels of KC do not change significantly with aging. Data are means ± SEM.

especially when constitutive promoters are being used. If these transgenic mice are analyzed solely at 90 days of age, a phase during which there are fewer infiltrating cells present, one could perhaps have assumed that the murine KC has little, if any, neutrophil-recruiting properties.

Expression of Chemokines KC and Monocyte Chemotactic Protein-1 in Central Nervous System

Many chemokines, such as KC and monocyte chemotactic protein-1 (MCP-1), are not constitutively expressed but can be rapidly induced in specific sets of cells. This pattern of expression most likely contributes to their proinflammatory activities *in vivo*. Hence, to understand the biological role of these molecules, it is important to use genetic elements that can reproduce some of these spatial and temporal patterns of expression. One such genetic element is the myelin basic protein (MBP) promoter. This promoter targets the expression of transgenes to oligodendrocytes in the CNS. Transgenes driven by this promoter are not significantly expressed during the embryonic period or during the first week of life. Starting at the second week of life, however, progressively higher levels of transgene expression can be detected throughout the entire CNS, with the expression being maximal between the third and fourth weeks in myelin-rich areas.[29] From that point on, the transgene levels fall to markedly lower levels.

We have used the MBP promoter to target the expression of chemokines to the CNS.[5,6] The chemokines used in these experiments were murine KC and MCP-1. MCP-1 belongs to the C-C family of chemokines and encodes in the mouse a microheterogeneous protein of M_r approximately 25,000.[30] Murine MCP-1 (also known as JE) is a potent chemoattractant for monocytes *in vitro*[31,32] but, when injected into mice, produces only a mild mixed inflammatory cell infiltrate.[30] To investigate whether MCP-1 and KC could recruit leukocytes *in vivo,* we programmed their expression to the CNS, a site where they are not normally expressed.

Two transgenes have been constructed: one containing KC (MBPKC) and the other containing mMCP-1 (MBPJE) (see Fig. 1B). These transgenes are constructed as follows. Murine MCP-1 and KC cDNAs are cloned into the *Xba*I–*Bam*HI site of the plasmid pMbP (a gift from Dr. R. A. Lazzarini,

[29] A. Gow, V. J. Friedrich, and R. A. Lazzarini, *J. Cell. Biol.* **119,** 605 (1992).

[30] C. A. Ernst, Y. J. Zhang, P. R. Hancock, B. J. Rutledge, C. L. Corless, and B. J. Rollins, *J. Immunol.* **152,** 3541 (1994).

[31] K. Matsushima, C. G. Larsen, G. C. DuBois, and J. J. Oppenheim, *J. Exp. Med.* **169,** 1485 (1989).

[32] T. Yoshimura, E. A. Robinson, S. Tanaka, E. Apella, J. I. Kuratzu, and E. J. Leonard, *J. Exp. Med.* **169,** 1449 (1989).

TABLE I

CELLS RECOVERED FROM BRAINS OF CONTROL
AND TRANSGENIC MICE EXPRESSING CHEMOKINES
KC (MBPKC) AND MCP-1 (MBPJE)[a]

Mice	Number of cells recovered/brain ($\times 10^{-3}$)
MBPKC	279 ± 12
MBPJE	476 ± 48
Control	0

[a] Data represent means ± SEM; $n = 5$ animals.

Mount Sinai Medical Center, New York, NY). This vector contains 1.9 kb (-1907 to $+36$ bp) of the promoter/enhancer of the myelin basic protein. Additional genomic sequences include splice and polyadenylation signals supplied by exon 6, intron 6, and exon 7 of the human proteolipid gene.[29] The transgenes are isolated from the vectors by restriction with NotI, followed by agarose gel electrophoresis. Fragments containing the transgenes are extracted from the agarose gel and purified as described before.

Several transgenic lines have been derived from injections of the transgenes. Transgenic expression of both KC and MCP-1 in the CNS results in specific, distinct, inflammatory infiltrates.[5,6] To better phenotype these infiltrates in the CNS, we have modified the method described by Renno et al.[33] The method basically consists of isolation of the infiltrating cells by Percoll gradient centrifugation and subsequent analysis of surface markers by flow cytometry. Briefly, brains are rapidly dissected, washed in PBS, and dissociated by gentle grinding through a metal sieve using a syringe plunger. The suspensions are diluted in 10 volumes of RPMI, 10% FBS, and centrifuged at 400 g for 10 min. The pellets are resuspended in 4 ml of 30% Percoll [3 parts isotonic Percoll, 7 parts Hanks' balanced salt solution (HBSS)] and overlaid on top of a discontinuous gradient containing 3.5 ml of 70% Percoll and 3.5 ml of 37% Percoll. Gradients are centrifuged for 20 min at 500 g. Cells are collected from the 37–70% interface and washed twice with HBSS, 10% FBS. The final pellet is resuspended in the same buffer for antibody labeling. This method produces viable leukocytes free of myelin and other cell debris. The viability of the recovered cells exceeds 85%. As shown in Table I, no leukocytes can be recovered from control brains, but more than 200,000 leukocytes can be readily isolated from the brains of animals expressing either KC or MCP-1 (age ~40 days, $n = 5$). Subsequent histological staining of the recovered cells demonstrates

[33] T. Renno, M. Krakowski, C. Piccirillo, J. Y. Lin, and T. Owens, *J. Immunol.* **154,** 944 (1995).

TABLE II
ANALYSIS OF SURFACE MARKERS ON CELLS
RECOVERED FROM BRAINS OF TRANSGENIC MICE
EXPRESSING CHEMOKINES KC (MBPKC) AND
MCP-1 (MBPJE)[a]

Antibody	MBPKC	MBPJE
Gr-1	84.8 ± 5.3	69.9 ± 1.6
F4/80	<5	75.1 ± 1.7
Mac-1	N.D.	76.6 ± 7.6
7/4	60.7 ± 4.3	65.4 ± 3.9
MOMA 2	N.D.	88.6 ± 3.2
CD3	<3	<3
TCR a/b	<3	<3

[a] Data represent means ± SEM percentage of total cells analyzed (n = 5).

that they are predominantly, if not exclusively, polymorphonuclear and mononuclear cells in the case of MBPKC and MBPJE mice, respectively.

Cell labeling is performed in HBSS, 10% FBS, at approximately 2.5×10^6 cells/ml for 30 min on ice. Most of the antibodies used in the flow cytometry experiments are obtained from Pharmingen (San Diego, CA), with the exception of F4/80 and 7/4 antibodies, which are obtained from Serotec. An average of 10,000 cells are assessed in each determination. Results from these experiments are presented in Table II. The infiltrate present in the brains of animals expressing KC is predominantly neutrophilic, as judged by the detection of the surface markers Gr-1 and 7/4 on the majority of the cells. Less than 5% of the cells recovered from the brains of transgenic animals express T-cell or macrophage markers. The infiltrate induced by the expression of MCP-1 is primarily composed of cells in the monocyte–macrophage lineage, as judged by the number of cells expressing the F4/80 and MOMA 2 markers. Interestingly, the vast majority of the cells expressing these markers, also coexpress granulocytic markers such as Gr-1 and 7/4 (data not shown). Additional ultrastructural studies have determined that these cells belong to the monocyte–macrophage lineage.[5] A small but insignificant number of T cells is also observed in the MBPJE infiltrates. The ease, speed, and sensitivity of flow cytometry make this method a very convenient tool to phenotype the infiltrates resulting from transgenic expression of chemokines.

Conclusion

Despite significant advances in understanding leukocyte trafficking under basal conditions and during inflammation, we still lack answers to many questions. What prompts specific subsets of cells to move into specific sites during hematopoietic development and inflammation? How is the recruitment of distinct subsets of leukocytes temporally structured during inflammation? Is there a combinatorial code dictating recruitment of leukocytes at different stages of development? If so, what is the molecular nature of these factors?

Results presented here and elsewhere clearly indicate that genetic tools can be used to address these questions and, it is hoped, provide us with answers. In this chapter, we describe methods to genetically control chemokine expression in specific organs and to analyze the composition of the inflammatory cell infiltrate caused by them. To study this process the following general strategies are used. (1) Transgenes are engineered to encode the chemokines without their 3'-end AU-rich regions, which have been shown to destabilize the mRNA. (2) Tissue-specific promoters are used. (3) Phenotypic analysis is performed using several transgenic lines, selected on the basis of expression. (4) Phenotypes are closely monitored in accordance with the spatial and temporal aspects of transgene expression.

Analysis of transgenic mice indicates that the expression of chemokines in specific tissues can initiate the cascade of events necessary for the transmigration of leukocytes from the vascular compartment into the parenchyma. Once in the parenchyma, the leukocytes migrate toward the cells producing the chemokines. Thus, these genetic experiments have established chemokines as an important group of molecules affecting leukocyte trafficking *in vivo*. Chemokines might affect trafficking of leukocytes by interfering with the expression of other molecules involved in controlling leukocyte trafficking, such as adhesion molecules, chemokine receptors, and presentation molecules, such as glycosaminoglycans. The mechanisms subserving this function can be further examined by a variety of biochemical and genetic approaches. A straightforward genetic approach to advance our understanding of this process would be to intercross chemokine-expressing animals with animals bearing mutations for adhesion molecules and other mediators. Such experiments will most likely contribute to defining the importance of each of these molecules to the process of leukocyte recruitment induced by chemokines.

The results presented here validate and extend our previous observations that the C-X-C chemokine KC and the C-C chemokine MCP-1 induce chemotaxis of specific subsets of leukocytes *in vivo,* while inducing little if any "inflammatory activation" of the recruited cells.[4-6] It is still unclear, however, whether the ability to promote recruitment without activation is a general property of KC, MCP-1, and/or other chemokines, or whether this effect results from the constitutive pattern of expression of the transgenes. The lack of significant inflammatory activation in the case of a more temporally controlled expression system (such as the MBP-driven transgenes) would argue in favor of a dissociation of the phenomena of recruitment and activation. The resolution of this issue will most probably require the use of conditional genetic approaches, with better control of the levels and timing of transgene expression.

Chronic, unregulated expression of KC is associated with an attenuation of the recruitment of neutrophils in thymus[4] and lung over time. The reduction in the number of recruited cells might reflect desensitization mechanisms taking place at one or several levels: leukocyte membranes, endothelium, and/or parenchyma. Alternatively, factors produced by the recruited cells, parenchyma, and/or endothelium might actively prevent further recruitment into the tissue producing the chemokine. Additional analysis of these transgenic animals will help us test some of these hypotheses, identify the specific mechanisms, and, it is hoped, contribute new insights into the biology of chronic inflammation.

In summary, gain-of-function approaches have been described to study the biological role of the chemokines KC and MCP-1. Results obtained in these experiments have established that these molecules are important regulators of leukocyte trafficking *in vivo,* and that genetic approaches are effective means to reproduce and understand important aspects of acute and chronic inflammation. It is likely that these and other genetic approaches will contribute to discern whether chemokines, in addition to their proinflammatory role, may also interfere with basal trafficking of leukocytes during homeostatic conditions.

Acknowledgments

The authors are grateful to P. Zalamea, M. Swerdel, and M. Burdick for technical contributions.

[21] Chemokines and Chemokine Receptors in Model Neurological Pathologies: Molecular and Immunocytochemical Approaches

By Richard M. Ransohoff, Marie Tani, Andrzej R. Glabinski, Ann Chernosky, Kimberly Krivacic, John W. Peterson, Hsiung-Fei Chien, and Bruce D. Trapp

Introduction

The central nervous system (CNS) is anatomically isolated from the periphery by the blood–brain barrier (BBB), which restricts traffic of both cells and macromolecules. Furthermore, the sites of antigen presentation and pathways of lymphoid drainage of the CNS have been uncertain. Consequently, the mechanisms that underlie immune-mediated inflammation of the CNS have been elusive. Studies using a small-animal model of inflammatory demyelination, experimental autoimmune encephalomyelitis (EAE), have begun to resolve the molecular participants in generating immunological tissue injury to the CNS. Pathological and clinical aspects of the EAE model are discussed elsewhere in this series.[1]

Autoreactive myelin-specific T lymphocytes are believed to be generated principally in the peripheral lymphoid compartment. These cells arise from immunization by the investigator in the case of mice with EAE. In humans with multiple sclerosis (MS), the origin of myelin-specific autoreactive T cells is a matter of speculation: virus-induced molecular mimickry is considered one likely culprit.[1a] To occupy the CNS, these T cells must be activated (either by superantigen or other means), cross the BBB, and undergo restimulation by myelin antigen engagement in the CNS perivascular space.[2] The subsequent step of inflammatory cell recruitment is essential for expression of clinical disease. Chemokines are proposed to play a role in CNS inflammatory disease at this stage of leukocyte recruitment. In this sense, chemokines exert an essential function in the development of immune-mediated CNS inflammation. This concept is supported by several studies of CNS chemokine expression during EAE, all of which indicate an intimate relationship between chemokine expression and clinical disease.

---

[1] A. R. Glabinski, M. Tani, V. K. Tuohy, and R. M. Ransohoff, *Methods Enzymol.* **288**, [13], (1997).

[1a] R. M. Ransohoff, P. V. Lehmann, and V. K. Tuohy, *Curr. Opin. Neurol.* **7**, 242 (1994).

[2] M. Tani and R. M. Ransohoff, *Brain Pathol.* **4**, 135 (1994).

Further, antibodies to one chemokine, macrophage inflammatory protein-1α (MIP-1α), reduces the severity of passive-transfer EAE.[3]

Inflammatory cells enter nervous system tissues in virtually all pathological states, with the character of the infiltrate differing widely, according to the pattern of tissue injury. We have documented chemokine production in the CNS during several models of nervous system pathology (Table I). These studies, in aggregate, begin to sketch an intriguing relationship between the preferential expression of individual chemokines and the recruitment of specific leukocyte cell types to the CNS or peripheral nervous system (PNS) compartment (Table I).

Quantitating Chemokine mRNA Expression

Chemokine expression is believed to be regulated largely at the level of transcription. Because chemokine protein concentrations in tissue usually reflect mRNA levels, it has been useful to monitor chemokine expression by determination of steady-state message. This concept of chemokine regulation is demonstrably valid for tissue culture cells or explants, and is believed to be correct *in vivo* as well. Message accumulation in response to stimuli is very rapid, occurring within minutes, and chemokine mRNAs transiently attain high levels. These features have resulted in the routine isolation of chemokine genes during differential hybridization experiments. Chemokine nomenclature has become extremely complex and uninformative, as many chemokines retain local clone designations.

The chemokine mRNAs are, as a general statement, relatively short-lived; timing of sample acquisition is an important consideration. Furthermore, amounts of available CNS or PNS tissues are frequently limited, so that very sensitive techniques have marked utility for studying neurological disease models. In addition, it is often essential to monitor simultaneous expression of multiple chemokines, placing a greater premium on sensitivity for the individual assays. Of the techniques available, RT-PCR (reverse transcription-polymerase chain reaction) dot-blot hybridization has proved particularly useful because of its flexibility, quantitative precision, and sensitivity. Alternative methods are mentioned briefly, followed by the RT-PCR dot-blot hybridization protocol and description of the steps needed to develop such an assay.

Northern Blot Hybridization Analysis

Northern blotting remains the standard technique for analyzing steady-state levels of mRNA species. The technique has been well described

---

[3] W. J. Karpus, N. W. Lukacs, B. L. McRae, R. M. Strieter, S. L. Kunkel, and S. D. Miller, *J. Immunol.* **155**, 5003 (1995).

TABLE I
CHEMOKINE EXPRESSION IN RODENT MODELS OF NEUROLOGICAL DISEASE[a]

Model	Infiltrate	Chemokines	Technique	Refs.
EAE	MØ > Ly	MCP-1, MIP-1α, GRO-α, RANTES, IP-10	Northern blot, RT-PCR, ISH, IC, ELISA	b–f
EAE/MOG	Ly > MØ	—	RT-PCR	g, h
Cortical stab/implant	MØ	MCP-1 ≫ MIP-1α, GRO-α, RANTES	RT-PCR, ELISA	i
Cortical cryolesion	MØ	MCP-1 ≫ IP-10	RT-PCR	j
Spinal cord: percussion trauma	MØ	MCP-1	RT-PCR	k
Axotomy	MØ	MCP-1 ≫ MIP-1α, GRO-α	RT-PCR, IC	l

[a] MØ, Macrophages; Ly, lymphocytes; MCP-1, monocyte chemoattractant protein-1; MIP-1α, macrophage inflammatory protein-1α; RANTES, regulated on activation, normal T cell expressed and secreted; IP-10, interferon γ-inducible protein-10 kilodaltons; RT-PCR, reverse transcription-polymerase chain reaction; ISH, *in situ* hybridization; IC, immunocytochemistry; ELISA, enzyme-linked immunosorbent assay; EAE/MOG, EAE induced by myelin oligodendroglial glycoprotein-specific T cells. ≫ indicates a difference in abundance of at least 10-fold.

[b] R. M. Ransohoff, T. A. Hamilton, M. Tani, M. H. Stoler, H. E. Shick, J. A. Major, M. L. Estes, D. M. Thomas, and V. K. Tuohy, *FASEB J.* **7,** 592 (1993).

[c] A. Glabinski, M. Tani, R. Strieter, V. Tuohy, and R. Ransohoff, *Am. J. Pathol.* **150,** 617 (1997).

[d] M. Tani, A. R. Glabinski, V. K. Tuohy, M. H. Stoler, M. L. Estes, and R. M. Ransohoff, *Am. J. Pathol.* **148,** 889 (1996).

[e] K. Hulkower, C. F. Brosnan, D. A. Aquino, W. Cammer, S. Kulshrestha, M. P. Guida, D. A. Rapoport, and J. W. Berman, *J. Immunol.* **150,** 2525 (1993).

[f] R. Godiska, D. Chantry, G. Dietsch, and P. Gray, *J. Neuroimmunol.* **58,** 167 (1995).

[g] C. Linington, T. Berger, L. Perry, S. Weerth, D. Hinze-Selch, Y. Zhang, H.-C. Lu, H. Lassmann, and H. Wekerle, *Eur. J. Immunol.* **23,** 1364 (1993).

[h] S. Lassmann, M. Tani, K. Krivacic, C. Linington, and R. Ransohoff, unpublished observations (1996).

[i] A. Glabinski, M. Tani, V. K. Tuohy, R. Tuthill, and R. M. Ransohoff, *Brain Behav. Immun.* **9,** 315 (1996).

[j] D. Grzybicki, S. Moore, R. Schelper, A. Glabinski, R. Ransohoff, and S. Murphy, *Arch. Neuropathol.* in press (1997).

[k] D. McTigue, B. Stokes, and R. Ransohoff, unpublished observations (1996).

[l] H.-F. Chien, J. Griffin, M. Tani, and R. Ransohoff, unpublished observations (1996).

in numerous manuals that are readily available. The procedure involves preparation of RNA, analysis of RNA on denaturing gels, transfer to nylon or nitrocellulose membranes, hybridization with probes that are typically radiolabeled, and detection, usually by autoradiography. Quantitation of the hybridization signal is achieved through densitometry, either with a laser densitometer or through the phosphor storage technique. One major advantage of Northern blotting is that formal quantitation of mRNA levels can be readily achieved, by determining the hybridization signal of the species of interest and normalizing for loading variation to an abundant and constitutively expressed species such as ribosomal RNA or house-keeping genes, such as actin or tubulin. A second advantage of Northern blotting is that the procedure allows for determination of the size of the mRNA under study, so that variant forms of specific mRNAs are readily detected. This attribute is important if the project requires detection of alternatively processed mRNA isoforms derived from a single transcript.

The principal shortcoming of Northern blotting involves its low sensitivity. Maximally, 30 μg of total RNA, representing 600 ng of mRNA, can be readily separated on denaturing agarose formaldehyde gels. Moderate-abundance mRNAs, present at levels of 10–50 pg, will not be easily detected. Preparation of the proper hybridization probe can markedly enhance sensitivity. Commonly, hybridization probes are prepared by radiolabeling cDNA fragments. In this instance the desired hybridization reaction is competed by the annealing of complementary strands of the probe. Single-stranded hybridization probes can be generated by *in vitro* transcription using bacteriophage polymerases such as SP6, T3, or T7, or by primer extension of single-stranded M13 bacteriophage templates.[4,5] Use of single-stranded hybridization probes augments the sensitivity of the Northern blot hybridization assay by a factor of 10- to 50-fold. To achieve adequate stringency, hybridization temperatures must be increased by 10° to 13° in the case of RNA–RNA duplexes, using conditions that are otherwise comparable to those established for formation of RNA–DNA duplexes. The sensitivity of the Northern blot technique can also be enhanced by preparation of poly(A)$^+$ RNA. Analysis of 6 μg of poly(A)$^+$ RNA represents a 10-fold increase in sensitivity over the maximum that can be achieved with total RNA. However, this step requires quantities of starting material (300 μg of total cellular RNA for this example) that are not readily available from CNS or PNS tissues.

Nuclease Protection Analysis

A further shortcoming of Northern blotting is that detection of multiple species requires repetitive stripping and rehybridization of filters, with uncertain retention of the RNA originally transferred. Nuclease protection

[4] R. M. Ransohoff, P. A. Maroney, D. P. Nayak, T. Chambers, and T. W. Nilsen, *J. Virol.* **56,** 1049 (1985).

[5] R. Ransohoff, C. Devajyothi, M. Estes, G. Babcock, R. Rudick, E. Frohman, and B. Barna, *J. Neuroimmunol.* **33,** 103 (1991).

assay provides a convenient means for detecting multiple RNA species. Furthermore, this solution hybridization method is directly quantitative and easily incorporates an internal control reference RNA. The technique relies on solution hybridization between target RNA and hybridization probe that is typically radiolabeled. After the hybridization reaction has reached equilibrium, the mixture is digested with nucleases that specifically degrade unpaired strands, leaving the hybrids. Undigested material is deproteinized by phenol extraction, recovered by ethanol precipitation, and analyzed on denaturing sequencing gels. Hybrids are then detected by autoradiography. The method was formerly limited in utility because hybridization probes were generated by radiolabeling double-stranded cDNAs. Each hybridization reaction needed to be carefully optimized to a stringency that favored formation of RNA–DNA (probe–target) hybrids over the reannealling reaction between the more abundant DNA–DNA strands of the probe.

Current methodology involves use of single-stranded cRNA probes, produced by *in vitro* transcription of cDNA templates with bacteriophage polymerases, whose recognition sites are incorporated in standard cloning vectors. This advance eliminates the need to optimize each individual hybridization assay, and it readily permist detection of multiple target species in one reaction.[6] Therefore, RNase protection assays are now widely used for determining steady-state levels of RNAs. The advantages over Northern blotting are greater sensitivity (10- to 100-fold more sensitive than Northern blotting of total cellular RNA) and the possibility of detecting multiple species conveniently in a single reaction. To realize this potential, it is necessary to generate cRNA probes yielding appropriate size protected fragments; hybrids ranging form 100 to 500 bp can conveniently be detected on a single autoradiogram. It is challenging at times to produce cRNA probes that generate readily distinguishable protected hybrid fragments from several genes of interest. The inherent limitation of nuclease protection for detecting chemokine mRNAs in rodent CNS or PNS samples is in the requirement for 25–50 μg of RNA starting material, an amount that cannot be obtained from a single mouse spinal cord or sciatic nerve sample. Nevertheless, for projects involving simultaneous quantitation of multiple mRNAs with high sensitivity, where large volumes of specimen are available, nuclease protection assay is the method of choice.

Reverse Transcription-Coupled Polymerase Chain Reaction Dot-Blot Hybridization Analysis: Introduction and RNA Preparation

Overview: Selection of RT-PCR Technique. Reverse transcription-coupled polymerase chain reaction (RT-PCR) is widely used for detecting

---

[6] M. Rani, D. Leaman, S. Leung, G. Foster, G. Stark, and R. Ransohoff, *J. Biol. Chem.* **271,** 22878 (1996).

mRNAs in biological specimens. The technique is relatively simple and extremely sensitive. Furthermore, it is flexible, as PCR primers for cloned genes of interest can be designed with assistance of computer-based programs, on the basis of information present in sequence databases.

The major pitfall in the use of RT-PCR lies in the requirement that each assay be individually calibrated to yield accurate quantitative results. Absent optimal conditions, the technique carries an onerous burden of error. Absolute quantitation of mRNA species by this technique requires the application of competitive RT-PCR. Competitive RT-PCR incorporates synthetic internal standards, with identical primer recognition sites, into each reaction. The synthetic internal RNA standard is usually created by deletion mutagenesis of the cDNA that encodes the gene of interest, followed by *in vitro* transcription with bacteriophage polymerase. The product generated by RT-PCR amplification of this template is distinguishable by size from the product generated from the endogenous transcript. Known amounts of synthetic template RNA are added to each sample, and the relationship between experimental and standard RNA is established by performing serial dilutions of the RT product before PCR. Each data point in each assay therefore consumes considerable resources of thermostable polymerase. This characteristic limits the applicability of competitive RT-PCR assay to examining RNA levels in settings where absolute quantitiation is required. Economically, it is not generally feasible to determine the abundance of numerous transcripts in a large variety of samples with competitive PCR. We have used a semiquantitative RT-PCR dot-blot hybridization assay to examine the abundance of chemokine transcripts in samples derived from rodent neural tissue, with high reproducibility.[7] These results correlate precisely with results obtained by *in situ* hybridization, a qualitative means of evaluating mRNA abundance.[8] Determination of corresponding protein levels by enzyme-linked immunosorbent assay (ELISA) has shown that protein levels reflect mRNA accumulation in these samples.[9,10] The detailed methods for quantitating chemokine levels in tissues by ELISA are discussed elsewhere in this volume.[10a]

---

[7] A. Glabinski, M. Tani, V. K. Tuohy, R. Tuthill, and R. M. Ransohoff, *Brain Behav. Immun.* **9,** 315 (1996).

[8] R. M. Ransohoff, T. A. Hamilton, M. Tani, M. H. Stoler, H. E. Shick, J. A. Major, M. L. Estes, D. M. Thomas, and V. K. Tuohy, *FASEB J.* **7,** 592 (1993).

[9] A. R. Glabinski, M. Tani, V. Balasingam, V. W. Yong, and R. M. Ransohoff, *J. Immunol.* **156,** 4363 (1996).

[10] A. Glabinski, M. Tani, R. Strieter, V. Tuohy, and R. Ransohoff, *Am. J. Pathol.* **150,** 617 (1997).

[10a] *Methods Enzymol.* **287,** 1997 (this volume).

RNA Preparation. Chemokine expression is typically focal and transient for most of the neurological pathologies under investigation in our studies.[11] Detecting chemokine RNAs in this setting places the investigator at a disadvantage, as the signal is diluted by a vast excess of noise represented by cells that are not expressing chemokines at any one time. It is therefore essential to have clear characterization of the disease model, so that entry of specific classes of leukocytes has been defined with regard to location and time. Samples should then be obtained before, during, and after the points at which leukocyte entry is observed, to facilitate relating chemokine expression to the biology of the disease model.

RNA preparation from brain and spinal cord is routinely performed with commercial reagents, such as Trizol (GIBCO-BRL, Gaithersburg, MD), according to manufacturer's instructions with a variation of the method of Chomczynski and Sacchi.[12] Other analogous products such as RNAzol can be substituted. This technique is readily applied to fresh tissue. It is important to remove and process tissue rapidly, to avoid post-mortem autolysis.

Animals are sacrificed by cervical dislocation. The skin overlying the skull is cut to expose the cranium. Carefully avoiding nicking the brain, the skull is cut from the foramen magnum forward at midline and laterally with a small, sharp scissors and the brain lifted out with blunt forceps. The brain is then divided with a midsagittal cut. One half is fixed and paraffin-embedded for *in situ* hybridization (ISH; see below) and RNA is extracted from the other half. The spinal column is isolated with sharp scissors, and divided in half if longer than 2–3 cm. The spinal cord is extruded with phosphate-buffered saline (PBS), through an 18-gauge needle attached to a 10-ml syringe. For extrusion, a seal is formed between the caudal, thinner end of the spinal column and the needle, by pinching between thumb and forefinger.

RNA is immediately prepared after tissue removal, using 1 ml of Trizol per 100 mg of tissue. We homogenize CNS tissues in Trizol with a Polytron (Kinematica, Lucerne, Switzerland) power homogenizer. Chloroform extraction, 2-propanol precipitation, and resuspension of the pellet are performed according to manufacturer's instructions. After suspension in water, RNAs are quantitated by absorbance at 260 nm on a spectrophotometer and analyzed on agarose gels to confirm integrity.

RNA PREPARATION FROM SCIATIC NERVES. Sciatic nerves are resected from anesthetized animals, snap frozen in autoclaved 1.5-ml Eppendorf tubes (Kimball) in liquid N_2, and held at $-80°$. After removal of a segment

[11] R. Ransohoff, A. Glabinski, and M. Tani, *Cytokine Growth Factor Rev.* **7,** 35 (1996).
[12] P. Chomczynski and N. Sacchi, *Anal. Biochem.* **62,** 156 (1987).

of fresh tissue, the animal is perfused with fixative, so that additional nerve samples can be optimally prepared for histology. To prepare RNA, the tubes are cooled in liquid N_2, and the nerve is crushed with a disposable Eppendorf pestle (Kimball). Trizol (500 μl) is added, and the tissue is ground further. Chloroform extraction is then performed, according to manufacturer's instructions. After centrifugation, the aqueous phase is carefully collected, and glycogen carrier (Boehringer-Mannheim, Indianapolis, IN) is added to a final concentration of 40 μg/ml. The RNA is precipitated with 2-propanol overnight at 4°, collected by centrifugation, washed with 80% ethanol, and the dried pellet resuspended in diethyl pyrocarbonate-treated (DEPC) water. Yields averaging 4–5 μg of total RNA per 5-mm mouse sciatic nerve segment are obtained, so that each nerve segment can be analyzed as an individual specimen.

STORAGE AND SHIPPING OF TISSUE SAMPLES FOR SUBSEQUENT RNA EXTRACTION. RNA can be protected against degradation by snap-freezing tissue and maintaining it at a temperature of −80° thereafter. Animals are sacrificed by cervical dislocation or decapitation, and tissues are harvested (see above), snap frozen in liquid N_2, and maintained at −80°; samples are shipped on dry ice. One should obtain advice from the courier service about regulations that pertain to the shipping of parcels containing dry ice, and preferably ship before midweek to avoid mishandling during weekends.

Developing Reverse Transcription-Coupled Polymerase Chain Reaction Dot-Blot Hybridization Assay

There are approximately 30 human chemokines, and a comparable number is present in mammalian species (mouse, rat, hamster, rabbit, dog) that are used in neurological disease models. Elucidating the precise role of an individual chemokine in each disorder is a significant challenge, and chemokine RT-PCR assays must be developed with sequence information that is isogenic with relation to the species under investigation. Assays for material from one species may not be applicable to another species, even if closely related. We have found, for example, that assays that have been optimized for mouse CNS or PNS tissues cannot be applied for quantitating rat chemokine expression. The investigator may well be in the position of developing RT-PCR assays to fit the need to monitor expression of a chemokine for which no previous assay has been reported, either due to novelty of the chemokine product or use of a species for which no assay has been described.

Selection of Primers and Primer Pairs. Many chemokine mRNA sequences are present in sequence databases. These sequences can be obtained by database search and downloaded to computers for analysis. Programs that assist design of PCR primers and primer pairs are readily available. Primer design functions are integrated into such general-purpose sequence analysis programs as MacVector and GeneWorks. These programs will require specification of product length and will request information about desired melting temperature, typically 52–60°. PCR primer pairs need to be well matched for melting temperature. Primer pairs should be designed to generate PCR products that are readily separated on 1.5% agarose gels; such products vary between 150 and 500 bp in the case of chemokines. Primers should also be designed to avoid intramolecular hairpin formation, which promotes mispriming. Pairs need to avoid self-hybridization or interprimer hybridization, either of which will reduce the effective primer concentration in the PCR reaction. Each of these problems will be addressed by primer-design programs. The final step in primer design is to revisit the eukaryote sequence database with each primer, using an algorithm such as BLAST, available through the National Center for Biotechnology Information (Bethesda, MD), to ensure that the primer sequence is uniquely represented in the gene of interest.[13]

It is frequently advisable to design primers that span an intron to detect genomic DNA contamination of samples. However, this PCR primer attribute is not essential for chemokine assays, as most experiments will incorporate numerous negative samples. Chemokine expression in normal neural tissue is essentially nil, so that detection of product in tissue from healthy animals should prompt concern about contamination.

Reverse Transcription. It is advantageous to prepare total cDNA as a PCR template, through the use of an oligo(dT) primer for the reverse transcriptase (RT) reaction. This alternative to using a gene-specific RT primer ensures an opportunity to reuse each sample multiple times and to reserve samples for future analysis for expression of additional mRNAs. First-strand synthesis conditions are used, with SuperScript II (GIBCO-BRL), the manufacturer's instructions, and RT buffer. Input RNA is 1 μg in water per reaction. The final step is to inactivate RT at 70° for 15 min. Reaction products may then be transferred directly to PCR reactions or stored at −20°.

Optimizing Polymerase Chain Reaction. Each PCR assay needs to be optimized in several respects. First, specific amplification of the sequence of interest must be assured. This step involves initially determining that PCR amplifications yield discrete products of the expected size on ethidium bromide-stained agarose gels. Next, these products must be shown to hybridize at high stringency to radiolabeled probes, after Southern transfer.

---

[13] S. F. Altschul, W. Gish, W. Miller, E. W. Myers, and D. J. Lipman, *J. Mol. Biol.* **215,** 403 (1990).

Finally, reaction products are subcloned and subjected to sequence analysis to achieve definitive characterization. Polymerase fidelity is governed primarily by Mg^{2+} concentration, which needs to be defined independently for each primer pair. The ideal PCR cycle number is defined by titration at the optimal Mg^{2+} concentration. Finally, the relationship of input RNA to output PCR product (monitored as hybridization signal) needs to be established.

A convenient way to begin optimizing each assay is to analyze a positive-control sample, either from a known tissue source or from cytokine-treated cells. The RT reaction mixture is diluted 1:25 in PCR reactions that are assembled with buffer supplied by the manufacturer (typically 15 mM MgCl$_2$ in 10× buffer) and subjected to a saturating number of cycles (35–40) of amplification. PCR conditions are routinely set as follows: first denaturation (94°, 3 min), with subsequent cycles of denaturation at 94° for 30 sec, primer-pair-specific annealing temperature for 60 sec, and extension at 72° for 60 sec. A final 2-min extension at 72° is frequently incorporated. Negative controls incorporate aliquots from mock cDNA synthesis reactions that omit RT or RNA. An aliquot of each reaction is analyzed on agarose gels to determine the presence of appropriate-size amplification product and then analyzed by Southern blotting, hybridizing with the appropriate radiolabeled probe to confirm a single expected product. Conditions that yield the appropriately sized product are used for magnesium titration.

MAGNESIUM TITRATION. Thermostable polymerases exhibit dependence on magnesium concentration for fidelity and efficiency. PCR buffer lacking Mg^{2+} is prepared and supplemented from a 0.5 M MgCl$_2$ stock to yield final [Mg^{2+}] ranging from 0.5 to 5.0 mM in 0.5 mM steps. PCR products from these reactions are analyzed on ethidium bromide-stained agarose gels and visualized under UV light. At suboptimal [Mg^{2+}], multiple bands arising from illegitimate priming are observed, and bands arising from genes of interest are faint (Fig. 1). At optimal concentrations, bands that represent the appropriate-size product from the target are prominent and extraneous bands are faint or absent (Fig. 1). MgCl$_2$ concentrations that are optimal for one primer pair are not reliably suitable for another pair (Fig. 1, compare top and bottom gels).

CYCLE NUMBER TITRATION. Polymerase chain reactions are saturable, and distinction between inputs is lost at cycle numbers that over- or under-amplify. Analysis of products from reactions that have not been optimized yields large errors in estimation of target mRNA concentration. Such errors usually bias toward underestimation of differences between samples.

Cycle-number titrations are conveniently performed by scaling PCR reactions to 100 μl and removing 10-μl aliquots at each step. Preparation of parallel reactions that are subjected to varying cycle numbers is an

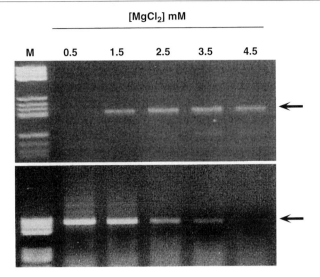

FIG. 1. Comparison of two primer pairs to demonstrate effects of magnesium concentration on efficiency and fidelity of thermostable DNA-dependent DNA polymerase in polymerase chain reaction. Primer pairs for two different chemokines were selected using MacVector 4.5.3 as described in the text. Following reverse transcription, PCR was carried out for 30 cycles of amplification at optimal annealing temperatures, at varying concentrations of MgCl₂ as indicated. Products were analyzed on 1.5% agarose gels, and visualized by UV transillumination. The expected size of the product is indicated (arrow). There are clear differences between the primer pairs produced by varying [MgCl₂], illustrating a requirement to determine the optimal [MgCl₂] for each primer pair.

adequate alternative. These two approaches yield identical results in our laboratory. After optimal [MgCl₂] is defined, RT products are subjected to cycle number titration, typically in two stages. Initially, cycles are increased in increments of 5 from 15 to 40 cycles (Fig. 2), producing a broad cycle-number titration for this primer pair. The relationship between cycle number and PCR product is then determined. In the example shown (Fig. 2), reaction products were quantitated by Southern transfer, hybridization, and densitometry on a PhosphorImager (Molecular Dynamics, Sunnyvale, CA). Alternatively, densitometry of ethidium bromide-stained agarose gels may be used. An optimal cycle number is then determined after a second round of titration across the exponential portion of the curve, using increments of 2–3 cycles at each step; the cycle number that represents the midexponential portion of the curve is selected. The cycle number that is defined in this fashion is then validated by cDNA input titration, using serial dilutions of the RT reaction product to program PCR amplifications, at the established optimal cycle number. Cycle number optima that are

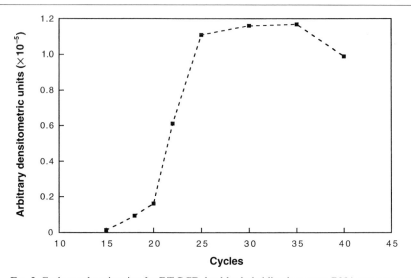

Fig. 2. Cycle number titration for RT-PCR dot-blot hybridization assay. RNA was prepared from rat spinal cord after percussive trauma and reversed-transcribed with an oligo(dT) primer. The RT products were amplified by PCR using primers directed against rat MCP-1 at optimal [MgCl$_2$] for varying numbers of cycles as indicated, transferred to nitrocellulose membranes on a dot-blot manifold, hybridized with a radiolabeled cDNA probe, and quantitated on a PhosphorImager. Midexponential amplification occurs at 22 cycles. Threshold effects are observed at 25–35 cycles of amplification.

practically useful will yield outputs that are linear with respect to input across 2 logs of serial cDNA input dilutions.

INPUT RNA TITRATION. The successful application of a predetermined optimal PCR cycle number relies on the assumption that varying levels of mRNA starting material will produce RT reaction products that directly correspond to input. This assumption is directly tested for each assay by RNA input titrations into the optimized assay. The output should remain linear across at least 1 log of input (Fig. 3).

Dot-Blot Hybridization. The PCR products can be quantitated by Southern transfer, hybridization, and densitometry. However, the transfer may vary in efficiency, and it consumes several days. One alternative is dot-blot hybridization. This technique has been directly compared in our laboratory with Southern transfer hybridization and provides identical data with a more time-efficient protocol. Dot-blot hybridization should only be applied after Southern transfer establishes that PCR products migrate as a single band at the expected molecular weight and sequence analysis identifies the target of amplification.

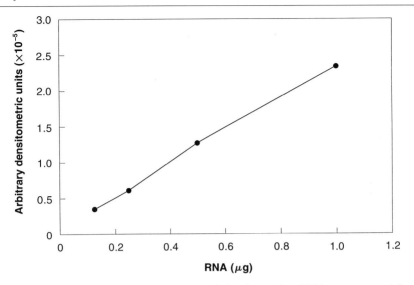

Fig. 3. RNA input titration at midexponential cycle number. RNA was prepared from rat spinal cord after percussive trauma. Varying amounts of RNA from 0.1 to 1.0 μg were used as template in RT reactions with oligo(dT) primer. Equal aliquots of each RT reaction were then used as input to the PCR, using primers directed against rat MCP-1. Products were analyzed by dot-blot hybridization and quantitated on a PhosphorImager. A linear relation between RNA input and densitometric output is observed.

The PCR products (5 μl) are denatured at 80° for 15 min in a reaction volume of 50 μl, containing 200 mM NaOH (freshly prepared) and 800 mM Tris-HCl (pH 7.6), and quenched on ice for 1 min. Each sample is adjusted to 3× standard saline citrate (SSC) by addition of an equal volume of 6× SSC. Any required serial dilutions are prepared similarly in 6× SSC. Nylon membrane and Whatman (Clifton, NJ) 3M paper are cut to size for the dot-blot manifold (Schleicher & Schuell, Keene, NH) and equilibrated in 2× SSC for 5 min. The dot-blot manifold is assembled with Whatman paper contacting the bottom half and membrane secured between the paper and top half. Vacuum is applied for 10–30 sec to aspirate excess buffer; samples are applied and allowed to remain in contact with membrane without suction for 30 min. Vacuum is applied for 30 sec; wells are washed twice with 200 μl of 6× SSC (each time with 30 sec vacuum). DNA is cross-linked to the membrane with UV irradiation (Stratalinker, Stratagene, LaJolla, CA). The membrane is air-dried and stored dry between sheets of Whatman filter paper. These membranes are suitable for hybridization for several months after storage in this fashion.

Membranes are prehybridized after equilibration in 2× SSC for 4–6 min. Prehybridization is performed at 42° for at least 3 hr, in 5× SSC, 5× Denhardt's, 50% formamide with 500 μg/ml sheared salmon sperm DNA. Hybridizations are performed in 5× SSC, 1× Denhardt's, 50% (v/v) formamide, 10% (v/v) dextran sulfate with 100 μg/ml sheared salmon sperm DNA at 42° for 14–16 h. Probes are generated from gel-purified cDNA inserts and labeled with ^{32}P by random priming or nick-translation; 2 × 10^6 cpm per milliliter of hybridization solution is used (1 × 10^7 cpm in 5 ml per 10 × 10 cm filter). Membranes are rinsed at room temperature for 10 min in 2× SSC, 0.1% (w/v) sodium dodecyl sulfate (SDS); washes in 2–3 changes of buffer are at 65° for 30 min in 0.1× SSC, 0.1% SDS. Filters are then air-dried for autoradiography.

Autoradiography and Quantitation by Storage Phosphor Technique. Autoradiography and densitometry are performed by the storage phosphor technique on a PhosphorImager. This instrument has an extended linear range, allowing for convenient quantitation without multiple exposures to X-ray film. It is important to use uniform exposure times for a single series of experiments. If samples from a series of animals are under examination over a long period, one sample is carried from assay to assay and serves as the normalizing standard for all values. Results from these determinations are displayed otherwise without normalization as arbitrary densitometric units.

Comment on RT-PCR Dot-Blot Analysis. The RT-PCR dot-blot hybridization technique has provided highly reproducible data from multiple models of nervous system pathology. Development of a new assay requires approximately 1 week. The correlations between results using this technique and ISH have been exact, indicating by a completely separate method of analysis that RNA levels defined by RT-PCR dot-blot hybridization were precise. Further corroboration has come from parallel determination of chemokine protein levels in parallel samples. The method does not control for integrity of RNA in all samples. For this purpose, amplification of tubulin or GAPDH mRNA in all samples is uniformly performed.

Central Nervous System Cellular Sources of Chemokines by *in Situ* Hybridization Analysis

Cellular sources of chemokines in varied pathologies have been extensively investigated. In contrast to the situation *in vitro*, where virtually any cell can express chemokines, the production of these products *in vivo* appears to be stringently limited. In model CNS pathologies, we detect a given chemokine mRNA in only a single cell type, despite the presence of multiple cell types that exhibit the ability to synthesize chemokines when explanted and grown *in vitro*. Chemokines are rapidly secreted from producer cells. The secreted proteins are (with few exceptions) highly basic and interact avidly with acidic extracellular matrix components. Further, the chemokine proteins possess half-lives between 8 and 24 hr. These characteristics render the detection of chemokines by immunocytochemistry (IC) relatively challenging. For these reasons, ISH has particular applicability to the identification of cells that produce secreted products such as chemokines *in vivo*.

Tissue Preparation

1. All tissue preparation is done with gloves to avoid RNase contamination.

2. Pour fixative into individual containers for each tissue. We use buffered 10% (w/v) formalin (Fisher, Pittsburgh, PA).

3. Animals are sacrificed by cervical dislocation; removal of brain and spinal cord tissues are described above (see RNA Preparation). For ISH, tissues are fixed by immersion in buffered 10% formalin for at least 5 days at room temperature. Alternatively, if fresh tissue is not needed, the animals may be perfused with fixative before tissues are removed.

4. Paraffin-embedded tissues are prepared by a histology service (housed either in a clinical pathology department or research department core facility). Conditions appropriate for animal CNS tissue need to be locally established. Tissues should be oriented so that midsagittal sections of brain and spinal cord can be cut in the same section. Paraffin-embedded tissues are sectioned at a thickness of 5–7 μm and mounted on Superfrost Plus (Fisher) microscope slides. Continue to wear gloves while handling slides, which are stored in a clean box at room temperature. One slide from each block is stained with hematoxylin and eosin, for histological correlation.

Glassware and Reagents

1. All glassware is baked individually wrapped in aluminum foil at 125–150° overnight.

2. Solutions can be made with autoclaved MilliQ or DEPC-treated water. Make DEPC-treated water by adding 0.1% (v/v) diethyl pyrocarbonate to distilled water and stirring overnight. Autoclave the solution the next morning to deactivate the DEPC.

3. Coverslips large enough to cover the tissue sections (typically 22 × 40 mm) are prepared for the hybridization step by dipping individual coverslips in Sigmacote (Sigma, St. Louis, MO), then 100% ethanol, then water. Coverslips are then placed in glass or porcelain coverslip racks (Fisher) and baked overnight at 125°.

In vitro Transcription to Generate Tritium-Labeled Riboprobes

1. Lyophilize ^{3}H-labeled nucleotides in an Eppendorf tube, using a SpeedVac (Savant, Holbrook, NY). For a final concentration of approximately 100 μM of each, lyophilize a quantity (in μl) equal to 2.5 times the specific activity (in Ci/mmol). For example, use 58 μl of [3H]CTP (Amersham, Arlington Heights, IL; ~23 Ci/mmol) or 108 μl of [3H]UTP (Amersham; ~43 Ci/mmol).

2. To the tube of dried nucleotides add the following:
5 μl of 5× transcription buffer (Stratagene)
1 μg of linearized DNA template
1 μl of 10 mM ATP
1 μl of 10 mM GTP
1 μl 0.75 M dithiothreitol (DTT)
1 μl (40 U) RNase inhibitor (Stratagene)
10 U (1 μl) diluted SP6, T3, or T7 RNA polymerase (Stratagene)
    DEPC-treated water to a final volume of 25 μl

3. Incubate at 37° for 30 min. Another microliter of RNA polymerase can be added and the mixtures incubated for another 30 min. For SP6 polymerase, add 20 U of enzyme, incubate at 40° for 30 min, add another 20 U, and incubate at 40° for 1 hr.

4. Add 1 μl of RNase-free DNase and incubate at 37° for 30 min.

5. Add 225 μl of 1× NETS (100 mM NaCl, 1 mM EDTA, 10 mM Tris, pH 7.6, 0.5% SDS) to each reaction and place on ice.

6. Add 250 μl of 1:1 (v/v) phenol:chloroform, invert several times, spin in microcentrifuge for 5 min, and remove upper aqueous phase into clean tube.

7. Extract with 225 μl of chloroform, centrifuge for 5 min, and remove upper aqueous phase into a clean tube.

8. Precipitate with 25 μl of 3 M sodium acetate, pH 5.2, and 690 μl of 100% ethanol at −80° for at least 30 min.

9. Pellet in a microcentrifuge at 4° for 15 min; carefully pipette off liquid.

10. Wash with ice-cold 80% ethanol, centrifuge for 10 min, and remove supernate.

11. Dry in SpeedVac.

12. Resuspend in 52 μl of DEPC-treated water.

13. Remove 1 μl of each transcript for gel electrophoresis and 1 μl to count in scintillation counter. Reserve 50 μl for hydrolysis.

14. Perform hydrolysis as follows. (a) Calculate the hydrolysis time for probes that are longer than 300 bp:

$$\text{Hydrolysis time (minutes)} = \frac{N(\text{kb}) - 0.3 \text{ kb}}{N(\text{kb})(0.3)(0.107)}$$

where N(kb) is the initial transcript length in kilobases. (b) Add 50 μl of mixed hydrolysis solution (200 μl of 200 mM NaHCO$_3$, 300 μl of 200 mM Na$_2$CO$_3$) to each transcript. (c) Incubate at 60° for calculated time. (d) Immediately add 8 μl stop solution (30 μl of 3 M sodium acetate, pH 5–6, 50 μl of 10% acetic acid) per hydrolysis reaction and place on ice. The mixture should bubble.

15. Add 1 μl of yeast tRNA (10 mg/ml) and 280 μl of ethanol; precipitate at −80° for at least 30 min.

16. Centrifuge at 4° for 15 min and remove supernatant.

17. Dry in a SpeedVac.

18. Add 23 μl of DEPC-treated water. Remove 1 μl for counts, 1 μl for gel electrophoresis, and 1 μl for acid-precipitable counts. Store remaining probe at −20°. The probes should include the following:

Unhydrolyzed probe
  1 μl for counts
  1 μl for gel electrophoresis
Hydrolyzed probe
  1 μl for counts
  1 μl for gel electrophoresis
  1 μl for acid-precipitable counts

19. For gel electrophoresis, prepare a 14-cm-long, 1.25% (w/v) agarose 3-[N-morpholino]propanesulfonic acid (MOPS)/formaldehyde denaturing horizontal gel (for 60 ml, boil 0.75 g agarose and 44 ml water; cool to 55°; add 10 ml of 37% (w/v) formaldehyde and 6 ml of 10× MOPS).

20. To prepare samples, use at least 5×10^5 cpm per sample. To 1 μl of sample, add 8 μl of FFM (1 ml formamide, 320 μl of 37% formaldehyde, 200 μl MOPS); heat to 60° for 15 min and then cool quickly on ice. Add 2 μl bromphenol blue/xylene cyanol tracking dye.

21. Electrophorese samples at 100 V for 2 hr.

22. Place in En^{3}Hance (DuPont, Boston, MA) for 1 hr. Wash gel in tap water for 15 min.

23. Dry gel on a gel dryer, place in an X-ray cassette, and expose to film overnight to visualize a minimum of 5×10^5 cpm per sample.

24. Determine incorporation into acid-precipitable counts as follows. (a) Boil salmon sperm DNA (10 mg/ml) for 10 min, quench briefly on ice, then add 100 μl to a 15-ml conical tube. (b) Add 1 μl of hydrolyzed probe. Add 5 ml of ice-cold 10% (w/v) trichloroacetic acid (TCA). Place tube on ice for 15 min. (c) Label GF/A filters (Whatman) with a pencil. Apply each probe to a filter by suction–filtration. (d) Wash filters with ice-cold 10% TCA. Wash with ice-cold 100% ethanol. Dry filters. Place in vials with scintillation fluid and count.

Hybridization

1. For deparaffinization, slides are placed vertically in a rack in a 60–65° oven for 15 min until the paraffin is melted. Slides are immediately placed in staining trays through two changes of fresh xylene (10 min each) with gentle mixing.

2. For rehydration, slides are then dipped through 100% ethanol twice, 95% ethanol twice, 85, 70, 50, and 30% ethanol, and water twice.

3. For proteinase K digestion, immerse slides in 100 mM Tris-HCl, pH 8.0, 50 mM EDTA, 1 μg/ml proteinase K (from a 20 mg/ml stock in DEPC-treated water) at 37° for 30 min. Gently agitate with slide rack then incubate.

4. For acetic anhydride treatment, while slides are in proteinase K, prepare 200 ml of 0.1 M triethanolamine (TRI), pH 8.0, as follows. Weigh TRI into a beaker (2.985 g for each 200 ml of final volume). Add about 3/4 of the total volume of water and adjust to pH 8.0 with 10 N HCl. Remember you will need two dishes of TRI for every rack of slides. Wash slides in DEPC-treated water then in TRI at room temperature for 10 min. While slides are in this wash, add 0.5 ml acetic anhydride to 200 ml TRI (0.25%, v/v) in a staining dish. Place slides in dish and dunk up and down for gentle mixing. Incubate the slides at room temperature for 10 min. Wash briefly in 2× SSC.

5. Dehydrate sections by passing through graded ethanol series ending with 100% ethanol twice. Allow slides to dry in dust-free chamber.

6. Perform hybridization as follows. (a) Prepare a sufficient amount of solution mix A (see below). (b) Prepare solution mix B separately (see below). Correct the final volume of mix B with DEPC-treated water and heat at 80° for 3 min. (c) Combine solutions and mix at room temperature (see below). Allow at least 45 μl of final hybridization solution for each tissue section to be covered by a 22 × 40 mm coverslip. (d) Hybridization solutions are prepared with the following recipes. Solutions A and B are combined to make the hybridization solution.

Reagent/final concentration	Stock concentration
Hybridization solution A components	
50% Formamide	100% Formamide
0.3 M NaCl	5 M NaCl
20 mM Tris-HCl, pH 8.0	1 M Tris-HCl, pH 8.0
1 mM EDTA	50 mM EDTA
10% Dextran sulfate	50% Dextran sulfate
1× Denhardt's	50× Denhardt's
DEPC-treated water	—

Hybridization solution B components	
0.5 mg/ml Yeast tRNA	10 mg/ml Yeast tRNA
0.001 M Tris-HCl, pH 8.0	0.1 M Tris-HCl, pH 8.0
Probe (see calculation below)	—
DEPC-treated water	—

For 400 μl hybridization mix, use the following amounts:

Reagent	Add
Solution A	
Formamide	200 μl
50% Dextran sulfate	80 μl
5 M NaCl	24 μl
1 M Tris, pH 8.0	8 μl
50 mM EDTA	8 μl
50× Denhardt's solution	8 μl
Total volume	328 μl
Solution B	
Probe	See calculation below
Yeast tRNA, 10 mg/ml	20 μl
0.1 M Tris, pH 8.0	4 μl
DEPC-treated water	To 72 μl
Total volume	72 μl

Probe calculation for 100% saturation:

Probe volume (in μl) per 100 μl of hybridization solution

$$= \frac{(\text{transcript size in kb})(0.3\ \mu g/ml/kb)(3 \times 10^7\ cpm/\mu g)(0.1\ ml)}{\text{acid-precipitable hydrolyzed probe counts (cpm/}\mu l)}$$

7. Add about 40 μl of hybridization mixture per slide and then cover each section with a siliconized coverslip. Gentle pressure on the slide should remove any air bubbles. Lay the slides flat in a covered container under a layer of mineral oil. The container can be reused multiple times.

8. Hybridize for approximately 16 hr at about 45° ($T_m - 25°$).

9. To remove the mineral oil, lift slides out from under the oil and allow the oil to drain off by resting the slide vertically against the container. Wipe excess oil from the ends of the slides. Place slides into glass carriers and pass through three changes of fresh chloroform for 15 to 20 sec each.

10. To remove coverslips, briefly blot excess chloroform and transfer slides to 4× SSC at room temperature. Pass slides through four changes of 4× SSC with vigorous shaking in the first wash until coverslips loosen and fall off. The remaining three washes should be at least 3 min each.

Posthybridization Washes

1. Make RNase buffer (500 ml per rack of slides) with the following recipe.

Stock solution	Volume added	Final concentration
5 M NaCl	50 ml	0.5 M NaCl
1 M Tris-HCl, pH 8.0	5 ml	0.01 M Tris-HCl, pH 8.0
0.5 M EDTA	1 ml	0.001 M EDTA
Distilled water	444 ml	—

2. Perform RNase digestion as follows. (a) Add 200 ml RNase buffer to each of two staining dishes. Allow the buffer to warm to 37°. Add 0.2 ml RNase A (stock 10 mg/ml) and 1.1 μl RNase T1 (stock 181,000 units/ml) to one dish. (b) Place the slide rack into the solution and agitate gently, being careful not to contaminate the water bath with RNase. Incubate at 37° for 30 min. (c) Wash slides in buffer without RNases at 37° for another 30 min.

3. Wash slides in 2 liters of 2× SSC, pH 7.0, at room temperature by putting racks in a large container covered with a lid and placing on a stir plate. Stir the wash solution with the slides for 30 min.

4. Wash the slides in 0.1× SSC at 62° ($T_m - 8°$) with gentle agitation for 30 min. Slides can be placed in a staining dish with 200 ml wash solution and incubated in a shaking water bath for this step.

5. Wash slides in 2 liters of 0.1× SSC as described above, at room temperature for 10 min. *Note:* Stop at this step if slides will be used for immunohistochemistry.

6. Dehydrate slides in a series of ethanol/300 mM ammonium acetate washes up to 95% ethanol.

Ethanol (%)	Volume 2 M ammonium acetate (ml)	Volume 100% ethanol (ml)	Volume distilled water (ml)
30	30	60	110
50	30	100	70
70	30	140	30
85	30	160	10
95	30	220	—

7. Rinse slides in two changes of 99% ethanol and then air dry overnight at room temperature.

Autoradiography

A darkroom with a properly positioned and filtered safelight is essential. Guidelines for appropriate filters and safelight distances are indicated by the photoemulsion manufacturer (Kodak, Rochester, NY). All other light sources in the darkroom should be covered. Also, avoid unplugging all electrical devices, which may cause sparks.

1. To prepare the NTB-2 photoemulsion (Kodak), melt the emulsion in the original container for 30 min in a 45° water bath. Dilute the emulsion 1:1 with 600 mM ammonium acetate, divide 20 ml into glass vials, cap, and completely cover in two pieces of foil. Store at 4° separated from any radioactive emission source.

2. Coat slides with photographic emulsion. In the darkroom, melt the emulsion in a 45° water bath for 30 min. Slowly pour the emulsion into a dipping chamber, avoiding bubbles. Allow emulsion to "rest" for 15 min at 45°. This removes more bubbles from the solution. Remove any remaining bubbles by dipping blank slides into the emulsion until slides come out clean. Slowly and evenly dip the hybridized slides into the emulsion using a steady motion at a rate of about 2 sec per slide. Blot the end of the slide on a paper towel and place vertically in a test tube rack.

3. Allow slides to dry for 1 h. The rack is then transferred to a tightly lidded container, humidified with damp paper towels lining the bottom, for 2.5 to 3 h. Remove slides from the humid chamber and allow to dry for another 30 min. This procedure aids in preventing the emulsion from cracking as it dries. Transfer the slides to a black, light-tight slide box along with a small amount of Drierite desiccant. Place the box into a plastic sandwich bag (e.g., ZipLoc) also with desiccant. Place the bag into a paper safe. Expose the slides for 4 to 8 weeks at 4°.

Note: Extreme care must be taken to ensure that the emulsion and slides are not exposed to any light source other than safelight until they are developed. Steps 1–3 are completed in a darkroom under a safelight.

Development and Counterstaining

1. Prepare fresh developer (Kodak D19 #146 4593, 31.3 g/200 ml water) and fixer (Kodak #197 1746, 35.8 g/200 ml). Store at 4° overnight.

2. In the darkroom, place staining dishes with D19 developer, 2% acetic acid, fixer, and water for rinsing into 15° water bath. All dishes and solutions should be fully equilibrated at 15°; if not, the emulsion may crack and flake.

3. All subsequent steps are carried out at 15°.

4. Under safelight conditions, remove black boxes from the light-tight

box and allow them to come to about 15° while still sealed (about 15 min at room temperature). This avoids condensation on the slides, which reduces signals.

5. Place slides in a glass slide carrier and immediately develop the emulsion for 2.5 min in D19 developer. Longer developing times and higher temperatures produce higher background.

6. Stop development in 2% acetic acid for 30 sec.

7. Fix in Kodak fixer for 5 min.

8. Rinse in distilled water for 15 min. You can turn on the lights at this point.

9. Rinse in cold running tap water for 15–30 min.

10. Counterstain with hematoxylin as follows. Filter hematoxylin and stain slides for 10 min. Wash in tap water until the water is clear. Place in Scot's tap water for 1–2 min. Wash with water. Pass slides through a graded ethanol series and then into several changes of fresh xylenes. Mount one slide at a time by draining excess xylene, adding a drop of Permount (Fisher) over the tissue, and then coverslipping, pressing gently to force out bubbles. Allow Permount to dry overnight. The next day, clean slides with 70% ethanol or glass cleaner to remove emulsion from the back of the slide.

Controls

It is essential to generate control hybridizations for each ISH experiment to facilitate interpretation of the data. Such controls adjust effectively for technical and biological variability. Technical controls are negative control sense-strand hybridizations, using probes of identical specific activity; and positive control ISH, using antisense probes directed against housekeeping genes such as β-actin. Biological controls are negative control antisense hybridizations using control tissues from healthy animals; and antisense hybridizations using control tissues from experimental animals that provided negative results, as monitored by RT-PCR dot-blot (Fig. 4). Initial screening of all slides is performed by an observer blinded to treatment or illness status of the animal from which the specimen was derived, probe identity, and probe polarity.

Colocalization Using in Situ Hybridization and Immunocytochemistry

After completing the ISH protocol up to (but not including) the ammonium acetate/ethanol dehydration series, proceed as follows.

1. Wash slides briefly in PBS, then block tissue in 3% normal goat serum (NGS), or other appropriate blocking solution, in PBS for 30 min,

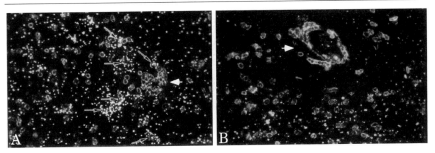

FIG. 4. Negative tissue controls for *in situ* hybridization. *In situ* hybridization (ISH) analysis with radiolabeled antisense MCP-1 probe was performed on CNS tissues from mice sacrificed on day 1 of EAE relapse (A) or the fourth day of remission (B). The dark-field images show hybridization signal as white grains. Arrows indicate hybridization-positive cells; arrowheads show inflammatory foci. ×100.

at room temperature, by laying the slides flat and covering the tissue with 100 to 500 μl of blocking solution in PBS. Follow this technique for the remainder of this protocol.

2. Incubate the tissue with primary antibody in 3% NGS in PBS for 1 hr. For GFAP (Dako, Carpinteria, CA), we use a 1 : 500 dilution.

3. Wash the slides three times in PBS at room temperature, 5 min per wash.

4. Incubate with secondary antibody in 3% NGS in PBS for 30 min. For GFAP immunohistochemistry, we use biotinylated goat anti-rabbit antibody (Vector Laboratories, Burlingame, CA) at a 1 : 1000 dilution. Make avidin–biotin complex (ABC) solution (Vector Elite kit, Vector Laboratories, Burlingame, CA) by mixing components A and B both at a 1 : 1000 dilution in PBS and leave at room temperature for 20 min prior to use.

5. Wash the slides three times in PBS at room temperature, 5 min per wash.

6. Incubate tissue with ABC solution for 1 hr at room temperature.

7. Wash slides three times in PBS at room temperature, 5 min per wash.

8. Make DAB solution by dissolving a single 10-mg tablet of 3,3′-diaminobenzidine tetrachloride (Sigma) in 15 ml PBS (some precipitate will form in the solution). Add 5 μl of 30% H_2O_2 and apply the solution to the tissue through a 0.45-μm syringe filter. Develop for a maximum of 8 min (we develop between 30 and 60 sec).

9. Rinse slides three times in PBS. Continue ISH protocol by dehydrating slides in ammonium acetate/ethanol series (see above).

Comment on in Situ Hybridization

We have applied [³H]uridine- and [³H]cytidine-labeled riboprobes for ISH analyses, because of the requirement for extremely low backgrounds.[14] The probes are subjected to limited base hydrolysis to approximately 300 bp, which increases probe complexity and enhances sensitivity.[15] As noted above, tissue sections are divided in half, with one half reserved for RT-PCR dot-blot assay and the remaining half subjected to ISH. This practice ensures that cellular sources of chemokine synthesis can be defined at the time points of maximal physiological relevance.

Immunocytochemistry

Tissue Fixation

Proper tissue fixation is essential to the success of most immunocytochemical techniques. In many instances, however, it is a double-edged sword. We desire stringently fixed tissue to optimally preserve the structures we want to visualize. However, the most stringent fixatives, such as the dialdehyde glutaraldehyde, produce such excessive cross-linking of tissue proteins that antibodies cannot penetrate the tissue and reach appropriate antigens. The statement that glutaraldehyde fixation destroys antigenic sites is usually incorrect. In fact, a popular means to produce antibodies in the past was to cross-link protein antigens with glutaraldehyde prior to immunization. Therefore, with some exceptions, glutaraldehyde does not necessarily destroy protein antigens, but instead inhibits penetration of immunoreagents to tissue antigens.

Paraformaldehyde or formalin fixation is an attractive choice for primary fixation of tissue for several reasons. It is routine fixation in most pathology departments and thus provides a wealth of banked diseased tissue. With appropriate pretreatment, immunoreagents will readily penetrate paraformaldehyde-fixed tissue sections. Tissue morphology following classic immunostaining procedures of paraformaldehyde-fixed tissue permits resolution of immunoprecipitate at the cellular and in many instances subcellular level.

The ideal tissue preparation would include intracardiac perfusion under high flow of perfusate. This is possible for animal studies but not feasible

---

[14] M. Tani, A. R. Glabinski, V. K. Tuohy, M. H. Stoler, M. L. Estes, and R. M. Ransohoff, *Am. J. Pathol.* **148**, 889 (1996).

[15] L. M. Angerer, M. H. Stoler, and R. C. Angerer, *in* "*In Situ* Hybridization: Applications to Neurobiology" (K. Valentine, J. Eberwine, and J. Barchas, eds.), p. 42. Oxford Univ. Press, Oxford and New York, 1987.

for analysis of human tissue. Fixation by immersion, however, can result in relatively well-preserved tissue. The time interval between death and fixation and the premortem condition of the patient are two variables that can affect the quality of fixation. Brains fixed within 12 hr of death generally provide adequate fixation. The best fixation occurs when tissue is sliced into segments no thicker than 2 cm. We routinely use 4% paraformaldehyde in 80 mM Sorenson's buffer.

8% Paraformaldehyde Stock Stolution. Dissolve 80 g of paraformaldehyde in 800 ml of distilled water. On a hot/stir plate heat the mixture and stir until dissolved. At 30° add 5–6 drops of concentrated HCl to help dissolve the paraformaldehyde. The paraformaldehyde should all be dissolved when the solution reaches 60°. Do not heat the solution over 65° as formic acid may form. Filter through Whatman filter paper and add distilled water for a final volume of 1 liter.

0.4 M Sorenson's Phosphate Buffer. Dissolve the following components separately:

7.16 g of $NaH_2PO_4 \cdot H_2O$ (sodium phosphate, monobasic, monohydrate) in 400 ml of distilled water

49.400 g of Na_2HPO_4 (sodium phosphate, dibasic, anhydrous) in 400 ml of distilled water

Once dissolved add the above solutions together and dilute to a final volume of 1 liter. The solution should be pH 7.6.

4% Paraformaldehyde Solution. Mix the following.

500 ml of 8% paraformaldehyde solution

200 ml of 0.4 M Sorenson's phosphate buffer

300 ml of distilled water

Although paraformaldehyde is the fixative of choice, some antigens do not survive paraformaldehyde fixation. If this appears to be the case, the next step is to try unfixed frozen tissue sections. Frozen tissue is mounted on the appropriate chuck with OCT (Miles, Elkhart, IN) compound (Tissue-Tek), and sections (10- to 20-μm thick) are cut on a cryostat, placed on glass slides, and immunostained. Short periods of fixation (15–30 min) and/or pretreatment with solvents such as ethanol and methanol may help.

Sectioning

There are many approaches for sectioning tissue blocks. In addition to the cryostat sections mentioned above, paraffin and free-floating sections are the most popular choice. Paraffin sections are the most common in routine pathology laboratories. They usually range between 8 and 20 μm in thickness. The tissue undergoes dehydration with ethanol and xylene treatment prior to embedding and rehydration of sections before immunostaining. These solvents can alter antigenicity in some instances. For this and other reasons described below, we prefer to use free-floating sections when possible.

To obtain free-floating sections, fixed tissue is cryoprotected in a solution containing 20% glycerol and 80 M Sorenson's buffer for 1–7 days. Once cryoprotected, the tissue block is frozen on the platform of the freezing, sliding microtome and sectioned at a thickness of 30–40 μm. The sections are placed in 24- or 48-well tissue culture plates in PBS. These sections can be stored for extended periods of time by placing them in a cryostorage solution [10 g polyvinylpyrrolidone (PVP-40), 500 ml of 0.2 M phosphate buffer, pH 7.4, 300 g sucrose, and 300 ml ethylene glycol] at $-20°$. Free-floating sections are immunostained in solution through the entire procedure and then placed on slides. They never enter solvents, and they provide the ideal section for three-dimensional analysis by confocal microscopy.

Pretreatment

A crucial step in immunostaining free-floating sections is pretreatment steps that permeabilize the tissue to permit penetration of immunoreagents. The most popular treatment reagent is the detergent Triton X-100. We generally use 1–10% (v/v) for 30 min. We also add 1–3% hydrogen peroxide (w/w) to the Triton X-100 if our procedure relies on peroxidase for immunodetection. This will inactivate endogenous peroxidase, which is most prominent in red blood cells.

The use of the microwave as a form of pretreatment has proved irreplaceable in producing successful immunocytochemistry results for many antigens. The rationale for using microwaves is not fully understood and is beyond the scope of this chapter. Suffice it to say, however, that a large number of antibodies work on an all-or-none basis with or without microwave pretreatment, and a significant number work at higher dilutions with greater sensitivity of detection after microwaving. The optimal time of microwave pretreatment can vary from section to section and is probably based on the water content and degree of fixation. We place 30- to 40-μm-thick, free-floating sections in 10 mM citrate buffer (pH 6.0) and microwave once or twice on high for 1–10 min; most sections work well with 1–2 min of microwave pretreatment. Cryostat and paraffin sections can also be microwaved prior to immunostaining.

Immunostaining Procedures

There are a number of methods for immunostaining tissue sections. We describe only two: the avidin–biotin complex (ABC) procedure using DAB as a chromogen and methods using fluorescently tagged secondary antibodies (see Fig. 5).

ABC Immunostaining. The ABC procedure provides a highly sensitive and permanent immunocytochemical detection method. In general, we follow the procedures included with the Elite kit provided by Vector Laboratories. All steps are done at room temperature on an orbital shaker unless otherwise specified.

1. Incubate sections in 1–20% normal serum for 30 min. This serves as a blocking step. The serum must be from a species that differs from the host species in which the primary antibody was made and is usually serum from the same host species in which the secondary antibody was made.

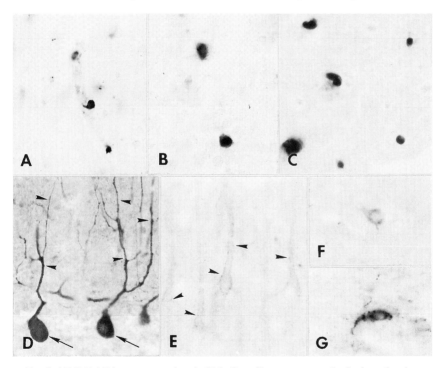

Fig. 5. ABC–DAB immunocytochemical labeling of human autopsy brain tissue for chemokine receptors. Sections were labeled for DARC (A, D, E), IL-8RB/CXC-R2 (B, F, G), and CC-R1 (C). (A–C) Staining of leukocytes associated with regions of inflammation in sections from multiple sclerosis brain. (D, E) Staining of Purkinje cells (C) and cortical neurons (E) with DARC antibodies. DARC immunoreactivity appears diffuse in Purkinje cell perikarya (D, arrows) and dendrites (D, arrowheads) and more particulate on the surface of cortical neurons (E, arrowheads). IL-8RB/CXC-R2 is detected in a small population of cortical neurons (F) and occasional neurons in the spinal cord (G). DARC, Duffy antigen receptor for chemokines; IL-8RB/CXC-R2, IL-8 receptor B; CC-R1, CC chemokine receptor 1. ×40.

2. Incubate in primary antibody in PBS with 1–3% normal serum for 18 hr to 7 days at 4°. Primary antibody dilutions must be titrated. In general, polyclonal antiserum should be diluted at least 1 : 250. Many polyclonal reagents work effectively at dilutions ranging from 1 : 1000 to 1 : 5000. Monoclonal antibody dilution depends on the concentration of immunoglobulin provided. This can vary extensively from supplier to supplier. Dilutions of monoclonal antibodies commonly used in our laboratory vary from 1 : 5 to 1 : 20,000. Most monoclonal reagents work in the 1 : 1000 to 1 : 2000 range.

3. Wash in PBS (three times, 5 min each time).

4. Incubate with a biotinylated secondary antibody at 1 : 500 to 1 : 1000 dilution in PBS with 1–3% normal serum made against the host species of the primary antibody used for 30–60 min.

5. Wash in PBS (three times, 5 min each time).

6. Incubate in ABC solution for 60 min. ABC is made using the ABC Elite kit from Vector Laboratories. Make the solution 20 min before use to allow the ABC complex to form.

7. Wash in PBS (three times, 5 min each time).

8. Incubate in DAB (Sigma) with 0.01% hydrogen peroxide for 5–10 min. To make DAB solution add one 10-mg tablet to 15 ml PBS in a 50-ml conical centrifuge tube and vortex to dissolve it. When ready to apply DAB solution add 5 μl of 30% H_2O_2, cap, and invert the tube to mix. Draw the solution into a syringe, remove the needle, replace with a syringe filter, and then dispense onto the sections.

9. Wash in PBS (three times, 5 min each time).

10. Enhance the DAB with 0.04% osmium tetroxide (OsO_4). Make a stock 4% OsO_4 solution with distilled water. Then dilute 1 : 100 with PBS in a scintillation vial to make enhancing solution. Dip the sections until DAB turns black-brown, which is usually within 30 sec (optimal time is determined empirically for each antibody).

11. Wash in PBS (three times, 5 min each time).

12. Transfer sections into 70% glycerol made with distilled water for 5 min.

13. Transfer sections into 100% glycerol, mount on slides, coverslip, and apply nail polish around the edges to seal.

Immunofluorescence. The use of fluorescence-tagged antibodies can provide many advantages for visualizing immunolabeling. One of the more popular and informative applications is double-labeling immunocytochemistry for conventional and/or confocal immunocytochemistry. The procedure we use for double-labeling free-floating sections is described below. We have utilized this technique for investigating aspects of CNS inflammation in MS brains.

1. Fixation and pretreatment of free-floating sections are performed as described above. Pretreatment with 0.08% osmium tetroxide for 30 sec will reduce autofluorescence in some instances.

2. Primary antibodies are both applied at the same time, and sections are incubated for 18 hr to 7 days at 4°. In general, primary antibodies are used at a higher concentration (25–50%) than used for the ABC–DAB procedure described above.

3. Wash in PBS (three times, 5 min each).

4. Incubate with fluorescent conjugated secondary antibodies (1:500 to 1:1000, Jackson ImmunoResearch Laboratories, Bar Harbor, ME) against the host species of the primary antibodies at room temperature for 60 min. Both secondary antibodies can be applied at the same time.

5. Wash in PBS (three times, 5 min each).

6. Mount the tissue with Vectashield (Vector Laboratories) mounting medium on microscope slides and coverslip. One modification we often employ to visualize the weaker labeling antigen is to use a biotinylated secondary antibody as in the ABC–DAB protocol (see ABC Immunostaining, step 4). Wash sections twice for 5 min each in PBS, and then for 5 min in 0.1 M sodium bicarbonate buffer at pH 8.5. Next we apply fluorochrome-conjugated avidin D (Vector Laboratories) at 1:500 to 1:1000 diluted in 0.1 M sodium bicarbonate buffer for 60 min at room temperature. Wash in sodium bicarbonate buffer (5 min, three times). Finally, mount the tissue with Vectashield.

Comments on Immunocytochemistry

Chemokines are soluble, highly basic proteins that interact with acidic extracellular matrix components. They become highly diffused within the neuropil, and they have half-lives between 8 and 24 hr. These properties render chemokine detection by immunocytochemistry rather difficult. As described earlier, *in situ* hybridization can be particularly useful for identifying cells that secrete chemokine products. Another informative approach for elucidating the functional role of chemokines is the identification of cells that have the ability to respond to chemokines by virtue of the presence of chemokine receptors on their surface. A number of chemokine receptors have been cloned and a battery of antibodies produced against their peptide sequences. The identification of chemokine receptors including CC-CR5 as an invasion coreceptor for human immunodeficiency virus (HIV) has resulted in a substantial interest in the cellular distribution of chemokine receptors. In addition, the detection of chemokine receptors on the neuronal cells has expanded the putative role of chemokines in the CNS beyond leukocyte attraction.

Chemokine receptors have many properties that facilitate their detection by immunocytochemical methods. They are cell surface transmembrane proteins that can be expressed at the cell surface at levels that are readily detectable. We have used a number of immunocytochemical techniques to localize adhesion molecules and receptors, including chemokine receptors, in tissue sections of human brain.[16,17] The approaches that have been productive for use are outlined above.

[16] L. Bö, J. W. Peterson, S. Mork, P. A. Hoffman, W. M. Gallatin, R. M. Ransohoff, and B. D. Trapp, *J. Neurol. Exp. Neuropathol.* **55,** 1066 (1996).
[17] L. Bö, S. Mork, P. Kong, H. Nyland, C. Pardo, and B. Trapp, *J. Neuroimmunol.* **51,** 135 (1994).

[22] Synthesis and Evaluation of Fluorescent Chemokines Labeled at the Amino Terminal

By Robin E. Offord, Hubert F. Gaertner, Timothy N. C. Wells, and Amanda E. I. Proudfoot

Introduction

Chemokines are a rapidly growing family of proinflammatory cytokines that are involved in cell recruitment and activation.[1] They are small 8- to 10-kDa proteins having a low level of amino acid sequence alignment, but they possess a four-Cys motif that allows their division into two subclasses depending on the spacing of the first two Cys residues: the α subclass has an amino residue between the two Cys (C-X-C), whereas in the β subclasses the two Cys are adjacent (C-C). Both subclasses of the chemokine family bind to seven-transmembrane receptors coupled to G proteins. As of June 1997, four human receptors have been identified that bind members of the α subclass, and eight that are activated by β-chemokines.[2-7] With the

[1] M. Baggiolini, B. Dewald, and B. Moser, *Adv. Immunol.* **55,** 97 (1994).
[2] C. A. Power and T. N. C. Wells, *Trends Pharmacol. Sci.* **17,** 209 (1996).
[3] C. C. Bleul, M. Farzan, H. Choe, C. Parolin, I. Clark-Lewis, J. Sodroski, and M. A. Springer, *Nature (London)* **382,** 829 (1996).
[4] M. Loetscher, B. Gerber, P. Loetscher, S. A. Jones, L. Piali, I. Clark-Lewis, M. Baggiolini, and B. Moser, *J. Exp. Med.* **184,** 963 (1996).
[5] M. Baba, T. Imai, M. Nishimura, M. Kakizaki, S. Takagi, K. Hieshima, H. Nomiyama, and O. Yoshie, *J. Biol. Chem.* **272,** 14893 (1997).
[6] G. J. Adema, F. Hartgers, R. Verstraten, E. de Vries, G. Marland, S. Monon, J. Foster, Y. Xu, P. Nooyen, T. McClanahan, K. B. Bacon, and C. G. Figdor, *Nature* **387,** 713 (1997).
[7] R. Yoshida, T. Imai, K. Hieshima, J. Kusuda, M. Baba, M. Kitaura, M. Nishimura, M. Kakizaki, H. Nomiyama, and O. Yoshie, *J. Biol. Chem.* **272,** 13803 (1997).

exception of CXCR1, which is specific for interleukin-8 (IL-8), all the receptors have been described as being promiscuous in that they bind more than one ligand with high affinity. Binding to purified leukocytes and to cells transfected with a recombinant receptor is generally performed with radiolabeled ligands.

We were interested in observing the direct binding of chemokines by techniques such as fluorescence activated cell sorting (FACS) analysis and confocal microscopy. Binding with fluorescent ligands has been limited to the commercially available phycoerythrin–IL-8 conjugate, which consists of the addition of a relatively large fluorescent moiety of 240 kDa attached to the 8-kDa protein through its side chains. Such a conjugate is many times larger than the chemokine itself, and because there are so many side chains available for conjugation, the product is likely to be a mixture of structural isomers.

We therefore sought a highly controlled, site-specific reaction to attach fluorophores to the polypeptide chain to ensure that each chain had one fluorophore, always at the same, chosen position in the chain. The chemical reactions involved had to be efficient and, so as not to compromise the biological behavior of the products, had to take place under mild, aqueous conditions. We chose a chemical reaction that can be limited to the terminal amino group only, without touching side-chain amino groups, or indeed any other side-chain functional group. The reaction used is the periodate oxidation of the 1,2-amino alcohol structure, which is found only in proteins that have an N-terminal serine or threonine. The reaction depends on the presence of a free amino group and an OH group on the neighboring carbon atom. It will therefore not occur with serine or threonine elsewhere in the sequence, because the amino group is not free. The new aldehyde-like functional group

$$H_2N-CH(CH_2OH)-CO-NH-protein \rightleftharpoons O=CH-CO-NH-protein$$

has a quite specific chemical reactivity,[8] clearly distinct from that of other groups normally found in proteins, and we show below that it readily lends itself to the subsequent attachment of a fluorophore. Many chemokines and cytokines have an appropriate N-terminus; if they do not, it is relatively easy by site-directed mutagenesis of recombinant proteins to introduce an N-terminal serine or threonine.

Because the sensitivity to periodate of a 1-amino 2-alcohol is thousands of times greater than that of the better known periodate oxidation of 1,2-diols (sugars, for example), the concentration of periodate can be kept very low (1 mM or less) and the reaction time can be kept short (3 to 10 min).

[8] H. B. F. Dixon, *J. Protein Chem.* **3**, 99 (1984).

This means that it is usually possible to avoid oxidative attack on otherwise sensitive side chains.

Any given polypeptide chain will have only one α-amino group irrespective of the number of side-chain groups, and so the ability to conduct reactions exclusively at the N-terminus has proved useful in a variety of ways.[8] We have previously used this approach to ligate unprotected protein chains to give large structures having a contiguous backbone chain[9] and to introduce noncoded elements within a protein sequence.[10] Geohegan and Stroh have used it for the introduction of fluorophores,[11] and a preliminary account of our work in this area has appeared.[12] Other, quite different methods exist to direct reactivity exclusively to the C terminus.[10] In addition, a recombinant method has been developed, for much the same purposes as we describe here, for placing a keto group in proteins.[13]

How can we use the aldehyde-like reactivity at the N terminus to introduce fluorophores? The most convenient approach is to functionalize the fluorophore with an aminooxy group, H_2N-O-, so that it will form an oxime with the oxidized protein. The aldehyde-like group might be

$$R-O-NH_2 + O=CH-CO-NH-\text{protein} \rightleftharpoons$$
$$R-O-N=CH-CO-NH-\text{protein}$$

able to form a Schiff base with an amino group elsewhere in the structure, but the aminooxy compound would rapidly break such a link, permitting the formation of an oxime without noticeable interference. The oxime, once formed, is highly stable under physiological conditions.

These amino terminally labeled chemokines are shown to retain their biological properties. We have shown that they are capable of binding to their receptors, and they are therefore useful for certain applications such as sorting cells with a high number of recombinant receptors.

Protein Expression and Purification

Reagents

Breakage buffer: 0.1 M Tris-HCl, pH 8.0, containing 1 mM dithiothreitol (DTT), 5 mM benzamidine hydrochloride, 1 mM phenyl-

[9] R. C. Werlen, M. Lankinen, K. Rose, D. Blakey, H. Shuttleworth, R. Melton, and R. E. Offord, *Bioconjugate Chem.* **5,** 411 (1994).
[10] H. F. Gaertner, R. E. Offord, R. Cotton, D. Timms, R. Camble, and K. Rose, *J. Biol. Chem.* **269,** 7224 (1994).
[11] K. F. Geohegan and J. G. Stroh, *Bioconjugate Chem.* **3,** 138 (1992).
[12] S. Alouani, H. F. Gaertner, J.-J. Mermod, C. A. Power, K. B. Bacon, T. N. C. Wells, and A. E. I. Proudfoot, *Eur. J. Biochem.* **227,** 328 (1995).
[13] V. W. Cornish, K. M. Hahn, and P. G. Schultz, *J. Am. Chem. Soc.* **118,** 8150 (1996).

methylsulfonyl fluoride (PMSF), 20 mg/liter DNase, and 2 mM MgCl$_2$

Endoproteinase Arg-C (Boehringer Mannheim, Germany)

Equipment

5-liter fermentor equipped with temperature, pH, and dissolved O$_2$ controls (LSL Biolafitte, France)

Cell breakage equipment such as French pressure cell (Amicon, Danvers, MA) or high pressure homogenizer (Gaulin) or equivalent.

Cooled centrifuge (4°, 10,000 g) for inclusion body preparation

Sephacryl HR S-200 column (5 cm diameter × 100 cm) (Pharmacia, Uppsala, Sweden)

HiLoad SP 26/10 and HiLoad Q 26/10 columns (Pharmacia)

FPLC (fast protein liquid chromatography) for protein chromatography (Pharmacia)

Procedure. The chemokines are purified from recombinant heterologous expression in the prokaryotic host *Escherichia coli*. The cDNA encoding for the mature form of IL-8 is expressed under the control of a Trp promoter as described.[12] RANTES and macrophage inflammatory protein-1α (MIP-1α) are expressed under the control of the T7 polymerase system as described.[14] In the case of MIP-1α the first alanine residue is removed by site-directed mutagenesis from the cDNA encoding the mature form of the protein so that the amino terminal residue will be Ser-2. Fermentations are carried out in 5-liter fermentors as described[15] with the following modifications. Induction of IL-8 expression is carried out by natural consumption by the cells of tryptophan in the medium, and the chemokines expressed under the control of the T7 polymerase promoter system are induced by the addition of 1 mM isopropylthiogalactoside (IPTG).

The *E. coli* cell pastes are suspended in a volume three times their wet weight of cell breakage buffer consisting of 0.1 M Tris-HCl, pH 8.0, containing 1 mM DTT, 5 mM benzamidine hydrochloride, 1 mM PMSF, 20 mg/liter DNase, and 2 mM MgCl$_2$. Cells are broken by three passages through a French pressure cell, with 30 sec of sonication on ice after each passage. The resulting solution is centrifuged for 60 min at 10,000 g at 4°. The recombinant proteins are purified from inclusion body pellets by solubilization in 0.1 M Tris-HCl buffer, pH 8.0, containing 1 mM DTT and 6 M guanidine hydrochloride followed by gel filtration on a Sephacryl HR S-200 column equilibrated in the same buffer. The proteins are renatured

[14] A. E. I. Proudfoot, C. A. Power, A. Hoogewerf, M.-O. Montjovent, F. Borlat, and T. N. C. Wells, *FEBS Lett.* **376**, 19 (1995).

[15] A. E. I. Proudfoot, D. Fattah, E. H. Kawashima, A. Bernard, and P. T. Wingfield, *Biochem. J.* **270**, 357 (1990).

by a 20-fold dilution into 0.1 M Tris-HCl buffer, pH 8.0, containing 1 mM oxidized and 0.1 mM reduced glutathione, and the solution is stirred overnight at 4°.

The renatured IL-8 and RANTES proteins are concentrated by applying the solution to a HiLoad S (26/10) column equilibrated in 50 mM sodium acetate, pH 4.5, and eluting the adsorbed protein with a linear 0.6–2 M NaCl gradient in the same buffer. The hexapeptide leader sequence is removed from the RANTES fusion construct by incubation in 50 mM Tris-HCl buffer, pH 8.0, with endoproteinase Arg-C (1 : 600, enzyme : substrate, w/w) overnight at 37°. The cleaved product is separated by cation-exchange chromatography as described above, except that 6 M urea is included in the buffers. The renatured MIP-1α is concentrated by anion-exchange chromatography on a HiLoad Q 26/10 column equilibrated in 20 mM Tris-HCl buffer, pH 8.0, and eluted with a 0–0.5 M NaCl gradient in the same buffer. The purified proteins are dialyzed extensively against 1% (w/w) acetic acid and then 0.1% (w/v) trifluoroacetic acid (TFA) and stored as lyophilized powders at −80°.

Preparation of Conjugates

The principal fluorescent chemokines that we have prepared are listed in Table I. Details of all the combinations of fluorophores and chemokines that these constructions represent are not given here, but we comment on each of the specific methods so that they can be adapted as required.

General Separation Methods

Reagents. All reagents are analytical grade, or better.

Equipment

   Chromabond C_{18} cartridges (Macherey-Nagel, Dueren, Germany)
   HPLC Nucleosil C_8 and C_{18} columns (Macherey-Nagel)
   Preparative PrepLC 25 mm Module (Waters Chromatography, Milford, MA)

Solid-Phase Extraction Cartridge: Double Chromabond Technique. For solid-phase extraction, a Chromabond C_{18} cartridge (1 g of resin, polypropylene construction) is opened, and the resin is removed and added to the resin in a second cartridge. The double cartridge is washed in acetonitrile and equilibrated with 0.1% aqueous TFA. A portion of the mixture to be separated, corresponding approximately to 50 mg of the product desired, is passed through the cartridge, which is then washed with 25 ml of 0.1% TFA/acetonitrile, 9 : 1 (v : v), or some other mixture as required by the protocol. The product is then eluted from the cartridge with 20 ml of 0.1%

TABLE I

FLUORESCENT CHEMOKINES PREPARED BY OXIMATION OF N-TERMINALLY OXIDIZED
CHEMOKINES WITH VARIOUS FORMS OF COMPOUND V^a

Compound (see Scheme 1)	Fluorophore[b]	Chemokine		
		IL-8	RANTES	MIP-1α^c
Va	FITC	8940.80 ± 0.8 (8941.36)	8406.67 ± 1.43 (8406.59)	8302.95 ± 0.45 (8302.23)
Vb	NBD	8714.02 ± 1.13 (8715.05)	8179.41 ± 0.21 (8180.28)	8075.22 ± 1.14 (8075.92)
Vc	Cy5	9190.93 ± 1.39 (9189.69)	8656.96 ± 0.63 (8654.92)	8552.01 ± 1.86 (8550.56)
Vd	Texas Red	9140.12 ± 1.53 (9140.71)	[d]	[d]

[a] Masses are in daltons as deduced by electrospray mass spectrometry. Theoretically expected values are given in parentheses.

[b] FITC is, more correctly, FTC (fluorescein thiocarbamyl); NBD is 7-nitrobenz-2-oxa-1,3-diazole-4-yl; Cy5 is a trade mark of Biological Detections Systems, Pittsburgh, PA, and Texas Red is a trade mark of Molecular Probes, Eugene, OR.

[c] Residue 1 of the natural sequence was removed by recombinant methods to expose an N-terminal Ser. Ala-9 was mutated to Thr for other reasons. We find that the Cy5 derivative of this protein gives better cell binding than the FITC and NBD derivatives.

[d] Conjugate not synthesized.

TFA/acetonitrile, $3:2$ (v/v), or some other mixture, as appropriate. The cartridge can be reequilibrated and reused as often as required. It is dried for storage after washing with 20 ml acetonitrile.

Analytical High-Performance Liquid Chromatography. For high-performance liquid chromatography (HPLC) separation, a gradient elution system is used between 0.1% TFA (solvent A) and acetonitrile/TFA/water, $900:1:100$ (v/w/v) (solvent B). If necessary, the reaction mixtures are carefully acidified (to pH 4.5 or lower) before chromatography.

Analytical separations are made on samples of approximately 10 μg, using linear gradients or isocratic elution, as indicated in the individual protocols. The columns are either Nucleosil C_8 for protein derivatives or Nucleosil C_{18} for fluorophores and the intermediates in their synthesis. Column dimensions are 4×250 mm.

Semipreparative High-Performance Liquid Chromatography. The columns, 10×250 mm, are eluted (isocratically or by a linear gradient) at a flow rate of 3.7 ml/min. The loading capacity of these columns in the solvent system used is about 5–7 mg of desired product.

Preparative High-Performance Liquid Chromatography. If an HPLC system is able to deliver, reliably, eluant at 20 ml/min, preparative-scale HPLC permits one to work with significantly larger quantities, or to avoid multiply repetitive solid-phase extractions or semipreparative HPLC. However, such equipment is not essential to any of the preparations described in this chapter.

Mixtures containing up to about 100–200 mg at a time can be loaded on a preparative HPLC equipped with a PrepLC 25-mm Module (Waters Chromatography) containing two PrepPak 25 × 100 mm cartridges filled with Nova-pak HR C_{18} 6 μm, 60 Å resin and a 25 × 10 mm guard Pak cartridge. Adsorption losses are prohibitive at loadings below 40 mg. The column is eluted at 20 ml/min using a linear gradient.

Recovery of Samples. The acetonitrile is removed from the sample solutions in a current of air, or by rotary evaporation at room temperature. The remaining liquid is then lyophilized.

Amino-Terminal Oxidation of Chemokines

Example: Amino-Terminal Oxidation of Interleukin-8. Interleukin-8 (2 mg) is dissolved in 1 ml of 1% NH_4HCO_3, pH 8.3, and 60 μl of 0.2 M methionine in water is added to give a 50-fold excess as scavenger. Then 48 μl of 50 mM sodium periodate in water (10-fold excess) is added, the solution incubated for 10 min in the dark, and the reaction stopped by the addition of 239 μl of 10 M ethylene glycol solution in water (10,000-fold excess). The solution is further incubated for 15 min and the product separated from reagents on a PD10 (Pharmacia) column equilibrated in 0.1 M sodium acetate, pH 4.6. The oxidized protein is stored frozen (up to 2–3 months at least) if not used at once.

Notes

1. Insoluble proteins, or those for which access to the amino terminus seems to be sterically hindered (e.g., RANTES), can be treated as above, but using a buffer that is 6 M in guanidinium chloride. We normally use 0.1 M sodium phosphate, pH 7.0, to which solid guanidinium chloride is added to provide a final concentration of 6 M. Account must be taken of the very considerable increase in volume produced by the dissolution of the solid guanidinium chloride.

2. The methionine is present to minimize oxidative attack on any unusually susceptible side chains that might exist in the protein. It is even possible, such is the specificity of periodate under these conditions, to have ethylene glycol (55 mM) present during the oxidation itself. The periodate attacks the susceptible N-terminus so rapidly that little of it is taken by reaction with the glycol, until the wanted reaction has finished.

3. The protein concentration of the eluate from the PD10 column can be adjusted using a centrifugal membrane concentration cell. Alternatively,

the reaction mixture after the quench with 10 M ethylene glycol can be dialyzed extensively against, for example, 1% (v/v) aqueous acetic acid, then freeze-dried and redissolved in the sodium acetate buffer.

4. The freeze-dried oxidized protein, like the solution, can be stored, if well sealed, at $-20°$ for many weeks. If guanidinium chloride has proved necessary for the oxidation, then it is best to add it to the stock solution of oxidized protein just before the oximation step (below). The storage time at $-20°$ is more severely limited once guanidinium chloride is present.

Site-Specific Conjugation of Aminooxy Fluorophores by Oximation

Formation of Fluorescent Chemokine. To 5 mg of oxidized protein in 1 ml of 0.1 M sodium acetate, pH 4.6, is added 0.23 ml of a 14 mM solution of N^α-aminooxyacetyllysine, carrying a fluorophore as substituent on the ε-amino group (for methods of preparation of such compounds, see below). The reaction is usually substantially complete overnight. The product is worked up by preparative or semipreparative HPLC. The products are stored at $-20°$ as lyophilized powders and, when needed for experiments, are taken up in distilled water before the addition of buffer.

Notes

1. Analytical HPLC on a small sample can be used to follow the course of the reaction. We leave some of the slower oximations (e.g., that of oxidized RANTES) for 2 days before working up.

2. There has so far never been any sign of the formation of Schiff bases between the aldehydic group and the ε-amino group of lysine residues. If such links are formed, they are presumably broken immediately by the much stronger nucleophile represented by the aminooxy compounds.

3. The desired product is usually the major component of the final reaction mixture, unless steric problems have been encountered, as indicated by low yields in preliminary experiments (10 μl of the protein solution with 2.3 μl of the aminooxy compound, time course by analytical HPLC of 2-μl samples each diluted into 200 μl of 0.1% TFA). In such cases we still obtain, normally, useful quantities of product, but yields can be significantly lower.

4. Guanidinium chloride does not interfere prohibitively with oximation, and we use it routinely when oximating a protein with which we have encountered steric problems, for example, RANTES.

5. A "one-pot" oxidation–oximation reaction is possible at around neutral or slightly alkaline pH. The pH must be sufficiently high to permit the oxidation step to take place but low enough to give reasonable kinetics of formation, and stability, of the oxime. Beyond demonstrating the possibility,

we have done little to optimize such procedures because the two-step procedure is quite adequately convenient already.

6. In theory, the final product could be further stabilized by reducing with, say, cyanoborohydride. We have not found this reaction at all easy to achieve (owing to very high concentrations of cyanoborohydride, long reaction times, and poor yields). Given the extreme stability of the oxime once formed, it is rarely necessary to reduce it.

7. Somewhat analogous conditions to the above can be used for the formation of hydrazones, or Schiff bases with aromatic amines. The hydrazones are still quite stable (typical conditions for breaking the bond are pH 2 in the presence of 1 mM CH_3ONH_2 for 2 days at 20°) and can be further stabilized, rather more readily than is the case for oximes, by cyanoborohydride reduction (0.2 M cyanoborohydride in 0.1 M sodium acetate solution, pH 4.6, 2 to 4 days at 20°[16]). Aromatic Schiff bases are readily stabilized by reduction (cyanoborohydride in the low millimolar range, pH 3.5–5, for a few minutes at 20°). In the case of aromatic Schiff base formation, such is the specificity of both periodate and cyanoborohydride that the oxidation and the reduction can take place simultaneously in the presence of both reagents, in a one-pot reaction.

8. An example of the formation of a hydrazone is afforded by the reaction between fluorescein thiosemicarbazide (Sigma, St. Louis, MO) and oxidized IL-8. The reaction is carried out as above, except that the thiosemicarbazide is dissolved to 14 mM in methylpyrrolidone/0.1 M sodium acetate, pH 4.6, 3:2 (v/v). Two hundred microliters of this solution is added to the 1 ml of the solution of oxidized protein, followed by 500 μl of methylpyrrolidone. Other examples of site-specific hydrazone formation with a fluorescent compound have been described.[8,10]

Preparation of Aminooxy Fluorophores

Although other bis-NH$_2$ compounds can be used, we normally build the fluorescent labels up on a lysine skeleton, with substitution of the α-HN$_2$ with the aminooxyacetyl group and substitution of the ε-NH$_2$ with the fluorophore. Most fluorophores are commercially available in the form of compounds designed to react easily with amino groups. Because, in our procedure, the amino-reactive reagents never see the polypeptide to be labeled, there is no danger of substitution of other side-chain groups of any type.

[16] H. F. Gaertner, K. Rose, R. Cotton, D. Timms, R. Camble, and R. E. Offord, *Bioconjugate Chem.* **3**, 262 (1992).

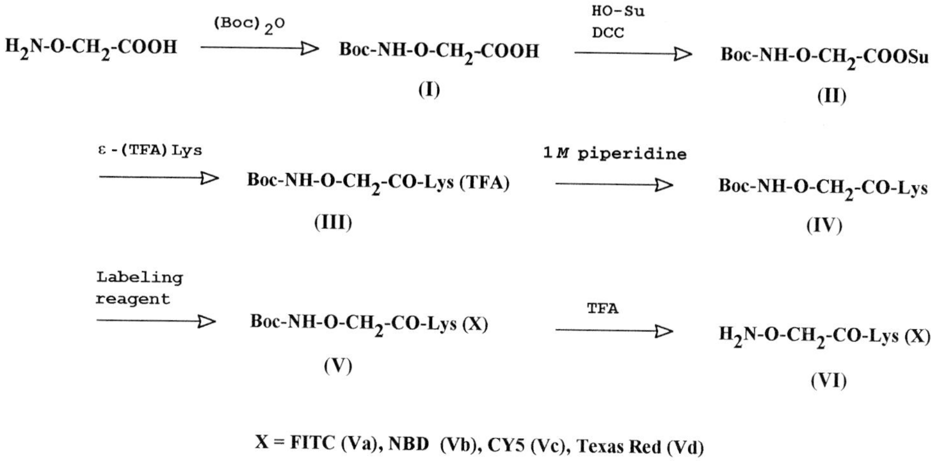

$H_2N-O-CH_2-COOH$ $\xrightarrow{\text{(Boc)}_2O}$ $Boc-NH-O-CH_2-COOH$ $\xrightarrow[\text{DCC}]{\text{HO-Su}}$ $Boc-NH-O-CH_2-COOSu$

(I)   (II)

$\xrightarrow{\epsilon-(TFA)Lys}$ $Boc-NH-O-CH_2-CO-Lys\,(TFA)$ $\xrightarrow{1\,M\text{ piperidine}}$ $Boc-NH-O-CH_2-CO-Lys$

(III)   (IV)

$\xrightarrow[\text{reagent}]{\text{Labeling}}$ $Boc-NH-O-CH_2-CO-Lys\,(X)$ $\xrightarrow{\text{TFA}}$ $H_2N-O-CH_2-CO-Lys\,(X)$

(V)   (VI)

X = FITC (Va), NBD (Vb), CY5 (Vc), Texas Red (Vd)

SCHEME 1

Scheme 1 shows the general series of reactions used. Although there are several steps, all are, with reasonable care, quite straightforward, and are based on the routine operations of peptide synthesis.

Boc-Aminooxyacetic Acid (I). A solution of 22.4 g KOH is prepared in a mixture of 400 ml methanol and 600 ml water. Ten grams (91.5 mmol) of aminooxyacetic acid hemihydrochloride is dissolved in 150 ml of the KOH solution in a round-bottomed flask. The apparent pH (glass electrode, uncorrected) is approximately pH 4.5, and the mixture is adjusted to pH 9.1 by adding the necessary volume (very approximately 100 ml) of the KOH solution.

To this solution is added 43.65 g (200 mmol) of *tert*-butyl pyrocarbonate [Boc-dicarbonate, $(Boc)_2O$], and the resulting emulsion is kept suspended with a magnetic stirrer. The apparent pH falls and is restored to pH 9.1 from time to time by the addition of more KOH solution. The flask is loosely covered with aluminum foil to permit the escape of the CO_2 released during the reaction. After the first 1 to 2 hr, and the addition of roughly 500 ml of the KOH solution, the mixture stabilizes at pH 9.1. The mixture is left, under agitation, overnight at 20°. The emulsion clarifies during the night. The apparent pH should remain above pH 8. The volatile organic substances and some water are removed by rotary evaporation under a water-pump vacuum until the volume of the remainder is approximately 100 ml. The solution is then mixed with an equivalent volume of ethyl acetate and cooled to 0°, and the pH is taken down to pH 3.3 by careful addition of small portions of 6 *M* HCl with vigorous agitation. The organic phase is separated and the aqueous fraction further extracted two times with 50 ml ethyl acetate. The combined organic layers are then dried over anhydrous Na_2SO_4 and finally taken to dryness in a rotary evaporator. Drying is completed by placing the flask in a desiccator under oil-pump vacuum. This product (**I**) is used without further purification. The yield is approximately 15 g.

Boc-Aminooxyacetyl-hydroxysuccinimido ester (II). To 1.91 g (10 mmol) of (**I**) in 30 ml of ethyl acetate is added a solution of 1.15 g (10 mmol) of *N*-hydroxysuccinimide in 30 ml of ethyl acetate. At room temperature, under constant mixing, a solution is added of 2.06 g (10 mmol) of *N,N*-dicyclohexylcarbodiimide (DCC), dissolved in 3 ml of ethyl acetate. After agitation for 5 hr at 20°, an abundant precipitate of the dicyclohexylurea forms. The complete disappearance of the carbodiimide is checked by thin-layer chromatography on silica gel of a 5-μl sample (chloroform/methanol, 1:1, v/v) developed with a spray of 1,3-dimethylbarbituric acid (500 mg in 9 ml pyridine plus 1 ml water[17]). A 5-μl spot of carbodiimide (30 mg in 1 ml of ethyl acetate) is run alongside for comparison and gives a blue-violet color. If some carbodiimide remains, 1.45 mg (5 mmol) of **I** is added and the reaction is continued until no carbodiimide is left. Once all the carbodiimide is consumed, the mixture is filtered on a number 3 glass frit, and the filtrate is partially dried down to obtain an oily solution. The material is then crystallized by the slow addition of one equivalent volume of diethyl ether and by cooling the solution for a few hours in an ice bath. The crystallized material is filtered on the glass frit, washed with previously cooled ether, and finally dried in a desiccator under oil-pump vacuum. The yield is 1.3 g. Material remaining in the filtrate can be dried down (1.2 g) and resolubilized in ethyl acetate for another crystallization.

N^{α}-(Boc-Aminooxyacetyl)-N^{ε}-(trifluoroacetyl)lysine (III). Compound **II** (570 mg, 2 mmol) is added to 300 mg (1.2 mmol) of N^{ε}-(TFA)-lysine (Novabiochem, Läufelfingen, Switzerland) suspended in 3 ml of dimethyl sulfoxide (DMSO). *N*-Ethylmorpholine (approximately 150 μl) is added until the externally indicated pH (moist narrow-range pH paper) is between approximately pH 8 and 9. By this time the mixture is a clear solution. After 15 hr at 20° and verification that the external pH is still alkaline, 5 ml of water is added, the external apparent pH is brought up if necessary to its original value with diisopropylethylamine, and the mixture left for 1 hr at 37° to hydrolyze any remaining active ester. The solution is then cooled to 0° (ice bath), and the apparent pH (glass electrode, uncorrected) is brought to pH 3.0 by careful addition of glacial acetic acid under constant agitation. Thirteen milliliters of cold water are then added, and the product

---

[17] J. Kohn and M. Wilchek, *J. Chromatogr.* **240,** 262 (1982).

is isolated either 50 mg at a time on a solid-phase extraction cartridge [equilibration in 0.1% TFA, wash in 0.1% TFA/acetonitrile 9:1 (v/v), elution in 0.1% TFA/acetonitrile 3:2 (v/v)] or, one-half at a time, by preparative-scale HPLC (gradient of 20% B to 50% B over 30 min), as described above. In the later case, the recovery is 380 mg.

Note: If the apparent pH is found to be acid the next morning, the reaction might not be complete. In such a case, the situation can be partially recovered by bringing the apparent pH back up to between pH 8 and 9 and allowing the reaction to continue for a few more hours. Such a fall in pH can be due to overcautious addition of N-ethylmorpholine at the start or (for small samples) inadequate sealing against atmospheric CO_2.

N^{α}-*(Boc-Aminooxyacetyl)lysine* **(IV)**. The dried **III** prepared as above is dissolved in 4 ml water, and 0.44 ml piperidine is added (final concentration 1 *M*). After 4 hr at 20° the mixture is adjusted to pH 3.0 at 0° as before and then diluted with 10 ml of 0.1% TFA. The deprotected material is then isolated, one-half at a time, by preparative HPLC using a linear gradient between 0 and 30% B over 30 min with a flow rate of 20 ml/min. The yield is 370 mg of the trifluoroacetyl salt. If the product is isolated on a double Chromabond column, the reaction mixture has to be diluted five times more in 0.1% TFA and only one-eighth of the sample is passed through per double cartridge.

Example: N^{α}-*(Boc-Aminooxyacetyl)-N^{ε}-(fluorescein thiocarbamoyl)lysine* **(Va)**. Compound **IV** (40 mg) is dissolved in 300 μl N,N-dimethylformamide, 40 mg fluorescein isothiocyanate (FITC) is added, and the mixture is adjusted to pH 8.0 with N-ethylmorpholine. After incubation for 15 hr in the dark, the product is purified by chromatography on a silica column [Kieselgel 60 (Fluka Chemie, Buchs, Switzerland), 1.5 × 20 cm] equilibrated in methanol/CH_3Cl (1:1, v/v). The excess FITC elutes in the flow-through fraction, and a second yellow fraction containing the expected product is eluted with methanol/CH_3Cl (4:1, v/v). The solvent is removed by rotary evaporation, and the product is characterized by electrospray ionization mass spectrometry (ESI-MS) (calculated M+H, m/z 709.6; found, m/z 709.4). Further purification is by semipreparative reversed-phase HPLC on the C_8 column, using a linear gradient between 30% B and 60% B over 30 min. The solvent is once again removed by rotary evaporation, followed by exposure to an oil-pump vacuum overnight. Alternatively, purification can wait until after the removal of the Boc group, see below.

Notes: The FITC derivatives normally elute as doublets on low-loading, analytical HPLC because of the use of the commercial mixture of fluorescein 5- and 6-isothiocyanate for the coupling to **IV**. This procedure is readily adaptable to other fluorophores (see Table II). The silica-gel step can be omitted for the other compounds, because it is only the FITC reagent that

TABLE II
PREPARATION OF OTHER VARIANTS OF COMPOUND **V**

Parameter	Fluorophore		
	NBD	Cy5	Texas Red
Reagent	NBD chloride	Cy5 hydroxysuccinimide ester	Sulfonyl chloride
Supplier	Sigma	Biological Detection Systems	Sigma
Solvent for reagent	Ethanol	Acetonitrile	Dimethylformamide[a]
Buffer for reaction with **IV**	0.5 M Borate (sodium), pH 8.0/ethanol, 1:1 (v/v)	0.1 M NaHCO$_3$	0.1 M Sodium bicarbonate, pH 9
Reagent amount (mg/mg of **IV**)	3.33	1.5	0.67
Final HPLC purification of derivative (before removal of Boc group)[b]	35% B isocratic	27% B isocratic	35% B to 100% B linear over 90 min
Solvent used for 14 mM solution of deprotected derivative when used for oximation	Water/acetonitrile, 4:1 (v/v)	0.1% TFA/acetonitrile, 1:1 (v/v)	Water/acetonitrile, 1:4 (v/v)

[a] The dimethylformamide must be freed of the amines that accumulate on storage: pure dimethylformamide can be stored at $-20°$, and "refreshed" by taking 5 ml (stock bottle to be allowed to reach room temperature in a desiccator), placing, without cooling, on the lyophilizer, and waiting until the volume has been reduced to about 3 ml. The flask should be rapidly sealed on removal from the lyophilizer and should remain closed until the contents (which cool due to evaporation) have reached room temperature. This operation should be carried out once only for any given 5-ml sample.

[b] These isocratic conditions should not be adopted without preliminary trials, since different lots of column fillings, and even column age, can have an effect on isocratic separations. These figures are given as a general indication of the elution behavior of these molecules.

contaminates HPLC columns and solid-phase cartridges to an unacceptable event.

N^α-(Aminooxyacetyl)-N^ε-(fluorescein thiocarbamyl)lysine (**VIa**). The dried product (**Va**) is dissolved in trifluoroacetic acid to a concentration not exceeding 20 mg/ml and left for 45 min at 20°. The majority of the trifluoroacetic acid is then evaporated in a stream of air, and drying is completed by lyophilization. If the **Va** has not been purified by reversed-phase HPLC as described above, then purification at the stage of **VIa** is equally possible on the semipreparative column, using 25–40% B over 15 min. The yield from 40 mg of **IV** is 18 mg. The m/z found by ESI-MS was 609.6 (M+H), calculated 609.6.

Note: It is in general best not to expose aminooxy compounds to chromatographic conditions in the unprotected state, since they are fairly reactive. In this instance, however, the relatively large quantities present probably mean that any small losses go unnoticed.

Membrane Preparation

Cell membranes are prepared from CHO (Chinese hamster ovary) cells expressing CCR1 by homogenization in an Ultra Turrax homogenizer using 50 mM HEPES (pH 7.4) buffer containing 1 mM EDTA, 10 mM MgCl$_2$, and a protease inhibitor cocktail (Boehringer Mannheim). An initial centrifugation at 500g for 30 min removes the cell debris. The supernatant is then spun at 48,000g for 30 min. The membrane pellet is resuspended in buffer and stored frozen at −80° in aliquots until required. The membranes are stable under these conditions for at least 6 months.

Competition Equilibrium Binding

Reagents

RPMI 1640 cell culture medium (GIBCO-BRL, Gaithersburg, MD)
^{125}Labeled chemokines (Amersham, Amersham, UK, specific activity 2000 Ci/mmol)
Ultima Gold scintillation fluid (Packard)

Equipment

96-well multiscreen filter plates (Millipore, Bedford, MA, MADV N6550)
Vacuum filtration apparatus for separating free and bound radioligand in 96-well filtration plates (Millipore MultiScreen System or equivalent)
Microbeta Scintillation Counter for counting 96-well plates (Wallac or other manufacturer)

Procedure. RANTES and MIP-1α binding assays are carried out on a scintillation proximity assay (SPA) on membranes from CHO cells expressing CCR1. Membranes are incubated at 2 μg/well with 25 μg/well wheat germ agglutinin SPA beads and 100 pM ^{125}1-MIP-1α for 4 hr, at 25°, on a shaking platform. Increasing concentrations of chemokine or fluorescent chemokine (in the range 10^{-13} to 10^{-6} M) are added to the incubation, to allow competition with the radiolabeled ligand for receptor binding. The plate is counted directly in a Wallac 1450 Microbeta plus counter (Turku, Finland) for 1 min per well.

IL-8 competition binding assays are carried out on neutrophils purified from healthy donors by sedimentation over dextran sulfate, centrifugation through Ficoll–Hypaque, and lysis of contaminating red blood cells by hypotonic shock. Cells are kept in RPMI 1640 medium, with 2% inactivated fetal calf serum (FCS), at 4° prior to assay. The IL-8 competition assay is carried out in 96-well multiscreen filter plates that have been pretreated for 2 hr with NaCl/phosphate buffer containing 10 mg/ml bovine serum albumin (BSA) and 0.02 mg/ml sodium azide (binding buffer). The assay is performed in a volume of 150 μl of binding buffer containing 3×10^5 neutrophils/well, 0.23 nM ^{125}I-labeled IL-8, and varying concentrations of competing chemokine. After 90 min of incubation at 4°, the cells are washed four times with 200 μl ice-cold NaCl/phosphate buffer, which is removed by aspiration. The filters are dried, 3.5 μl Ultima Gold scintillation fluid is added, and filters are counted on a Microbeta Scintillation Counter. The assays are performed in triplicate, and the data are fitted directly to the equation: $B = B_{max} [L]/K_d + [L]$, where [L] is the molar concentration of ligand. For graphing purposes, these are shown in transformations as Scatchard plots.

Assays are performed in triplicate and the data are analyzed with Excel 5.0 and Grafit 3.01 software[18] (Erithacus Software, Staines, UK). Results are fitted to the equation $B = B_0/(1 + [L]/IC_{50})$, where B_0 is the amount of radioligand binding in the absence of competitor, [L] is the concentration of cold competitor added, and IC_{50} is the concentration of cold competitor required to reduce the radioactive binding to 50%. For a single-site competitive interaction, $IC_{50} = K_d + [L^*]$, where $[L^*]$ is the concentration of radiolabel added; therefore, under the conditions of the assay the IC_{50} approximates the K_d.

As shown in Fig. 1, the fluorescent chemokines are able to compete for binding of the radiolabeled ligand. The addition of the fluorescent group at the amino terminus of MIP-1α results in a small decrease of affinity as shown by the IC_{50} for Cy5-MIP-1α of 55 nM compared to that observed for MIP-1α of 15 nM in this assay. Similarly, there is an increase in the IC_{50} for FITC–IL-8 (12 nM compared to 1 nM observed for IL-8 binding to neutrophils). The addition of the NBD group to RANTES causes the largest effect on binding, where the IC_{50} for NBD–RANTES displacement

---

[18] R. J. Leatherbarrow, GraFit Version 3.01. Erithicus Software Ltd., Staines, UK, (1992).

A

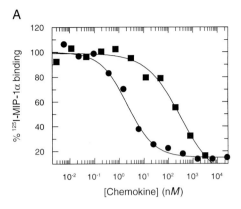

B

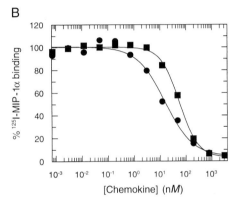

C

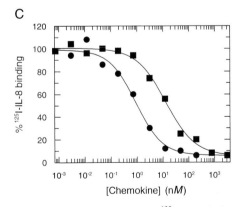

FIG. 1. Competition equilibrium binding assays of ^{125}I-labeled chemokine by fluorescent chemokines. (A) Competition of ^{125}I-MIP-1α in an SPA assay on CHO membranes containing the recombinant CCR1 receptor by RANTES (●) and NBD–RANTES (■). (B) Competition of ^{125}I-MIP-1α in an SPA assay on CHO membranes containing the recombinant CCR1 receptor by MIP-1α (●) and Cy5–MIP-1α (■). (C) Competition of ^{125}I-IL-8 on neutrophils by IL-8 (●) and FITC–IL-8 (■).

of ^{125}I-MIP-1α is 300 nM compared to 2 nM for the displacement of this radiolabeled ligand by RANTES.

Flow Cytometry Analysis

Equipment

FACStar plus and FACScan (Becton Dickinson, Evembodegem, Belgium)

Procedure. Cells (2×10^5) are suspended in 200 μl PBS containing 10 mg/ml BSA and the fluorescent chemokine at the concentrations indicated in either the presence or absence of varying concentrations of unlabeled ligand. The cells are incubated at 4° for 2 hr for the IL-8 conjugates and for 4 hr for the RANTES and MIP-1α conjugates, after which the cells are washed twice with phosphate-buffered saline (PBS) for IL-8 and MIP-1α, and with 50 mM HEPES buffer, pH 7.2, containing 1 mM CaCl$_2$, 5 mM MgCl$_2$, 0.5% BSA, and 0.5 M NaCl for RANTES. Prior to analysis on a FACScan the cells are resuspended in 100 μl PBS containing 10 mg/ml BSA. The binding of the fluorescent chemokines to cells expressing the relevant receptors is easily monitored by FACS analysis. Binding of 1 μM NBD–RANTES to CHO cells expressing the CCR1 receptor is shown in Fig. 2, and the specificity of this binding is demonstrated by the fact that it can be blocked competitively by RANTES in a dose–dependent manner.

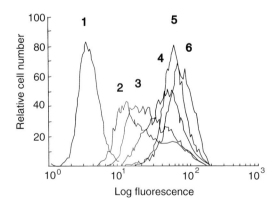

Fig. 2. FACS analysis of the binding of NBD–RANTES to CHO cells expressing CCR1 and its displacement by RANTES. CHO cells were incubated alone (peak 1, background autofluorescence) or with 1 μM NBD–RANTES for 4 hr at 4° (peak 6) or with 1 μM NBD–RANTES and 50 μM RANTES (peak 2), 10 μM RANTES (peak 3), 5 μM RANTES (peak 4), or 2 μM RANTES (peak 5).

MIP-1α, which also binds to this receptor, can similarly compete with the binding of NBD–RANTES (results not shown). The binding of NBD–RANTES to HEK cells expressing CCR-5 is shown in Fig. 3.

Recombinant cells are sorted on a FACStar plus gated to select cells expressing high levels of receptor.

Confocal Microscopy

Cells are grown in plastic microscope chambers overnight at 37° to allow them to adhere to the slide. The medium is removed, and the cells are washed twice with ice-cold binding buffer. The fluorescent chemokine is dissolved first in water and then diluted into ice-cold binding buffer at the required concentration. Two hundred fifty microliters of the fluorescent chemokine solutions is added to the chambers, and the cells are incubated on ice in a cold room in the absence of light. To observe plasma membrane binding, the chemokine solution is removed, and the cells are fixed with 3.7% (w/w) formaldehyde in PBS for 15 min at room temperature. The slides are then mounted using Citifluor antifade solution. To observe internalization the fluorescent chemokine solution is removed after the 4-hr incubation, and the cells are washed twice with binding buffer and incubated

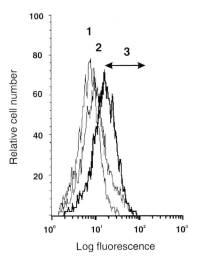

FIG. 3. FACS analysis of the binding of NBD–RANTES to HEK cells expressing murine CCR5. The cells expressing the highest number of receptors were pooled as indicated by the arrow. Peak 1, Cells were incubated alone (background autofluorescence); peak 2, cells incubated with 0.2 μM NBD–RANTES and 2 μM RANTES; peak 3, cells incubated with 0.2 μM NBD–RANTES.

for the required time at 37°, in the absence of light, prior to fixing and mounting as described.

The fluorescent binding can be competitively blocked with a 100-fold excess of MIP1-α and a 500-fold excess of RANTES. A time course of binding shows that strong fluorescence is observed only after 3 to 4 hr of incubation. The optimal concentration for binding to the CCR1 receptor expressed in CHO cells is found to be between 250 and 500 nM. The plasma membrane binding using 500 nM NBD–RANTES after 4 hr of incubation is shown in Fig. 4A, and the internalization of 500 nM Cy5-MIP1-α after binding on ice for 4 hr followed by a 30 min incubation at 37° is shown in Fig. 4B.

Chemotaxis

Equipment

96-well Boyden microchambers (Neuro-Probe, Cabin John, MD)
48-well Boyden microchambers (Neuro-Probe)
Cell Titer 96 Nonradioactive Cell Proliferation Assay (Promega, Madison, WI)
Thermomax microtiter plate reader (Molecular Devices, Palo Alto, CA)

A   B

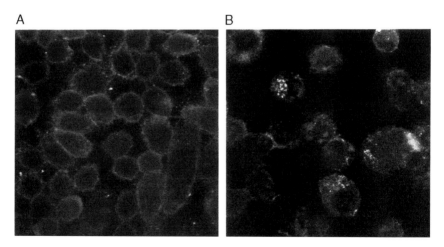

Fig. 4. Confocal microscopy analysis of (A) plasma membrane binding of NBD–RANTES to CHO cells expressing the CCR1 receptor after 4 hr at 4° and (B) internalization of Cy5–MIP-1α after a 30-min warm up at 37° following 4 hr of binding at 4°.

Procedure. Assay of *in vitro* chemotaxis of human polymorphonuclear leukocytes (PMNs) induced by IL-8 and its conjugates is carried out on neutrophils purified from fresh blood. Fresh blood (50 ml) is collected into 15 ml 0.1 M EDTA, 3% dextran, and 3% glucose to prevent aggregation. The solution is allowed to sediment for 1 hr at 37°, after which the plasma is layered onto Ficoll (plasma:Ficoll, 2:1, v/v), and centrifuged for 20 min at 296g at 25° with the brake off. The PMNs form a pellet, and the lymphocytes are located at the interface of the Ficoll and the plasma. Contaminating erythrocytes are removed from the PMNs (mainly neutrophils) by hypertonic lysis. The residual leukocytes are washed and resuspended at a concentration of 10^6 leukocytes/ml of RPMI 1640 medium.

Monocyte chemotaxis induced by MIP-1α and conjugates is carried out on monocytes purified from buffy coats using the following isolation procedure. One hundred milliliters of buffy coat solution is diluted with 100 ml PBS and layered onto Ficoll and centrifuged as above to separate the lymphocytes from the PMN pellet. The lymphocytes forming the interface are washed twice with PBS, collected by centrifugation at 296g for 10 min, and resuspended at a concentration of 10^6 leukocytes/ml in RPMI 1640 medium. Then 40–50 × 10^6 monocytes/ml are purified from the lymphocyte fraction by adding 10^6 sheep red blood cells/ml and rosetting overnight at 4°, followed by a further Ficoll gradient centrifugation. The monocytes are washed in PBS buffer and resuspended at 2 × 10^6 cells/ml in RPMI 1640 medium.

The chemotaxis assays are carried out in 48-well Boyden microchambers. Twenty-five microliters of the chemokine solutions, previously diluted in RPMI medium containing 2 nM glutamine, 25 mM HEPES, and 10% inactivated FCS, is placed in the lower chamber. The lower chambers are covered with a polyvinylpyrrolidone-free polycarbonate membrane with pore sizes of 3 μm for neutrophils and 5 μm for monocytes. Fifty microliters of solution containing 10^6 cells/ml is added to the top chambers. The assay plates are incubated at 37° for 20 min for neutrophils and 30 min for monocytes. The cells on the upper surface of the membranes are then washed with PBS, and the cells on the lower surface are fixed in methanol before staining with a mixture of Field's A and B stains (Bender and Hobein, Zurich, Switzerland) and air-dried. The cells on the undersurface of the membranes are counted using a microscope fitted with the IBAS image analyzer software.

RANTES-induced chemotaxis of the promonocytic cell line THP-1 is carried out using 96-well Boyden microchambers fitted with 5-μm filters. Approximately 5.6 × 10^5 cells in 200 μl of medium (RPMI 1640 containing 10 mM HEPES, 10% heat-inactivated FCS (v/v), 2 mM L-glutamine, and 0.005% (w/v) gentamicin) are placed in the upper chamber. Then 370 μl of the medium described above, but without FCS and containing the ligand, and appropriate dilutions of NBD–RANTES or Met–RANTES are placed in the lower chamber. The agonist concentration is fixed at five times the EC_{50} value required to induce the response. After 60 min of incubation at 37° under 5% CO_2, the cells are removed from the upper wells, and 200 μl PBS containing 20 μM EDTA is added to detach the cells bound to the filter. After 30 min of incubation at 4°, the plate is centrifuged at 1800g for 10 min and the supernatants removed from the lower wells. The number of cells that have migrated are measured by the Cell Titer 96 nonradioactive cell proliferation assay (Promega), which monitors the conversion of tetrazolium blue into its formazan product. The absorbance at 590 nM is read

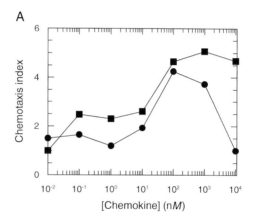

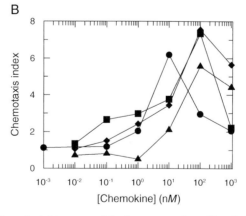

FIG. 5. *In vitro* chemotaxis bioassays of the fluorescent chemokine derivatives. (A) Monocyte chemotaxis induced by MIP-1α (●) and Cy5–MIP-1α (■). (B) Neutrophil chemotaxis induced by IL-8 (●), Cy5–IL-8 (■), FITC–IL-8 (▲), and NBD–IL-8 (◆).

with a microtiter plate reader. The antagonism of THP-1 cell chemo-taxis is carried out by adding various concentrations of the antagonist and a fixed concentration of the agonist in the lower wall.

The data obtained are fitted using Grafit 3.01 software[15] to a four-parameter logistic equation. The EC_{50} and IC_{50} values are defined as the concentrations of agonist and antagonist giving one-half the maximal re-sponse. The fluorescent derivatives of IL-8 and MIP-1α, in accordance with their ability to mobilize calcium (demonstrating that they retain their biological activities), are also active in this *in vitro* bioassay, as shown in Fig. 5. The drop in affinity in receptor binding is mirrored in these responses. NBD–RANTES shows no agonist activity in inducing THP-1 chemotaxis but retains the ability to antagonize the response to RANTES. Met–RANTES is a potent antagonist, inhibiting the chemotaxis of THP-1 cells induced by 3.5 nM RANTES with an IC_{50} of 8 nM, whereas NBD–RANTES inhibited the response with an IC_{50} of 150 nM, again agreeing very closely to the affinity observed in the receptor binding assay.

Acknowledgments

We thank Brigitte Dufour and Fréderic Borlat for skilled technical assistance, M. Pierre-Olivier Regamey for mass spectrometric analyses, Sami Alouani for the binding assays, Rapha-elle Buser for the chemotaxis assays, Christine Power for chemokine and receptor cloning, J.-P. Aubry for FACS analyses and confocal microscopy, and R. Solari for scientific discussion. H. F. G. is supported financially by Gryphon Sciences, South San Francisco, CA. R. E. O. thanks the Swiss National Science Foundation for financing background technical facilities.

[23] Solid-Phase Binding Assay to Study Interaction of Chemokines with Glycosaminoglycans

By Gabriele S. V. Kuschert, Rod E. Hubbard, Christine A. Power, Timothy N. C. Wells, and Arlene J. Hoogewerf

Introduction

Cell surface glycosaminoglycans interact with numerous proteins includ-ing acidic and basic growth factors,[1,2] antithrombin III,[3] and lipoprotein

[1] O. Saksela, D. Moscatelli, A. Sommer, and D. B. Rifkin, *J. Cell Biol.* **107**, 743 (1988).
[2] H. Mach, D. B. Volkin, C. J. Burke, and C. R. Middaugh, *Biochemistry* **32**, 5480 (1993).
[3] U. Lindahl, G. Backstom, L. Thunberg, and I. G. Leder, *Proc. Natl. Acad. Sci. U.S.A.* **77**, 6651 (1980).

lipase,[4] serving diverse functions. The binding of heparin to fibroblast growth factors results in ligand and receptor dimerization and subsequent cellular activation.[5,6] The binding of heparin to antithrombin III results in an acceleration of its ability to inactivate thrombin (reviewed by Griffith[7]), and the binding of heparan sulfate to lipoprotein lipase serves to anchor the enzyme in the vascular endothelium where it will have access to its lipoprotein substrates and the products of hydrolysis can easily be taken up for utilization or storage.[8]

Heparin also interacts with chemokines such as platelet factor 4, resulting in a neutralization of the anticoagulant property of heparin.[9] Other chemokines also bind to heparin,[10] but the nature and role of these interactions are not yet clearly understood. The binding of chemokines to cell surface glycosaminoglycans may serve to immobilize leukocyte-activating chemokines in the vascular endothelium so that they can be presented to leukocytes during the course of inflammatory cell recruitment.[10] In this case, injection of heparin would be expected to displace the chemokines from the cell surface and prevent chemokine-mediated leukocyte activation at the appropriate vascular site. Intravenous injection of heparin has already been shown to reduce the number of neutrophils in the peritoneal lavage following thioglycolate injection in mice,[11] and it has also reduced eosinophil recruitment in the airways of immunized rabbits.[12] This anti-inflammatory effect of heparin was suggested to be partially dependent on the ability of heparin to block L- and P-selectin, but the interaction of heparin with other molecules involved in the inflammatory cascade may also be important. To better understand the nature of the interaction of chemokines with glycosaminoglycans, we have developed a solid-phase binding assay, using immobilized heparin, radiolabeled chemokines, and various competitor molecules. This assay not only may be useful for studying the interaction of chemokines with a variety of different glycosaminoglycans, but could

[4] C.-F. Cheng, G. M. Oosta, A. Bensadoun, and R. D. Rosenberg, J. Biol. Chem. 256, 12893 (1981).
[5] T. Spivak-Kroizman, M. A. Lemmon, I. Dikic, J. E. Ladbury, D. Pinchasi, J. Huang, M. Jaye, G. Crumley, J. Schlessinger, and I. Lax, Cell (Cambridge, Mass.) 79, 1015 (1994).
[6] M. W. Pantoliano, R. A. Horlick, B. A. Springer, D. E. Van Dyk, T. Tobery, D. R. Wetmore, J. D. Lear, A. T. Nahapetian, J. D. Bradley, and W. P. Sisk, Biochemistry 33, 10229 (1994).
[7] M. J. Griffith, New Compr. Biochem. 13, 259 (1986).
[8] M. S. Bosner, T. Gulick, D. J. S. Riley, C. A. Spilburg, and L. G. Lange III, Proc. Natl. Acad. Sci. U.S.A. 85, 7438 (1988).
[9] D. A. Lane, J. Denton, A. M. Flynn, L. Thunberg, and U. Lindahl, Biochem. J. 218, 725 (1984).
[10] D. P. Witt and A. D. Lander, Curr. Biol. 4, 394 (1994).
[11] R. M. Nelson, O. Cecconi, W. G. Roberts, A. Aruffo, R. J. Linhardt, and M. P. Bevilacqua, Blood 82, 3253 (1993).
[12] M. Sasaki, C. M. Herd, and C. P. Page, Br. J. Pharmacol. 110, 107 (1993).

also be applied to the study of the interaction of glycosaminoglycans with other proteins as well.

General Considerations

A 96-well format competitive binding assay for the study of the interactions of chemokines with glycosaminoglycans is simple to perform and allows the simultaneous analysis of the binding of multiple radiolabeled ligands to immobilized heparin under a variety of conditions and in the presence of various competitor molecules. Competition of the binding by an unlabeled ligand, competitor glycosaminoglycans, or other small molecule competitors yields 50% inhibition constants (IC_{50} values). Comparison of the IC_{50} values obtained with various competitor molecules allows one to determine the relative potency and selectivity of the competitors.

The study of the binding of radiolabeled proteins with immobilized glycosaminoglycans will obviously depend on several factors, including the time of incubation, the concentration of both the ligand and the immobilized glycosaminoglycan, the temperature, and the buffer composition, including ionic strength. Because all of these factors may influence the measurement of the relative affinity between chemokines and glycosaminoglycans, they must be carefully defined to obtain meaningful and reproducible data.

The basic strategy for performing binding assays to study the interaction between proteins and glycosaminoglycans is to (1) choose initial conditions, including the concentration of ligand and immobilized heparin; (2) optimize and validate the binding; and (3) use the optimized conditions to study the interactions in the presence of various competitor molecules. The protocol for the basic procedure, the choosing of initial conditions, and assay optimization are addressed in this chapter.

Solid-Phase Assay for Studying Glycosaminoglycan–Chemokine Interactions

Reagents

Heparin-Sepharose (Pharmacia, Uppsala, Sweden); alternatively, heparin-acrylic beads, heparin-agarose, or heparin-agarose with terminal aldehyde coupling (all from Sigma, St. Louis, MO)

Sepharose beads (Pharmacia)

96-Well filter plates (Millipore, Bedford, MA, MultiScreen MADVN6510)

Soluble glycosaminoglycan competitors: Heparin (Sigma), unlabeled chemokines (Peprotech, Rocky Hill, NJ; R & D, Minneapolis, MN), or other proteins

^{125}I-Labeled chemokines (Amersham, Amersham, UK), specific activity 2000 Ci/mmol

Buffer reagents: Bovine serum albumin (BSA, Sigma), HEPES buffer (GIBCO, Grand Island, NY), $MgCl_2$, $CaCl_2$, NaCl (reagent grade)

Scintillation fluid (Wallac Optiphase Supermix, Turku, Finland).

Equipment

Vacuum filtration apparatus for separating free and bound radioligand in 96-well filtration plates (Millipore MultiScreen System or equivalent)

Microbeta Scintillation Counter for counting 96-well plates (Wallac or other manufacturer)

Shaker to accommodate 96-well plates

Basic Procedure

1. Rehydrate a defined amount of heparin-Sepharose, wash according to the manufacturer's suggestions, and remove the fines. Resuspend the wet gel in binding buffer (1 mM $CaCl_2$, 5 mM $MgCl_2$, 0.5% BSA, and 50 mM HEPES, pH 7.2) at a stock concentration of 0.25 ml wet gel per milliliter total volume (for heparin-Sepharose from Pharmacia, 1 g dry weight $\approx$ 4 ml wet gel, and the concentration of heparin is 2 mg/ml of wet gel, giving a concentration of heparin of 0.5 mg/ml). In this chapter, all concentrations of the immobilized heparin refer to the mass of the heparin, rather than the mass of dry gel or volume of wet gel.

2. Dilute the heparin-Sepharose stock solution to the appropriate concentration to deliver 0.015 to 1.5 μg/well immobilized heparin in a volume of 50 μl. Add heparin-Sepharose to the wells (50 μl/well) of a Millipore filtration plate.

3. Add competitor (other proteins, competitive ligands, or glycosaminoglycans) to the filtration plate in a volume of 30 μl, preparing duplicate to quadruplicate wells for each point. For the glycosaminoglycan competitors used in our studies, the concentration of glycosaminoglycans are 0–2 mg/ml, spanning four orders of magnitude with 10–12 points.

4. Add a constant amount of radioiodinated ligand in a volume of 20 μl. For the studies described in this chapter, we are using 5 nM unlabeled interleukin-8 (IL-8), with a tracer concentration of 0.25–0.5 nM ^{125}I-labeled IL-8 (specific activity 2200 Ci/mmol).

5. Incubate the plates for 30 min to 8 hr at room temperature with shaking.

6. Separate free and bound radioiodinated ligand by vacuum filtration.

7. Wash the heparin-Sepharose three times with 200 μl wash buffer (0.15 M NaCl, 0.5 M NaCl in binding buffer).

8. Allow the plates to dry for 30 min at room temperature.

9. Add 30 μl of scintillation fluid to each well.

10. Measure the radioactivity with a calibrated Microbeta Scintillation Counter.

11. Analyze the data using GraFit Software[13] or other appropriate curve-fitting software to calculate the 50% inhibition concentration of the competitor. We use an equation based on a single-site model: $B/B_{max}^{app} = 1/[1 + ([L]/IC_{50})]$, where B is cpm bound, B_{max}^{app} is cpm bound in the absence of competing ligand, [L] is the molar concentration of competing ligand, and the IC_{50} is [radioligand] + K_d (Ref. 14).

Choosing Initial Conditions

In binding assays, various parameters will affect the interaction between the chemokine and glycosaminoglycan being studied. Initial studies should be performed to choose the optimum conditions for the binding studies. Because at the outset of the experiment the affinity of the specific chemokine–glycosaminoglycan is not known, it is best to examine several concentrations of immobilized heparin and incubation periods. An example is shown in Fig. 1, in which the binding of ^{125}I-labeled IL-8 to four different concentrations of immobilized heparin was studied as a function of time. The first observation from this kinetic study was that the approach to equilibrium was faster at the lower concentrations of immobilized heparin, indicating that if the higher concentrations of immobilized heparin were chosen for the binding studies, sufficient time would be needed to reach equilibrium. The second observation is that at the two higher concentrations of immobilized heparin, 30–60% of the total radiolabeled ligand was bound, indicating that ligand depletion had occurred. A high concentration of receptors causes a significant reduction in the concentration of free ligand, resulting in a complicated association kinetic that shifts the equilibrium competition curve to the right. Hence, when optimizing the conditions of the binding assay, it is important to keep the receptor concentration low to avoid ligand depletion, while maintaining a suitable signal-to-noise ratio. In this example, we used lower concentrations of immobilized heparin to avoid both ligand depletion and a slow approach to equilibrium. For a

[13] R. J. Leatherbarrow, GraFit Version 3.01, Erithicus Software, Staines, UK, 1992.
[14] Y. Cheng and W. H. Prusoff, *Biochem. Pharmacol.* **22**, 3099 (1973).

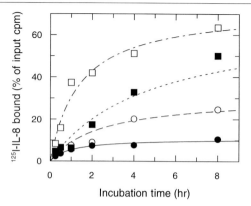

FIG. 1. Effect of increased concentrations of immobilized heparin on ligand depletion and equilibrium kinetics. ^{125}I-Labeled IL-8 was incubated at room temperature with 0.005 (●), 0.015 (○), 0.05 (■), or 0.5 μg/well (□) immobilized heparin for 30 min to 8 hr in binding buffer (1 mM CaCl$_2$, 5 mM MgCl$_2$, 0.5% BSA, and 50 mM HEPES, pH 7.2). The free and bound radioactivity was separated by vacuum filtration, and the immobilized heparin was washed three times with 200 μl binding buffer containing 0.15 M NaCl. The radioactivity was measured with a Microbeta scintillation counter. Each data point represents the mean of duplicate or triplicate samples, and the variation between samples was less than 10%. The total radioactivity added was 25,000 cpm. The data shown are from a single experiment representative of three experiments.

more comprehensive review of receptor–ligand interactions and ligand depletion kinetics, the reader is referred elsewhere.[15]

Optimizing and Validating Binding Assay

Using basic conditions determined from the initial binding experiment, the time course of the binding of a radiolabeled ligand to the immobilized heparin should also be determined in the presence of competitor ligands. This is to determine if the competitor IC$_{50}$ values are time-dependent during the course of the assay. An example is given in Fig. 2. The binding of ^{125}I-IL-8 to a single concentration of immobilized heparin was competed by increasing concentrations of soluble heparin for time periods between 30 min and 8 hr. Figure 2 shows that greater amounts of radioactivity were bound with increasing times, which leveled off between 4 and 8 hr. The IC$_{50}$ values for the competition by soluble heparin were 6.0 ± 0.3 μg/ml for the 8-, 4-, 2-, and 1-hr times, indicating that the IC$_{50}$ values for competition of the binding of this particular ligand, IL-8, did not vary with the time of

---

[15] E. C. Hulme and N. J. M. Birdsall, *in* "Receptor–Ligand Interactions" (E. C. Hulme, ed.), p. 63. Oxford Univ. Press, Oxford and New York, 1992.

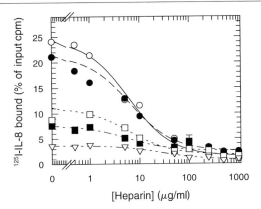

FIG. 2. Kinetics of soluble heparin competition for the binding of ^{125}I-labeled IL-8 to immobilized heparin. ^{125}I-Labeled IL-8 was incubated for 30 min (▽), 1 hr (■), 2 hr (□), 4 hr (●), or 8 hr (○) with 0.015 μg/well immobilized heparin and increasing concentrations of soluble glycosaminoglycans. The radioactivity was measured following the separation of free and bound radioligand and three washes by vacuum filtration. Each data point represents the mean of duplicate or triplicate samples, and the variation between replicates was less than 10%. The total radioactivity added was 25,000 to 30,000 cpm. The data shown are from a single experiment representative of four experiments.

assay. This is in contrast to the dependence of the IC_{50} value on the concentration of immobilized heparin, which was determined in similar kinetic competition experiments with various concentrations of immobilized heparin and several different chemokine ligands. In general, 10-fold increases in immobilized heparin concentration resulted in 2- to 5-fold increases in the IC_{50} values (Table I and data not shown).[16]

Buffer conditions, including variations in ionic strength and divalent cation composition, may also affect the IC_{50} measurement. Because the studies of the interaction of immobilized heparin with various ligands does not involve cells, very simple buffers can be used. We use 50 mM HEPES at pH 7.2, and include the divalent cations Mg^{2+} and Ca^{2+}. This buffer can also be supplemented with NaCl to increase the ionic strength. The chemokine–glycosaminoglycan interactions are partially charge-dependent, and increasing the ionic strength causes a decrease in the apparent affinity of the interaction. Figure 3 shows the effect of ionic strength on the binding of ^{125}I-labeled IL-8 to immobilized heparin and the competition of the binding by soluble heparin. The binding of IL-8 to 0.015 μg/well immobilized heparin (Fig. 3A) was severely reduced (by 80%) in the presence of 0.10 M NaCl, and binding could not be detected in the presence

[16] G. S. V. Kuschert and A. J. Hoogewerf, unpublished observations (1996).

TABLE I

IC$_{50}$ Values for Heparin Competition of Binding of
^{125}I-Labeled IL-8 to Immobilized Heparin

[NaCl] (M)	IC$_{50}$ Values[a] for immobilized heparin concentration	
	0.015 μg/well	0.15 μg/well
0.0	3.3 μg/ml	11 μg/ml
0.05	7.5 μg/ml	16 μg/ml
0.10	23 μg/ml	53 μg/ml
0.15	No detectable binding	69 μg/ml

[a] Competitive binding assays were performed with ^{125}I-
IL-8, 0.015 or 0.15 μg/well immobilized heparin, and
increasing concentrations of heparin competitor for 4
hr at room temperature. The IC$_{50}$ values were determined as described in the Basic Procedure. The data
shown are from one experiment representative of three
experiments for IL-8. Similar results were also obtained
with three other radiolabeled chemokines.[16]

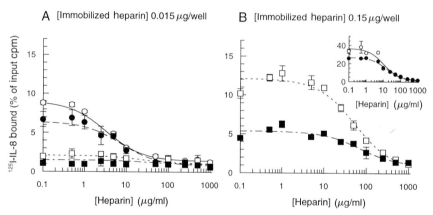

Fig. 3. Effect of ionic strength on the binding of ^{125}I-labeled IL-8 to immobilized heparin
and on the competition by soluble heparin. ^{125}I-Labeled IL-8 was incubated for 4 hr at room
temperature with either 0.015 (A) or 0.15 μg/well immobilized heparin (B) in binding buffer
supplemented with 0.15 (■), 0.10 (□), or 0.05 M NaCl (●), or without NaCl (○). The binding
was competed by increasing concentrations of soluble heparin. The bound radioactivity was
measured, the data were analyzed, and IC$_{50}$ values were determined as described in the Basic
Procedure. Each point represents the mean ± SEM of triplicate determinations, and the data
shown are from one experiment representative of two. Total radioactivity was 23,000 cpm
for (A), and 10,000 cpm for (B). So that the binding in the presence of 0.10 and 0.15 M NaCl
could be visually compared for the two concentrations of immobilized heparin, the higher
values for the binding in the presence of 0.05 M NaCl (●) or without NaCl (○) were plotted
on the inset graph for 0.15 μg/well immobilized heparin (B, inset).

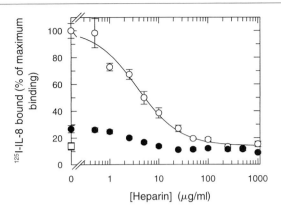

FIG. 4. Determination of the level of nonspecific binding of ^{125}I-labeled IL-8. Competition binding experiments were performed with ^{125}I-IL-8, increasing concentrations of soluble heparin competitors, and either 0.015 μg/well heparin-Sepharose (○), an equivalent concentration of Sepharose (●), or a 10^4-fold excess of unlabeled IL-8 (□). The binding was carried out for 4 hr at room temperature. The radioactivity was measured as described in the Basic Procedure. Each point represents the mean ± SEM of triplicate determinations, and the data shown are from one experiment representative of two. Maximum binding equals 5800 cpm, which represented 26% of the total added radioactivity.

of 0.15 M NaCl. Similar to the muscarinic system, the IC$_{50}$ values increased 2- to 7-fold with increasing ionic strength.[15] The same experiment was performed using a 10-fold higher concentration of immobilized heparin (Fig. 3B). Significant binding was noted at all ionic strengths using 0.15 μg/well immobilized heparin, and the effect of increasing ionic strength to increase the IC$_{50}$ values was also observed. The effect of ionic strength and immobilized heparin concentration on the IC$_{50}$ values for heparin competition of ^{125}I-labeled IL-8 binding are presented in Table I. The dependence of IC$_{50}$ values on these two parameters was also noted for other chemokines in addition to IL-8.[16] In addition, the effect of divalent cations or other buffer components on the interaction of chemokines with glycosaminoglycans may also be examined.

Another optimization/validation study that should be performed is to examine the nature of the nonspecific binding in the system. Traditionally, this is performed by measuring the binding in both the presence and absence of a large excess of unlabeled ligand. Figure 4 shows the binding of ^{125}I-IL-8 to immobilized heparin and the competition of the binding by soluble heparin. In this example, the nonspecific binding was assessed by two methods. The first method was to measure ^{125}I-IL-8 binding in the presence of 50 μM unlabeled IL-8 (a 10^4-fold excess). In the presence of excess unlabeled Il-8, 14 ± 3% of the radioactivity was bound, compared to the binding in the absence of the unlabeled IL-8. The second method of analyzing the nonspecific binding was to duplicate the soluble heparin competition experiment using unsubstituted Sepharose beads instead of the heparin-Sepharose beads. Using this method, we have compared ^{125}I-IL-8 binding to immobilized heparin to the nonspecific ^{125}I-IL-8 binding to other components of the system, including the Sepharose beads and the filter plates. The low level of binding in the presence of the Sepharose beads clearly demonstrates that the IL-8 is binding to heparin rather than to other components of the system.

Conclusions

We have developed a solid-phase binding assay for chemokine–glycosaminoglycan interactions. Optimization and validation studies were performed so that we could recognize and avoid significant ligand depletion, retain a significant signal-to-noise ratio, explore the effects of buffer composition, and examine nonspecific binding. Our results show that variations in immobilized heparin concentration and ionic strength will affect the IC_{50} values in competition experiments. The results from this type of assay are relative, rather than absolute measurements of the affinity between chemokines and immobilized heparin. Considering this limitation, the assay is suitable for making comparisons between the relative potencies of different glycosaminoglycans or other small molecule competitors, and it may be useful in identifying molecules that can selectively inhibit the interactions between chemokines and glycosaminoglycans. Molecules identified by this assay could be further tested in other systems to characterize their suitability as anti-inflammatory molecules.

[24] Biological Activity of C-X-C and C-C Chemokines on Leukocyte Subpopulations in Human Whole Blood

By Maryrose J. Conklyn and Henry J. Showell

Introduction

The CD11b/CD18 complex is a β2-integrin that is found both on the surface and in intracellular organelles in leukocytes. After exposure of leukocytes to stimuli such as chemokines, complement fragment C5a, leukotriene B_4, and certain cytokines, for example, tumor necrosis factor-α (TNF-α), granulocyte colony-stimulating factor (G-CSF), and granulocyte–

METHODS IN ENZYMOLOGY, VOL. 287

macrophage colony-stimulating factor (GM-CSF), CD11b/CD18 is rapidly mobilized to the surface of specific leukocyte subsets.[1] These activated leukocytes then attach more firmly to the vascular endothelium,[2] an important step prior to diapedesis and migration into inflammation foci.

There are several advantages to studying the responses of leukocytes in whole blood as opposed to isolated cell preparations. Among these are (1) avoiding cell activation by the isolation procedure itself,[3] (2) maintaining a relatively more physiological condition, and (3) simultaneously monitoring the responses of different cell populations in the same sample. In whole blood assays, each sample can contain its own positive (responding cells) and negative (nonresponding) controls.

Principle

CD11b expression on a cell surface is monitored by immunofluorescent techniques using a fluorescein isothiocyanate (FITC)-conjugated antibody to CD11b and a flow cytometer. Activation of cells such as neutrophils is demonstrated by an increase in the fluorescent intensity of cells exposed to a stimulus when compared to that seen in neutrophils incubated with buffer. The different leukocyte subpopulations are identified by two parameters: their light scatter properties and either autofluorescence (eosinophils[4]), labeling by anti-CD14 phycoerythrin (PE) (monocytes), or neither (neutrophils). Because CD14 clones that bind to neutrophils will obscure the autofluorescence of the eosinophils, the selection of an anti-CD14 clone that has little to no binding to neutrophils is important.

The assay is divided into in four parts: (1) activation, (2) fluorescent labeling, (3) quantitation, and (4) analyses.

Materials

The materials used in this assay along with suggested vendors are listed in Table I. There are other whole blood lysing reagents and protocols that may be substituted. Nunc immuno Minisorp tubes (12×75 mm) and caps are purchased from Marsh Biomedical Products (Rochester, NY). A flow cytometer capable of four-parameter analysis (forward scatter, side scatter, FL1, and FL2) and equipped with a histogram analysis program is needed.

[1] E. C. Butcher, *Cell* (*Cambridge, Mass.*) **67,** 1033 (1991).
[2] T. A. Springer, *Nature* (*London*) **346,** 425 (1990).
[3] M. G. Macey, D. A. McCarthy, S. Vordermeier, A. C. Newland, and K. A. Brown, *J. Immunol. Methods* **181,** 211 (1995).
[4] G. J. Weil and T. M. Chused, *Blood* **57,** 1099 (1981).

TABLE I
REAGENTS AND SUPPLIERS

Reagent	Supplier and location
CD11b FITC (Clone Bear-1)	Caltag Lab., San Francisco, CA
CD14 PE (Clone M5E2)	PharMingen, San Diego, CA
Mouse IgG$_1$ FITC	Caltag Lab., San Francisco, CA
Mouse IgG$_{2a}$ PE	Caltag Lab., San Francisco, CA
Bovine Serum Albumin (low endotoxin)	Sigma, St. Louis, MO
Dulbecco's phosphate-buffered saline (10×)	JRH Biosciences, Lenexa, KS
Heat-inactivated fetal bovine serum	GIBCO-BRL, Grand Island, NY
Sodium azide	Fisher Scientific, Springfield, NJ
Human IL-8 (72 amino acids)	PeproTech, Rocky Hill, NJ
Human eotaxin	PeproTech, Rocky Hill, NJ
Human MCP-1 (MCAF)	PeproTech, Rocky Hill, NJ
Ethylenediaminetetraacetic acid (EDTA)	Sigma, St. Louis, MO
Dimethyl sulfoxide (DMSO)	Pierce, Sigma, Rockford, IL
FACS lysing solution (10×)	Becton Dickinson, San Jose, CA

Stock Solutions

7.5% EDTA. The anticoagulant, 7.5% ethylenediaminetetraacetic acid (EDTA), is prepared as follows: 7.5 g of EDTA is placed in a solution containing 10 ml of 10× Dulbecco's phosphate-buffered saline (PBS) and 60 ml of distilled water. The pH of the solution is adjusted to pH 7.25 with 10 N NaOH. (The EDTA is solubilized with the addition of the NaOH.) The final volume is brought to 100 ml with distilled water. The solution is sterilized by filtration, divided into aliquots in sterile 12 × 75 polystyrene tubes (2.5 ml EDTA/tube), and stored at room temperature.

PBS Wash. The wash buffer consists of 100 ml of 10× PBS, 20 ml heat-inactivated fetal calf serum (FCS), and 2 g sodium azide in a final volume of 1 liter. EDTA (final concentration of 10 mM, pH 7.25) may be included, if desired. PBS wash is sterile filtered and stored at 4°.

Chemokines. The primary stock solutions of chemokines are prepared by dissolving the chemokines in PBS to a final concentration of 100 μM and storing as aliquots at −20°. The 10× stocks are made by 1:10 serial dilutions in PBS containing 2 mg/ml bovine serum albumin (BSA) (low endotoxin) in Nunc Minisorp tubes. These dilutions are prepared and used on the day of assay.

Heat-Aggregated IgG. Human immunoglobulin G (IgG) (10 mg/ml in PBS) is heated to 63° for 20 min. It is stored at 4°.

Procedures

Blood Collection

Human whole blood is collected by venipuncture into 50-ml syringes containing 1 ml 7.5% EDTA. Heparin may be substituted for EDTA for most studies.[5] The blood is kept at room temperature and is used as soon as possible after phlebotomy. Do not place blood in ice[6] until after the chemokine stimulation step. It should be noted that the percentage of eosinophils in human blood from different donors has a wide range (less than 1% to greater than 6%). Whole blood containing 3–5% eosinophils is optimal for both consistency of response and reasonable accumulation times on the flow cytometer.

Assay

Activation

1. Add 20 μl of increasing concentrations of $10\times$ stocks of chemokines to 12×75 Nunc Minisorp tubes (see Table II). Place in a 37° water bath. For evaluation of baseline CD11b expression, substitute PBS–BSA for chemokine.

2. Warm the whole blood in a 50-ml polypropylene tube in a 37° water bath for 5 min.

3. Add 200 μl whole blood per chemokine-containing tube. Incubate in a 37° water bath for 15 min (optimal incubation time for maximal CD11b up-regulation). The reaction is terminated by the addition of 1 ml of cold PBS wash.

Fluorescent Labeling

4. Gently centrifuge samples at 200 *g* for 7 min at 4°. Remove the supernatant fluid by aspiration as much as possible. Resuspend the pellet by gentle shaking.

5. Add 10 μl of 1 mg/ml heat-aggregated IgG and incubate for 10 min at 4°. To each assay tube, add antibodies as listed in Table II. CD11b FITC and CD14 PE should be titrated to ensure that the antibody concentration

---

[5] M. J. Conklyn, K. Neote, and H. J. Showell, *Cytokines* **8,** 762 (1996).
[6] K. D. Forsyth and R. J. Levinsky, *J. Immunol. Methods* **128,** 159 (1990).

TABLE II
ASSAY SETUP

| | Tube number | | | | | | | | | |
Stock reagent	1	2	3	4	5	6	7	8	9	10
Part 1										
PBS–BSA	X	X	X	X	—	—	—	—	—	—
1E-10 M C-C[a]	—	—	—	—	X	—	—	—	—	—
1E-9 M C-C	—	—	—	—	—	X	—	—	—	—
1E-8 M C-C	—	—	—	—	—	—	X	—	—	—
1E-7 M C-C	—	—	—	—	—	—	—	X	—	—
1E-6 M C-C	—	—	—	—	—	—	—	—	X	—
1E-5 M C-C	—	—	—	—	—	—	—	—	—	X
Whole blood	X	X	X	X	X	X	X	X	X	X
Part 2: Antibody binding[b]										
CD11b FITC	—	—	—	X	X	X	X	X	X	X
CD14 PE	—	—	X	X	X	X	X	X	X	X
M IgG$_1$ PE	—	X	X	—	—	—	—	—	—	—
M IgG$_{2a}$ PE	—	X	—	—	—	—	—	—	—	—

[a] C-C, Chemokine; M, molar.

[b] Include three extra samples (CD11b alone, CD14 alone, and no antibody added) in the assay setup for adjusting the compensation for flow cytometry.

is at a saturating concentration. Incubate for 20 min at 4°. Add 1 ml PBS wash and pellet cells as described in step 4.

6. This step is to be performed according to established guidelines for the safe handling of biological materials. Resuspend cell pellet by gentle shaking. Add 1.5 ml of 1× fluorescence activated cell sorting (FACS) lysing buffer. Cap the tubes. In a biological safety hood, gently vortex the samples. Incubate 5 min at room temperature. Vortex again. Incubate at room temperature for 5 min. Gently centrifuge samples at 200 g for 7 min at 4°. Aspirate supernatant fluid. Resuspend cells in 1 ml cold PBS wash.

7. Centrifuge as before. Repeat the wash step.

8. The cells are resuspended in 250 μl of PBS wash and kept on ice until analyzed.

Quantitation

Data Acquisition. The cellular response to the chemokines is quantitated by standard flow cytometry. In the following examples, a Becton Dickinson (Lincoln Park, NJ) FACScan with a 488-nm argon laser and three detectors for 530 nm (FL1, FITC), 585 nm (FL2, PE), and >650 nm (FL3, Red) was used. CellQuest software was used for both acquisition and analysis. For

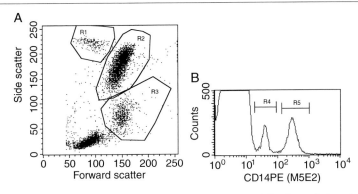

FIG. 1. Identification of monocyte, neutrophil, and eosinophil subpopulations in human whole blood. In (A), eosinophils (R1), neutrophils (R2), and monocytes (R3) are defined by their forward and right-angle light scattering properties. In (B) eosinophils (R4) are defined by their autofluorescence peak in the FL2 channel and monocytes (R5) by the binding of CD14 PE.

data collection, a live gate is drawn around the eosinophil population identified by the light scatter profile, and 2500 events are counted. All the data are stored in list mode.

Analysis. Leukocyte subsets are defined by the logical gating technique employing two parameters to define a cell population. First, the three leukocyte subpopulations are defined by their light scattering properties as shown in Fig. 1A. Then the three subpopulations are defined by the pattern of fluorescence in the FL2 channel (Fig. 1B). The logical gates are defined as follows: eosinophils are defined as cells found in both R1 and R4, monocytes are defined as cells occurring in both R3 and R5, and neutrophils are those cells found in R2 but not R5. By using a two-parameter gate, the effects of intrusion of one cell type into the region of another is minimized. This is especially important for the least numerous leukocyte subpopulation, the eosinophils.

CD11b expression is quantitated by histogram analysis using the gates as defined above. Fluorescence intensity (*FI*) is expressed as mean channel fluorescence for neutrophils and median channel fluorescence for monocytes. In Fig. 2, histograms with overlay curves show the shift in CD11b FITC fluorescence on stimulation of whole blood with interleukin-8 (IL-8), eotaxin, or macrophage chemotactic protein-1 (MCP-1).

For comparing the potency of different stimuli, the data are expressed as percent baseline and are calculated as follows in Eq. (1):

$$\% \text{ Baseline} = \frac{FI(\text{chemokine}) - FI(\text{isotype control})}{FI(\text{PBS–BSA}) - FI(\text{isotype control})} \times 100 \qquad (1)$$

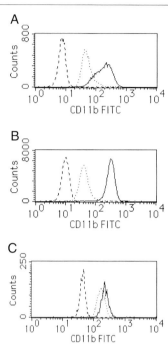

FIG. 2. Histograms of CD11b expression in human whole blood monocytes, neutrophils, and eosinophils. The CD11b expression patterns in stimulated (——) and unstimulated (·····) monocytes (A), neutrophils (B), and eosinophils (C) are shown. Also presented are isotype controls (– – –), which were incubated with mouse IgG$_1$ FITC.

The *FI* of the isotype control (tubes 2 and 3 from Table II) should be equal to that of cellular autofluorescence (tube 1). In Fig. 3, typical dose–response curves for neutrophils (IL-8[7]), monocytes (MCP-1[8]), and eosinophils (eotaxin[9]) are shown. Of the three leukocyte populations examined, eosinophils will have the lowest maximal increase in CD11b expression. These are usually 1.5 to 2-fold in magnitude. The maximal increase in CD11b expression in monocytes is usually in the 2- to 4-fold range. Neutrophils have the most dramatic increases, which can range from 4- to 10-fold. The maximal increase(-fold) in CD11b in a given cell type will vary according to the choice of stimulus.

    Possible sources of leukocyte activation that are not chemokine-specific

---

[7] P. A. Detmers, D. E. Powell, A. Walz, I. Clark-Lewis, M. Baggiolini, and Z. A. Cohn, *J. Immunol.* **147**, 4211 (1991).

[8] K. Vaddi and R. C. Newton, *J. Immunol.* **153**, 4721 (1994).

[9] K. Tenscher, B. Metzner, E. Schopf, J. Norgauer, and W. Czech, *Blood* **88**, 3195 (1996).

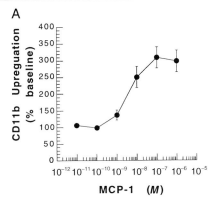

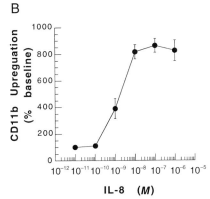

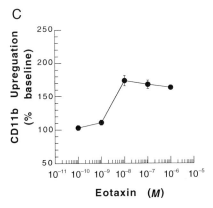

FIG. 3. Concentration-dependent increase in CD11b expression of whole blood leukocytes stimulated with either MCP-1 (monocytes), IL-8 (neutrophils), or eotaxin (eosinophils). Human whole blood was incubated with the indicated concentration of chemokine as described in the text, and the response of monocytes (A), neutrophils (B), and eosinophils (C) ($n = 3$) were determined.

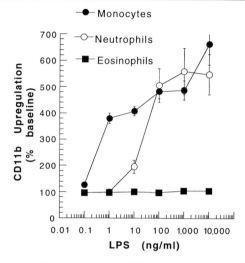

FIG. 4. Lipopolysaccharide up-regulates CD11b on neutrophils and monocytes but not eosinophils in human whole blood. LPS was added to human whole blood and incubated for 15 min, 37°. Changes in the expression of CD11b were determined as described in the text ($n = 3$).

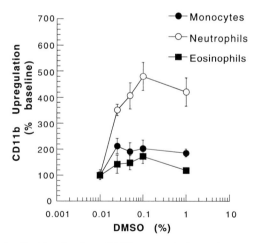

FIG. 5. Dimethyl sulfoxide stimulates CD11b up-regulation on human whole blood leukocytes. Human whole blood was incubated with DMSO, and the effect on CD11b expression of monocytes, neutrophils, and eosinophils ($n = 3$) was determined.

are lipopolysaccharide[10] (LPS) and dimethyl sulfoxide (DMSO). As shown in Fig. 4, CD11b up-regulation occurs in monocytes and neutrophils but not eosinophils in the presence of LPS. Monocytes ($EC_{50} = 0.47 \pm 0.05$ ng LPS/ml) are most sensitive to LPS; neutrophils ($EC_{50} = 26.9 \pm 0.9$ ng LPS/ml) are about 10-fold less sensitive to LPS than monocytes. Eosinophils do not respond to LPS in this assay system. DMSO, a universal solvent used in many inhibitor studies, can activate all three cell types (Fig. 5). Concentrations less than 0.01%, however, have little or no effect on monocytes, neutrophils, or eosinophils. DMSO obtained from different sources vary in ability to stimulate cells, and therefore the reagent should be assayed prior to use. We have found DMSO from the two sources listed in Table I to be the most reliable in this regard.

Summary

We have described an assay that monitors the activating effects of a variety of chemokines on leukocyte subsets in human whole blood. This procedure has the following advantages: (1) minimal manipulation of the cells, (2) maintenance of more physiological conditions, and (3) simultaneous monitoring of the responses of monocytes, neutrophils, and eosinophils.

---

[10] R. Weingarten, L. A. Sklar, J. C. Mathison, S. Omidi, T. Ainsworth, S. Simon, R. J. Ulevitch, and P. S. Tobias, *J. Leukocyte Biol.* **53,** 518 (1993).

Author Index

Numbers in parentheses are footnote reference numbers and indicate that an author's work is referred to although the name is not cited in the text.

H

M

Subject Index

A

Actin polymerization, *see* Chemotaxis assay
Affinity chromatography, *see* Heparin affinity chromatography; Immunoaffinity chromatography
Alanine scan mutagenesis, *see* Site-directed mutagenesis
Allergy, RANTES role, 293–294

B

Basophil, isolation from human blood, 174–175
BB-10010, *see* Macrophage inflammatory protein-1α

C

Calcium flux
  dose-response curve compared to chemotaxis, 179
  fluorescent probes
    calcium concentration by ratio fluorescence, 123, 181–182
    loading of cells, 179–180
    types, 123, 179–180
  monocyte assay in response to monocyte chemotactic proteins, 123, 125, 179–182
CD, *see* Circular dichroism
CD11a/CD18, chemokine upregulation assays
  monocytes, 185
  neutrophils, 72–74, 76
CD11b
  mobilizers, 378–379
  whole blood upregulation assay by flow cytometry

activation, 381
data acquisition, 382–383
data analysis, 383–384, 387
fluorescence labeling, 381–382
materials, 379
principle, 379
sample collection, 381
stock solution preparation, 380–381
Chemokine classification, 59, 89, 186, 206, 267, 348
Chemotaxis assay
  actin polymerization and cell polarization, 182–183
  fluorescent chemokine assay, 366–369
  granulocyte chemotactic protein-2, 27–28
  hippocampal injection of chemokines in animal models, 268–269
  human peripheral blood lymphocyte– severe combined immunodeficiency mouse model
    chemokine administration, 271–272, 275
    flow cytometric analysis of cell types, 273, 287
    graft-versus-host disease in model, 283, 287
    growth hormone
      administration, 272
      engraftment effects, 270, 281–283
    immunostaining of lymphocytes, 272– 273, 277
    leukocyte trafficking effects
      adhesion molecule binding, 289–290
      blocking by chemokine antibodies, 278–279, 284
      interleukin-8, 273, 279, 288
      IP-10, 273–274, 278, 288, 291
      macrophage inflammatory protein- 1α, 270, 273–274, 276–277, 290

W